Dynamical Modelling and Estimation in Wastewater Treatment Processes

Dynamical Modelling and Estimation in Wastewater Treatment Processes

Denis Dochain
CESAME
Université Catholique de Louvain
Louvain-La-Neuve, Belgium

and

Peter Vanrolleghem
BIOMATH
Universiteit Gent
Gent, Belgium

Publishing

Published by IWA Publishing, Alliance House, 12 Caxton Street, London SW1H 0QS, UK
Telephone: +44 (0) 20 7654 5500; Fax: +44 (0) 20 7654 5555; Email: publications@iwap.co.uk
Web: www.iwapublishing.com

First published 2001
© 2001 IWA Publishing

Printed by TJ International (Ltd), Padstow, Cornwall, UK

British Library Cataloguing in Publication Data
A CIP catalogue record for this book is available from the British Library

Library of Congress Cataloging- in-Publication Data
A catalog record for this book is available from the Library of Congress

ISBN: 1 900222 50 7

Contents

Preface

The present book has a rather long history that dates back to a meeting of the European COST action 682 in San Sebastian, Espana in September 1994. At that time, the action was entitled "Optimizing the Design and Operation of Biological Wastewater Treatment Plants Through the Use of Computer Programs Based on a Dynamic Modelling of the Process". Later on, it has been renamed with a more comprehensive title: "Integrated Wastewater Management". The first title was given by our hosts in San Sebastian, i.e. Jaime Garcia-Heras and his group at the CEIT who were at the origin of this COST action. This particular meeting is certainly memorable for those who were lucky enough to participate. But for us, it was also the occasion to launch the idea of writing a book on identification in wastewater treatment processes. Indeed we had the feeling that this matter was a key topic in wastewater treatment for many reasons: the models of WWTP are complex; their dynamics are often badly grasped, a characteristic shared with the other biological processes but with the specific aspect that both the kinetics and the hydrodynamics may be badly defined; the data necessary for model identification suffer from a lot of drawbacks, they are basically too rare in terms of the number of measured variables and in terms of time distribution, and they are often not informative enough to perform a reliable model identification. And we had realised that there was no monograph that could fit to the needs of the practitioners and to the students in WWTP engineering. Available textbooks were too mathematically involved and/or not dedicated to the specific problems encountered in the identification of WWTP, or they were too much practice oriented and gave insufficient insight in the quantitative methods used. What was left was a large body of detailed papers and a few good review papers. We all know how difficult it is to get a comprehensive understanding of a topic as wide as WWTP modelling and estimation.

To start the writing we had basically already some good material coming from our individual and mutual scientific experience. We already had the chance to test the basic ideas of the book in a course that we had been asked to offer at a general meeting of the same COST action a few months before in Kollekolle, Denmark.

After six years of intensive writing, we are now able to deliver a manuscript that aims at gathering the main aspects of dynamical modelling, model building and selection, experiment design, parameter and state estimation in a comprehensive and pedagogical presentation. This is at least our intention. We have tried to select at each step (W)WTP examples that are drawn from our experience in these fields.

Obviously the scientific experience that we have gained in these fields is due to many people with whom we have interacted. These include scientific collaborators and students. From the interaction with each of them we have been able to gather the scientific expertise compiled in the present manuscript. We hope that we have not forgotten anyone in the list below: Spyros Agathos, Jens Alex, Kokou Avowlanou, Eduardo Ayesa, Jean-Pierre Babary, Georges Bastin, Danielle Baetens, Willy Bauwens, Bruce Beck, Ahmed Benhammou, Olivier Bernard, Herwig Bogaert, Geert Boeije, Jean-Francois Béteau, Benjamin Bonnet, Cédric Bonvillain, Bouchaib Bouaziz, Sylvie Bourrel, Juan Canals, Bernard Capdeville, Bengt Carlsson, Ronald Carrette, Jacob Carstensen, Gilda Carvalho, Chantal Cenens, Filip Claeys, Filip Coen, John Copp, Diedert Debusscher, Bob De Clercq, Bart De heyder, Carl Demuynck, Dirk De Pauw, Martijn Devisscher, Jeremy Dudley, Maxime Estaben, Amaya "Cindirella" Franco, Chantal Fronteau, Jaime Garcia-Heras Antoine Genovesi, Patrick Gérain, Krist Gernaey, Sylvie Gillot, Jean-Luc Gouzé, Sandra Grégoire, Koen Grijspeerdt, Serge Guiot, Willi Gujer, Zakaria Hadj-Sadok, Jérôme Harmand, Poul Harremoës, Sef Heijnen, Chris Hellinga, Mogens Henze, Lisa Hopkins, Steven Isaacs, Ulf Jeppsson, Sabine Julien, Asma Karama, Karel Keesman, Bas Kops, Peter Krebs, Patrick Labat, Luis Larrea, Juan Lema, Paul Lessard, Carl-Fredrik Lindberg, Stefano Marsili-Libelli, Jurgen Meirlaen, Henri Naveau, Fatiha Nejjari, Bart Nicolai, Ingmar Nopens, Jacques Nyns, Gustaf Olsson, Etienne Paul, André Pauss, Marco Pengov, Davide Perez Alvarino, Michel Perrier, Britta Petersen, Martin Pléau, Monique Polit, Marie-Noëlle Pons, Ana Punal, Isabelle Queinnec, Wolfgang Rauch, Peter Reichert, Enrique Roca, Alberto Rozzi, Manfred Schuetze, Ivan Simeonov, Laszlo Somlyody, Henri Spanjers, Jan Spriet, Dirk Stevens, Jean-Philippe Steyer, Hans Stigter, Imre Takacs, Nadia Tali-Maamar, Olivier Thas, Vincent Van Breusegem, Marc Van Daele, Bart Vanderhaegen, Alexis Vanderhasselt, Marijke Van de Steene, Hans Vangheluwe, Henk Vanhooren, Jan Van Impe, Mark van Loosdrecht, Lieven Van Vooren, Frederik Verdonck, Lode

Vermeersch, Willy Verstraete, Karina Versyck, Stefan Weijers, Aris Witteborg, Zhiguo Yuan and all the ones we have forgotten to mention here.

Several people have carefully read early versions of (parts of) the book: in particular we would like to thank the sometimes tedious efforts of Jean-Philippe Steyer and Michel Perrier who read unachieved bits of text.

Another person has played an important role in the writing of the present monograph, in particular via numerous discussions on the topic and his sensible comments: Paul Lessard. The first time we met him was at the already mentioned COST action meeting in San Sebastian. By that time, Paul was on sabbatical leave at the INSA of Toulouse in France. Since then, both of us have developed intense, friendly relations with Paul. We have both been impressed by his scientific rigour and his comments have always been highly welcomed.

Denis Dochain and Peter Vanrolleghem
Montréal and Gent, 1 May 2001

To our parents,
to Annick, Chantal, Kevin, Simon, Arnaud and Nicolas.

1

Dynamical Modelling

1.1 Introduction

As for any industrial process, models in biological wastewater treatment have always played an important role in the process design and in the determination of optimal operating conditions. Modelling of biological wastewater treatment processes is certainly a very active research area today. The models, originally developed for characterising the carbon removal in Continuous Stirred Tank Reactors (CSTRs), are constantly reviewed and revised in order to reflect the improvements achieved in the knowledge about the mechanistic behaviour of the involved processes. However, the lack of understanding of the underlying biochemical processes has always been a major limitation for the optimisation and effluent quality control in biological wastewater treatment processes. Hydrodynamics also play an essential role in the dynamics of wastewater treatment processes. In simple situations like CSTRs, these are well known and characterised. Yet in many instances, the behaviour of wastewater treatment processes deviates from the ideal simple CSTR. For instance, the complete mixing assumption may fail to be fulfilled in practical operation: the stirring system may not be sufficiently efficient to have a homogeneous medium in the tank. Also, the large dimensions of the plants imply that the matter may have a long way to go between the plant input and output: then transport delays cannot be neglected anymore. The hydrodynamics become an even more complex matter in other reactors than the CSTR (fixed or fluidised bed reactors, airlift loop reactors, settlers,...). This advanced reactor technology is increasingly operated for wastewater treatment plants.

The above means that better understanding of the complex biological processes (like nitrification, denitrification or phosphorus removal in activated sludge processes or the role of hydrogen in anaerobic digestion) as well as of the hydrodynamics of the reactors in which the processes are operated, appear to be essential to improve the operation and control of wastewater treatment plants. The objective of this book, and of Chapter 2 in particular, is to introduce some modelling concepts and, when possible (for some classes of processes), systematic model formalisms which will be useful in the context of parameter identification and monitoring of biological wastewater treatment processes.

Mankind is using models in nearly every aspect of human life as the principal vehicle to describe reality. In wastewater treatment, they are currently used in manifold applications, e.g.

1. to increase the understanding of the underlying bioprocess mechanisms;
2. in the design of (full-scale) processes, and of control and operating strategies;
3. for the training of operators and process engineers.

In these examples the mathematical model can only be successfully applied if it is a proper description, in terms of both model structure and model parameters, of the underlying process. Theoretical (deductive) modelling is most often not enough. There is also a need for identification of the model from experimental data (induction).

The goal of model identification is to find and calibrate a model for the system under investigation that is adequate for its intended purpose.

Although mathematical models can be classified according to many different criteria, an interesting classification is to distinguish between *models for understanding* and *models for prediction/forecasting* [53].

Models applied for understanding aim to increase knowledge of system behaviour. The objective is to develop a simple, though universal model of the system under consideration that gives an adequate description of reality as it was observed [213]. The use of models for the purpose of understanding is most frequent in research and education. These models may not reach the goal their name implies: full understanding is often too ambitious. These models should therefore be considered no more than a (dense) description of hypotheses or conjectures that are acceptable explanations of process behaviour. Such models will hold as long as there are no significant deviations between model predictions and measurements as the model is put into jeopardy under different conditions [41]. In other words, a model for understanding can never be validated, it can only be disproved or confirmed [121]. The confidence in the model assumptions (and the mechanisms they encapsulate), however, increases as the model passes more and more severe tests. The bottom line is that accepting a hypothesis never means that the hypothesis is proved, but only that it was not possible to contradict the hypothesis with the aid of the available data.

The prediction of either future or hypothetical system behaviour is one of the basic tasks in practice. Models applied for prediction aim at providing an accurate and fast image of a real system's behaviour under different conditions than the ones prevailing during model building. The model used can either aim at forecasting future states of the system (simulation with new inputs) or at predicting system behaviour under hypothetical scenarios (simulation with new parameter values). The latter mentioned application is most frequent in design. It is important to stress that prediction, e.g. in control applications, is possible with calibrated so-called *black box* models without understanding the basic mechanisms of the system. This approach is sometimes even preferable as the development time of such models is substantially lower.

Before going any further, let us draw the attention of the reader to the following key points:

1. The models that we shall derive are *dynamical* models, i.e. models aimed at describing the time evolution of the processes. The *static* or *steady state* models will then be deduced as a particular case (when there is no time variation, or more precisely the time derivatives are equal to zero). This approach is essential for us, especially in the context of system analysis, and its use for process monitoring and control: static models are a particular case of dynamical models, and not the opposite. A typical example of this "traditional" attitude is the dynamical modelling of settlers (see Chapter 2): from the solid flux theory a basically static model has been derived without any a priori reference to dynamics. It is only in a second step that the results have been incorporated in dynamical models for settlers. Unfortunately, because of the above argument, this can possibly make the applicability of such models for representing the time evolution of the process questionable.

2. Because the use of physical laws ("first principles") is a priori the best reference for deriving reliable models, we shall start the modelling by introducing the notion of mass balances. As we shall see, a fundamental aspect of any reactor dynamical model is that its dynamics is composed basically of two terms, transport dynamics (including hydrodynamics and mass transfer), and conversion (i.e. the specific influence of the biological reactions).

Before going into the mathematical derivation of the dynamical models, let us first introduce a classification of mathematical models that will (can) be used for parameter and state estimation (Section 1.2), and consider the question of model building (Section 1.3). The basic ideas of optimal experiment design, i.e.

how to design experiments so as to obtain the best data possible to solve properly
the model building exercise, will be introduced in Section 1.4. Finally, because
people active within the field of water quality management have very different
backgrounds and jargon [51], a glossary with terminology typically used in this
discipline has been put in the Glossary.

1.2   Classification of Mathematical Models

The abstract representation of a real system by the ideas on its constituents and
functional relationships is called a **conceptual model**. Mathematical formulation
of these ideas leads to a mathematical model that can be used to give quantitative answers to questions on its behaviour under given external conditions. Such
mathematical models are referred to simply as "models" in this book. Because environmental systems are much too complicated to be described in detail, models
must be drastically simplified descriptions of reality. Since the aspect of a system
that is relevant depends on the question to be answered, a unique model for an environmental system does not exist, but different models must be used for different
purposes, and even in a given context several adequate descriptions are possible.

1.2.1   *Model Constituents*

Ultimately, a model is a 'machine' that transforms inputs (u) to outputs (y) by defined relations [52], [60], where u and y are, when discretised, sequences of either
scalars or vectors. (In continuous time formulation inputs u may be also a vector of
forcing functions from outside forces.) The features of the input-output relations
determine the basic structure type of the model, which is either an input/output
or a state-space description [52]. The outputs are these variables the model user
is interested in. The inputs of a model consist of disturbances and manipulated
variables that affect the outputs.

An **input-output model** is, strictly speaking, only the set of **transfer functions**
(g) that transform the inputs u directly to outputs y. An example of such a model
(in discrete time formulation) where the output y at time t_k depends on past and
present inputs $u(t_i)$, is:

$$y(t_k) = \frac{B(q)}{A(q)} u(t_k) \tag{1.1}$$

where $A(q)$ and $B(q)$ are polynomials in the backward shift operator q, i.e.

$$q^{-j}(y(t_i)) = y(t_{i-j}) \tag{1.2}$$

$$A(q) = 1 + a_1 q^{-1} + a_2 q^{-2} + \ldots + a_n q^{-n} \tag{1.3}$$

$$B(q) = b_0 + b_1 q^{-1} + b_2 q^{-2} + \ldots + b_m q^{-m} \tag{1.4}$$

The order of the polynomials n and m and the model parameters a_i and b_i are to
be estimated from a set of input-output data to make the model complete.

Alternative black box models are artificial neural network models that have been shown to be capable to describe any nonlinear input-output mapping. This feature has attracted a lot of attention and intense research is going on for their use in biotechnological applications. More details are given in te Braake *et al.* [246].

The most important feature of a **state-space model** is the introduction of **state variables** (vector x) which act as mediators between the inputs and the outputs. These state variables are additional model constituents and the system description is consequently addressed as internal. It is characterised by the fact that x is obtained from present and past x and u by means of the **state-transition equation** and y is generated from x by means of the **observation equation**.

The **state** of the system (as described by the model) is defined as the values of the state variables at any instant of time. Note that by definition it is neither required that state variables are measurable nor that they are meaningful in terms of natural science (although they frequently are the latter). It should also be stressed that the equations of a model can be formulated as either algebraic or differential equations.

In many cases, the dynamics of the state variables x considered important for the adequate description of the process can be described by the following state-space model:

$$\frac{dx}{dt} = Ax + Bu, \quad x(t = 0) = x_0 \tag{1.5}$$

and the output observations y are given by

$$y = Cx \tag{1.6}$$

In this model A, B and C are matrices containing the characteristic parameters of the system.

Nonlinearities of the bioprocesses, however, often ask for a different representation than the linear one given above. A more general model is:

$$\frac{dx}{dt} = f(x, u, t, \theta), \quad x(t = 0) = x_0 \tag{1.7}$$

$$y = h(x, u, t, \theta) \tag{1.8}$$

One can observe the nonlinear relations f and h between the state variables, inputs and outputs and the model parameters θ.

Whereas the dynamics of stirred tank reactors (STRs) are characterised by ordinary differential equations (ODEs)(like (1.7)), the dynamics of non completely mixed reactors (fixed bed reactors, settlers...) are generally speaking characterised by a presence of spatial gradients, i.e. a dependence of the key variables (typically the process component concentrations) not only on time but also on the spatial position in the tank: their dynamics are described by partial differential equations

(PDEs)(see Section 2.5). The systems described by ODEs are called "lumped parameter systems", while those described by PDEs are called "distributed parameter systems". We shall discuss PDEs in detail later, but let us anticipate with a simple illustrative example: the mass balance equation in a plug flow reactor of the concentration of a reactant x involved in one reaction with first order kinetics. The dynamics are then described by the following PDE:

$$\frac{\partial x}{\partial t} = -v \frac{\partial x}{\partial z} - k_0 x \tag{1.9}$$

$$x(t = 0, z) = x_0(z) \tag{1.10}$$

$$x(t, z = 0) = x_{in}(t) \tag{1.11}$$

where z is the space variable ($0 \leq z \leq$ L, with L the reactor length), v is the fluid superficial velocity, k_0 is the kinetic constant, and x_{in} is the inlet reactant concentration. Note that:

- x does not only depend on time t but also on space z (otherwise the partial derivatives would be meaningless);
- the model needs boundary conditions to be complete (here only one is necessary since the derivative with respect to z is of order 1).

The picture of mathematical models is only complete if one also considers *grey box* models. These are mainly considered as models in which part of the model structure is based on a priori knowledge of the process while another part consists of black box descriptions such as empirical rules. Such hybrid models combine the advantages of both approaches: identifiability and extrapolative power [50] [248].

Fuzzy models may also be considered grey box models though in a different sense. This is because these models incorporate insights in the internal working of the processes under study in a qualitative way. Therefore these models are particularly useful for description of systems where the process mechanisms are not (yet) completely understood. The mathematics of fuzzy set theory and some applications are introduced in te Braake *et al.* [246].

Irrespective of the **model structure**, the mathematical equations that relate inputs to outputs contain three types of constituents, that of **variables, constants** and **parameters**. Inputs, outputs and states are seen as variables in the equations. The difference between constants and parameters is less evident and gave rise to some confusion in modelling terminology. In the following we denote all model constituents that never change their value throughout all possible applications of the model as constants (e.g π, e, g,...).

A parameter, on the other hand, is a model constituent whose value may vary with the circumstance of its application. Hence, the value needs to be determined for each particular application of the model.

In some cases this strict definition is not followed for ease-of-use and communication reasons. Indeed, the value of some of those parameters may need to be modified during a specific application, e.g. adaptive control. Also, some 'time-varying' models exist in which a parameter is introduced in the model which is a priori known to be time-varying due to a dependency, e.g. on temperature. However, for the sake of clarity, this dependency is not explicitly included in the equations, but it is hidden elsewhere, leading to an apparent time-varying parameter. For some applications a parameter can be replaced by a function describing its time and space dependency but in fact that should be seen as a model extension. Henze *et al.* [120], [121] incorrectly use the term constant for parameter in the widely used IWA Activated Sludge Model No. 1 and 2 as they are application dependent.

1.2.2 *Model Attributes*

The model constituents describe the fundamental elements of a model, while the characteristics of a model can be addressed by a number of descriptive modelling terms. These terms are called model attributes. Some attributes have a clear and stringent definition (e.g. linear versus nonlinear), while others have a weak or relative definition (e.g. phenomenological versus mechanistic). In the following the model attributes are referred to as being strong or weak depending on the stringency of their definition. It is obvious that the stronger model attributes are applied, the better a model is characterised.

Strong model attributes

The model attributes **linear** and **nonlinear** relate to the structure of the model equations. The model may be linear with respect to the variables or to the parameters. Thus, a model can be nonlinear in the parameters but linear in the variables and vice versa. Linearity is a basic characteristic of a model that has quite some impact on the properties of solution, e.g. linear models are frequently used, because the analytical solution can be found. For nonlinear models numerical solutions are predominant.

The evaluation of the linearity of a model is conveniently performed by differentiating the function with respect to the variable or parameter of interest and evaluating whether the derivative is still function of the same variable or parameter. If this is the case, the model is nonlinear in the variable or parameter.

In wastewater treatment processes models are often characterised as **dynamic** in the sense that the variables evolve over time. A model which is not **dynamic** is called **static** or **steady state**. Thus, dynamic relates to a time dependency in the model which can be formulated as dynamic input variables and/or state variables. The output of a dynamic model is often called time series (e.g. [159]).

If the model parameters are constant in time the model is characterised as being **time-invariant**. Strictly speaking any model therefore is time-invariant. However, as discussed above (section 1.2.1), sometimes parameters are conveniently taken as time varying to increase clarity in the model description, e.g. the temperature dependency of the growth rates is well-known and induces a time-varying growth rate but this is not reflected in the specification of the growth rate equation. A model can have a space-dependency (e.g. a clarifier model). Such models are referred to as **distributed parameter** models. Currently, in wastewater treatment modelling only time- and space-derivatives are concerned, and dynamic distributed models are normally formulated as partial differential equations. However, in other biotechnological applications (and expected to become applied in wastewater treatment modelling in the near future), another type of partial differential equation formulations is used. In these models time is one independent variable, but the age of the biomass population is also considered and seen as an independent variable [127]. This leads to another type of distributed parameter model as the state variables have a time- and age-dependency. In pure culture fermentation biotechnology, these models have been termed **segregated models** because biomass is segregated into different age classes. Similarly particle sizing may be used to segregate the biomass and lead to PDEs.

Models can be termed discrete or continuous, and in most cases these attributes relate to the model formulation of difference/differential equations with respect to time, i.e. the correct terminology should be **discrete-time** or **continuous-time** model. However, **discrete-space** or **continuous-space** are two other model attributes describing a space relationship for up to three dimensions in the model formulation. A continuous-time differential equation is either solved analytically or discretised into a discrete-time difference equation (according to an Euler, Runge-Kutta, etc. scheme) which is solved numerically. It should be stressed that most computer programs apply a discretisation to the continuous-time and continuous-space differential equations, and the discretised equations are solved as algebraic equations.

If a model contains elements of randomness, it is called **stochastic** otherwise **deterministic**. The uncertainty encapsulated in any model is due to a combination ([189], [21], [23]) of

1. uncertainty in input variables,

2. uncertainty in parameter values, and

3. uncertainty in model structure.

Measurement uncertainty is embedded in these three different forms of uncertainty. In case all uncertainty aspects are neglected, the model is deterministic and the output is determined uniquely by input and initial conditions. The output of a stochastic model can be described as a probability density function. The terms stochastic and statistical are occasionally confused, and the use of statistical as model attribute should be avoided, since the term statistical is referring to methods of analysing data. Likewise, the term deterministic is occasionally confused with mechanistic, physical or white-box implying that deterministic models are always based on physical, chemical and biological laws. This is not true, however, because a black-box model may be deterministic as well, e.g. a neural network or a spline.

Stochastic input to a model is denoted as innovations, realisations, disturbances or perturbations.

The strong model attributes are highly recommended for characterising models as the meaning of the terms is well defined. However, the use of additional adjectives such as purely, totally or completely (e.g. 'purely deterministic nor purely stochastic model' in Harremoës and Cartensen [112] has no meaning when the stringent definitions of the strong model attributes are referred to – in fact, these combinations should be avoided as they are confusing. The majority of models for wastewater treatment so far are formulated as nonlinear dynamic continuous-time deterministic models [219], [137].

Weak model attributes

These terms have less clarity in their interpretation and may in the lack of strong model attributes potentially lead to confusion. However, provided that the terms are used correctly these attributes also signify the background of the model. To a large extent many of the weak model attributes describe almost the same model property, i.e. the degree of conceptualism, basis in physical, chemical and biological laws, simplification level, etc.

Words like **mechanistic**, **physical** and **white-box** are used to describe that the model's structure is based on physical, chemical and biological laws. A priori knowledge is the only information source during the creation of such a model, i.e. a deductive modelling strategy is adopted. The attribute **transparent** has the same interpretation as **white-box**, which means that every detail in the model has a mechanistic explanation. The extreme is **reductionist** models (should not be confused with reduced order models, see below) that are based on the attempt to include as many details as possible. The term causal is also used with the same meaning as mechanistic, but in some scientific fields a causal model is strictly defined as a model which only depends on past observations and inputs. Thus, the use of causal as model attribute should be avoided. **Phenomenological, black-box**, **empirical** and **heuristic** (by rules of thumb) are used as model attributes to describe that the model is based on empiricism rather than laws. It is data-driven, i.e. inductive modelling is adopted. A black-box model has not necessarily a structure compatible with the underlying physical, chemical or biological reality [252]. The essential feature of black-box models is that they assume no knowledge of physical or internal relationships between the system inputs and outputs other than that the inputs should produce observable responses in the output. Hence, the system is considered 'black box' and no use is made of the available a priori knowledge (Casti [52] denotes it as an external system description), i.e. it omits the consideration of the mechanism by which inputs are related to outputs.

A combination of the mechanistic and phenomenological approach is frequently called **grey-box** modelling. Holst *et al.* [128] refer to grey-box models as reflecting a priori knowledge as well as black-box parts, while in Carstensen *et al.* [49], grey-box models are given two virtues – the properties of parameter interpretability and parameter identifiability. Ljung [159] refers to this approach as **semi-physical** modelling.

Another issue for characterising models is the degree of simplicity or complexity in the model. A **simple** model is characterised by few equations and parameters while a **complex** model has many equations and parameters, but it remains unclear when a given model should be termed simple or complex. As a rule of thumb, mechanistic and phenomenological models are normally formulated with a high and low degree of complexity, respectively. However, an artificial neural network is considered to be a black-box model but may at the same time have a high level of complexity [122]. The terms simple or complex for characterising a model indicates that the model is derived from a basis model or compared to another model. The confusion occurs when this reference model is not given or just assumed to be well-known. The level of model complexity/simplicity can also be addressed with the attributes **aggregated/segregated**. Jeppsson and Olsson [134] use a **reduced order** model to describe the model's derivation as being **lumped** or aggregated. A model is lumped or aggregated compared to an original or base model when model variables or equations are united in a simplified description. At this stage it is good to note that lumping of variables inevitably results in modelling errors. These errors end up in the simplified model, typically in the parameters thereof. This aspect will be dealt with to some detail in Chapter 2.

Even though it is obvious to some people that a model is complex based on the state-of-the-art in modelling today, this fact is very likely unclear to researchers unfamiliar with the specific topic or researchers within the field 10 years from now.

1.3   The Model Building Exercise

Using the 'story' of a model building exercise, a number of terms involved in this activity will be introduced within their appropriate context. At the same time a short review is given on the current state-of-the-art in modelling methodology. The diagram in Figure 1.1 summarises the aspects of model building which are described in detail below. Once the steps in the figure have been fulfilled successfully, the model can be applied for its intended purpose. These applications typically involve **simulation** that may be regarded as virtual experimentation with the virtual reality of the model.

Problem formulation

An often forgotten task in a model building exercise is the clear formulation of the goal of the model that is to be constructed. While in most cases this task is rather intuitively performed by the modeller in case he is also the problem supplier, problem formulation or goal incorporation is much more difficult in case these people are not the same. In this case, an important effort must be spent to answer such questions as desired accuracy of the results, degree of uncertainty in the provided

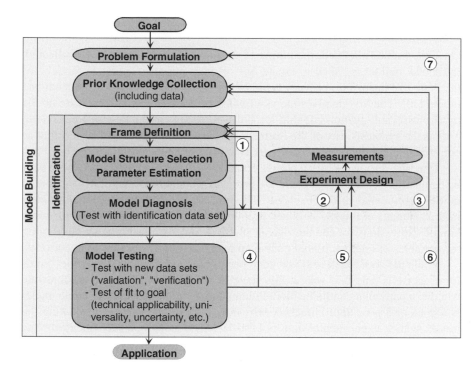

FIG. 1.1. The model building exercise [51].

answers, time scale of the solution, system boundaries, important variables, environmental conditions under which questions will be asked for which the model must give an answer, etc.

Prior knowledge collection

The next task is to collect the available, relevant a priori knowledge from literature and experts or from a model building environment that supports re-use of model-encapsulated knowledge [256]. At this (early) stage of the exercise, some experiments may be conducted or some data collected during previous experiments may be retrieved and stored in the experimental database.

Frame definition

As soon as these two tasks have been performed once, a first iteration of the model building can start. The **frame definition** phase aims to delineate the conditions under which a model will be used (e.g. temperature), to choose the class of models that seems fit for the task (time series, state-space, distributed parameter, stochastic...), to specify the variables that seem important to find a solution to the formulated problem (inputs, outputs, states), the range of time constants that need to be covered by the model, etc.

Candidate model set creation

Once the frame is defined, the purposes set and the a priori knowledge collected, the model builder is asked to create one or more possible candidate models for the system at hand. Two primary types of reasoning are used to create a candidate model [20]. The first involves the assembling of all prior hypotheses made on the mechanisms and phenomena that are believed to govern the system's behaviour. This is the synthesis part of the exercise. The analysis part consists of refuting or confirming these hypotheses on the basis of a set of field data.

Typically during the initial phases of model building, linguistic models are built using qualitative reasoning. There are however no methods or advices available to guide the modeller in this exercise. The translation of these conceptual models into mathematical models is better supported. Textbooks provide plenty of basic kinetic expressions for the biological, physical and chemical interactions among the variables of interest. It is of course up to the analyst to extract the proper ones.

While the above task is the creative one of the model builder, the model building exercise is only half way at this stage. It is equally important to demonstrate whether a particular candidate model (among many proposals that are being made) is a good or a poor approximation of reality. This second stage of evaluating the model against experimental data is best described as system **identification** (the term used in control theory). Simply stated, system identification may be considered as model **calibration**. However, as stated in Beck [20],

the word calibration suggests an instrument already well-designed and in need only of small, fine-tuning adjustments. Such a view weakens the role of experimental observation, relegating it to a minor test on what is the more or less inevitable path to applying the model with which the model-building procedure was first begun. It encourages the attitude that there is little to be learned from the field data that was not known before. Our preferred view of system identification is quite the contrary: it is an integral part of the process of developing scientific theories about the behaviour of a system; it, too, is about knowledge acquisition and hypothesis generation.

This statement brings us back to the fact that modelling is an iterative cycle in which experiments play the role of indicating areas of model deficiency that are subsequently tackled in a new hypothesis generating step. The statement clearly originated from an environmental systems analyst. Indeed, for most physical and chemical applications, the a priori knowledge is of such high quality that the system framework and most of the model structure can be deduced from it. The modelling methodology developed for these systems is adequate to estimate the parameters and solve the minor uncertainties in the model structure by using final validation experiments and eventually iterating a small number of times through the procedure.

In contrast with this, the inherent characteristics of bioprocesses, i.e. their non-linearity and nonstationarity, coupled with the lack of adequate measuring techniques, is such that this mathematical modelling methodology cannot be applied without modification [267]: more emphasis must be given to inductive reasoning to infer a larger part of the model structure from the scarce (or harder to obtain) experimental data. Consequently, structure characterisation (or model selection) methods become a more important tool, because the chance of obtaining an invalid model is much larger and, hence, the number of modelling iterations may increase substantially.

Model structure selection

The goal of **classical model structure selection** (**model structure characterisation**) is to select a unique model structure according to the principles of a quality of fit and parcimony [238], [113]. However, it is also possible to select a set of models that are attributed different weights reflecting their probability of appropriateness [79], [214]. Most model structure selection criteria require the parameter values to be estimated, e.g information criteria such as AIC [6] and BIC [228]. However, structural selection criteria that only require basic data analysis also exist for particular applications [238], [261]. A detailed discussion of this model building task will be given in Chapter 3.

Parameter estimation

Parameter estimation is typically based on the maximisation or minimisation of a goodness-of-fit criterion such as Least Squares, Weighted Least Squares, Maximum Likelihood, etc. and aims to provide values for the parameters in the model and, in case a state space representation is adopted, also values for the initial (and boundary) conditions of the state variables. Although several powerful estimation (nonlinear **regression**) algorithms are available (both for **recursive** and **batch estimation** depending on the number of data points used for a parameter update, for details, see Chapters 6 and 7), their success is highly dependent on the experimental dataset available. The identifiability analysis performed prior to the parameter estimation can provide answers to the key question whether, given a set of measured variables, unique parameter values can be obtained. Two types of answers can be given depending on the applied method. In case **structural** (also termed **theoretical** or **a priori**) **identifiability** [100] [185] is evaluated the answer is either yes or no, respectively meaning that the parameters can be given unique values or not at all [74]. However, it is not ensured that the data always contain sufficient information to provide reliable and unique estimates. Methods for the **practical** (also termed as **numerical** or **posteriori**) **identifiability** study allow to evaluate the information content of the dataset intended for parameter estimation [258]. The basis of these methods is also underlying methods that can provide a solution to a practical identifiability problem, i.e. **optimal experiment design.** This design procedure uses the model for which unique parameters are to be found to calculate experimental conditions such that sufficient information is contained in the data. One should note that a structural identifiability problem encountered cannot be solved without altering the candidate model or frame definition (e.g. include other variables in the system description). **Model reduction** can lead to models that become less 'data-hungry' and hence their identifiability properties may improve [134].

Model diagnosis

Once the parameters are estimated it remains to investigate whether the identified model violates the assumptions made in the frame definition. For instance, statistical tests of systematic deviations between model results and measurements (residuals) and their distribution are frequently used [38], [232]. Also, evaluation whether non-sense parameter values (e.g. negative affinity constants) or initial or boundary values are obtained can allow diagnosis of potential violation of the experimental frame.

Model testing

Fitness of a model can be evaluated by comparing its performance with data obtained under different conditions than the ones prevailing at the time of the data collection performed for model identification. This process of putting the model in jeopardy [41] or, in other words, straining the model to its limits, may reveal model inadequacies that may be sufficient to conclude that the model is no longer 'fit' for the purpose it was intended for. Hence, the whole model building process may have to start all over. Sometimes this may even lead to a reformulation of the problem as the modelling exercise may have provided considerable insight in the system under study and its behaviour. This process of putting the model into jeopardy by confronting it with new data is most often called **model validation**, but serious arguments are put forward against this term. As a model only describes part of the reality (the one defined in the frame) in a simplified manner, it is obvious that a model never can describe reality completely. Therefore, there will always exist experimental conditions for which the model is not valid. Hence, validation of a model is utopian! A completely other approach is to term this process of jeopardising the model a model **falsification** step [53] [212], which if answered negatively, provides more confidence in the selected model. However, the term falsification appears too negative and one has therefore looked for other terminology that is less pronounced (quantitative) as validation but still gives a qualitative insight in the level of confidence one has in a selected identified model. The terms put forth for this are **corroboration** and **confirmation** [200]. Finally, the term **verification** is frequently interchanged with validation, but in the quite related field of simulation methodology, verification has a different, but specific meaning. In this field, verification is the evaluation whether the coding of a model has been performed correctly, i.e. whether the implemented model equals the created model [156]. From this discussion it is advocated to use the term validation as it is the most widespread used and has no dubious meaning.

1.4 Optimal Experiment Design: Basic Ideas

This book devotes a lot of attention to the main tasks of a model building exercise, model selection and parameter estimation. However, these tasks can only be performed successfully if the sources of information, i.e. goal, a priori knowledge and experimental data, are of sufficiently high quality. Probably the definition of the objective is not the most difficult task albeit sometimes overlooked and not stated in a quantitative way. The analysis of the a priori knowledge leading to the construction of the set of candidate models may take some more time as knowledge must be acquired from the literature, experts and so forth. The main efforts however are, certainly for bioprocesses, devoted to the collection of data. The design of the experiments that will provide the data is an increasingly important task for the model builder for the following reasons:

- the limited resources available, e.g. experimentation time and expenses;
- the increased degrees of freedom that have to be exploited in an optimal way;
- the trade-off that has to be made between the information content of the data and the pursued goal;
- the requirement of good information content for the data for obtaining reliable models.

Goals pursued during an experiment design procedure can be categorised in the following areas:

- experiment design for a reliable selection of an adequate mathematical model structure of the process;
- the design of experiments for precise estimation of model parameters;
- the dual problem of structure characterisation and parameter estimation.

These will be considered separately in the rest of the book.

Each of these goals is associated with one or another quantitative function which will determine the focus of the experimental efforts. Some authors also suggest to separately consider [159]:

- the optimal sampling frequency;
- the best location of sensors and sampling points in a reactor system.

However, these items should be regarded as a part of the degrees of freedom available and the constraints to which one has to adhere during the experimental design process. An experiment design typically consists of three steps:

1. definition of an objective function that is the mathematical translation of the pursued goal of data usage;
2. enumeration of the available degrees of freedom and experimental constraints;
3. extremisation of the objective function by appropriate variations of the degrees of freedom without violation of the applicable constraints.

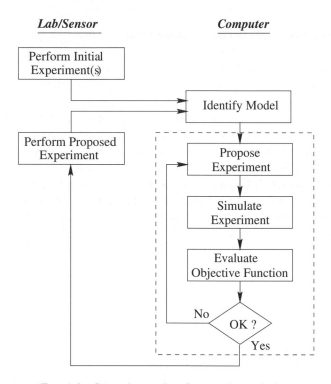

FIG. 1.2. General procedure for experiment design.

Figure 1.2 schematises the procedure used for (automated) optimisation of the experimental conditions with relation to (w.r.t.) the defined objective function. An essential feature of the procedure is that a preliminary model is available that is identified on the basis of previously acquired data. Essentially, the algorithm performing the design performs "simulated experiments" to quantify the potential effect of proposed experimental conditions on the objective function. Using a (non-linear) optimisation algorithm that is able to take into account the experimental constraints, an "optimal" experiment is proposed that is subsequently applied in real-life. Real-life may be a specific laboratory experiment or a full-scale process on which some degrees of freedom are available that can be exploited to gather informative data.

Since the experiment design procedure depends to a large extent on a model of the process under study, it is evident that a prerequisite to good design is the availability of a good process model.

Questions that may be of guidance when initiating an experiment design study are the following ones.

What to measure? This question stresses the importance of a good system definition (input, output and state variables...).

Where to measure? This addresses the problem of sensor location.

When to measure? Here the sampling strategy is paid attention to.

Which manipulations? In many cases this is the most important question and requires creative work related to the design of excitation signals that will act on the process in such a way that highly qualitative information can be gathered. Examples are sine-waves, pulses, PRBS (pseudo random binary sequences [163]), etc.

What data treatment? Though most of the time not considered part of the experiment design, pretreatment of the data is often applied before model identification starts. It is therefore part of the collection of data for model identification purposes. Typical data pretreatment includes noise rejection, outlier detection/removal and elimination of "uninteresting" dynamics.

1.5 Conclusion

The objective of this chapter was to introduce the basic concepts of dynamical modelling. An important issue is the notion of dynamics, necessary to present the time evolution of processes. In Section 1.2 we have introduced a classification of mathematical models in terms of model constitutents (variables, constants and parameters) and of model attributes (in particular, linear vs nonlinear, dynamic vs static, lumped parameter vs distributed parameter, discrete-time vs continuous-time, deterministic vs stochastic, white-box, grey-box, or black-box). Subsequently, we have introduced the different elements of the model building exercise, from problem formulation to model testing. Finally the basic ideas of optimal experiment design, i.e. how to design experiments so as to obtain the best data possible to solve the model building exercise properly, have been introduced in Section 1.4.

2

Dynamical Mass Balance Model Building and Analysis

2.1 Introduction

The preceding chapter has introduced the basic concepts of dynamical modelling. In this chapter we shall go deeper into the subject of dynamical modelling. Because the use of physical laws ("first principles") is a priori the best reference for deriving reliable models, we shall concentrate in this chapter on the notion of mass balances and related derivation and dynamical analysis of dynamical models. It is also worth noting that the mass balance models will play a key role in the rest of the book.

Section 2.2 will introduce the basic concepts of mass balances ("first principle"), and apply them to some simple bioprocess dynamics situations. The end of Section 2.2 will be dedicated to kinetics with a particular emphasis on mathematical models of the specific growth rate. In Section 2.3, we shall introduce several process examples, and their dynamical mass balance models in stirred tank reactors (STRs). In Section 2.4, we shall extend and generalise the mass balance dynamical model developed in Section 2.2 and applied in Section 2.3 by introducing a General Dynamical Model framework for completely mixed reactors (STRs). This will be further extended to non completely mixed reactors (plug flow reactors, fixed bed reactors, settlers) in Section 2.5. In Section 2.6, we will discuss the respective notions of linear and nonlinear systems a little further. In Section 2.7, we shall concentrate on the notion of equilibrium points, on the linearisation of nonlinear models around equilibrium points and on their stability properties. The last sections of the chapter will concentrate on the properties and the analysis of a model: a state transformation (Section 2.8), a model reduction method based on singular perturbation theory (Section 2.9), and the use of Laplace transformation in models described by PDEs (i.e. non completely mixed reactors).

2.2 The Notion of Mass Balances

2.2.1 Stirred Tank Reactors

Basically the reactor in which the wastewater treatment process is operated is a tank in which one or several biological reactions occur simultaneously. Let us first consider a stirred tank reactor, as schematised in Figure 2.1. The tank is characterised by a liquid medium volume V, and inflow and outflow rate, F_{in} and F_{out}, respectively. Let us consider that one (bio)chemical reaction is taking place in the tank. The mass balance of a component involved in the reaction, and characterised by a concentration C in the reactor can be formalised as follows:

$$\begin{pmatrix} Time\ variation \\ of\ the\ mass\ of \\ the\ component \\ in\ the\ tank \end{pmatrix} = \begin{pmatrix} Mass\ of\ the \\ component \\ that\ goes\ in \\ the\ tank \end{pmatrix} - \begin{pmatrix} Mass\ of\ the \\ component \\ that\ goes\ out \\ of\ the\ tank \end{pmatrix}$$

$$\pm \begin{pmatrix} Mass\ of\ the \\ component \\ produced\ or \\ consumed\ by \\ the\ reaction \end{pmatrix} \qquad (2.1)$$

The assumption of perfect mixing in the tank implies that the concentration C is homogeneous in the tank, and in particular its value at the reactor output is equal to the one anywhere in the tank. We now have all the elements to write the mass balance equation for the component:

$$\frac{d(VC)}{dt} = F_{in}C_{in} - F_{out}C \pm V\phi \qquad (2.2)$$

where C_{in} is the component concentration in the influent, and ϕ the rate (per unit volume!) at which the component is produced ($+\phi$) or consumed ($-\phi$), i.e. the conversion rate of C. As we shall see later, the conversion rate ϕ is the product of a (positive) yield coefficient (ideally, a stoichiometric coefficient) with the reaction rate ρ, which generally speaking is a positive function of different process variables, e.g. the reactant concentrations, the product and biomass concentrations, or

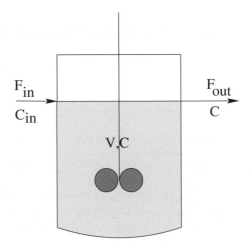

FIG. 2.1. Schematic view of a stirred tank reactor.

the physico-chemical conditions like the temperature or the pH. Let us rewrite the mass balance equation (2.2) by expanding the left hand side:

$$V\frac{dC}{dt} + C\frac{dV}{dt} = F_{in}C_{in} - F_{out}C \pm V\phi \qquad (2.3)$$

and by dividing both sides by V and moving the second term of the left hand side to the right hand side, we obtain the differential equation describing the concentration dynamics:

$$\frac{dC}{dt} = \frac{F_{in}}{V}C_{in} - \frac{F_{out}}{V}C - \frac{C}{V}\frac{dV}{dt} \pm \phi \qquad (2.4)$$

An important comment at this point is that the dynamics in a reactor is composed of two types of terms:

- transport dynamics ($\frac{F_{in}}{V}C_{in} - \frac{F_{out}}{V}C + \frac{C}{V}\frac{dV}{dt}$);
- conversion ($\pm\phi$).

By defining the dilution rate D:

$$D = \frac{F_{in}}{V} \qquad (2.5)$$

a useful alternative formulation of equation (2.4) can be written as follows:

$$\frac{dC}{dt} = DC_{in} - DC \pm \phi \qquad (2.6)$$

which covers the following important type of operations of STRs: batch, fedbatch (which may be encountered e.g. in Sequential Batch Reactors (SBRs)), and continuous (without and with volume variation). These are shortly described below.

Batch reactors.   In batch reactors, there is no inflow nor outflow:

$$F_{in} = F_{out} = 0 \qquad (2.7)$$

and equation (2.4) reduces to:

$$\frac{dC}{dt} = \pm\phi \qquad (2.8)$$

which can also be obtained from (2.6) by setting D to zero ($D = 0$).

Fedbatch reactors.   In fedbatch reactors, the tank is initially filled with some amount of reactants and catalysts, and is progressively filled with reactants (this is what happens in SBRs when the filling occurs, or when additional substrate is introduced during a denitrification phase): this means that there is an inflow but no outflow, and the time variation of the liquid volume V is equal to the inflow rate F_{in} (total mass balance):

$$F_{out} = 0, \ \frac{dV}{dt} = F_{in} \ (= DV) \qquad (2.9)$$

If we add the above volume equation in (2.4), the mass balance equation (2.6) fits also to fedbatch operation.

Continuous reactors.   In continuous tanks, the reactor is "continuously" fed with the reactants. Since the tank is filled, the inflow and outflow rates are equal, and the volume is constant:

$$F_{in} = F_{out}, \ \frac{dV}{dt} = 0 \qquad (2.10)$$

and once again the mass balance in continuous stirred tanks fits into equation (2.6).

Continuous reactors with volume variation.   The above argument corresponds to the classical continuous stirred tank (CSTR). But it may happen in some process configurations (like in Figure 2.2 which shows three types of overflow weir designs (see [188])) that the reactor is basically a continuous reactor (with inflow and outflow), but for which during transients (e.g. variations of the influent flow rate F_{in}), the volume is varying, and the outflow rate F_{out} is different from the inflow rate F_{in}. Then the above equations (2.10) have to be replaced by the following ones:

$$\frac{dV}{dt} = F_{in} - F_{out} \qquad (2.11)$$

More precisely, we can have the following relationship between the (out)flow rate and the height h:

$$F_{out} = c + Nah^b \qquad (2.12)$$

where N is the number of weirs, a and c are functions of weir type or width (c is only different from zero for the Sutro weir), and b is a function of the weir design

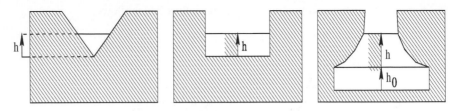

FIG. 2.2. V-notch, rectangular and Sutro weirs.

($b = 2.5$, 1.5 and 1 for the V-notch, the rectangular and the Sutro weir, respectively, (see [231], [187]). Besides, note that:

$$V = Az = A(z_0 + h) \qquad (2.13)$$

where A is the reactor section, z is the total depth, and z_0 the (constant) distance between the bottom of the tank and the lower edge of the weir. Then if we consider a tank with a constant section A, the mass balance equation (2.11) becomes:

$$\frac{dh}{dt} = -\frac{Na}{A}h^b - \frac{c}{A} + \frac{F_{in}}{A} \qquad (2.14)$$

As was pointed out in [188](and as suggested by the above equation (2.14)), the weir designs may have an important influence on the propagation of hydraulics variations and disturbances in the plant.

The models mentioned above and their parameters a, b, c can be estimated from relatively simple experiments in which the pump flow rates are changed in a step-wise manner. In Figure 2.3, the simulation results are given of a variable volume model that is fitted to experimental data that were collected by measuring the effluent flow rate of an activated sludge system subjected to a sequence of step changes to influent and recycle flow rate [64].

2.2.2   A Simple Biological Reactor Model

Let us now consider an example more directly linked to biological reactors. Assume that the following microbial growth reaction is taking place in the reactor:

$$S \longrightarrow X \qquad (2.15)$$

where S is the limiting substrate and X is the biomass. Note that the biomass X plays the role of an autocatalyst (it is both a product and catalyst) in the above reaction (sometimes a feedback arrow is added on top of the reaction arrow to emphasise this, see [14]). It is usually assumed (but this is not always correct, as mentioned by Henze [118], important amounts of biomass can be present in the influent) that only substrate is present in the reactor incoming flow: let us

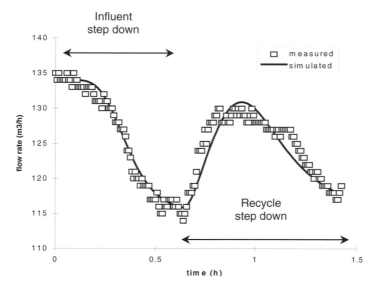

FIG. 2.3. Hydraulic propagation experiment [64].

denote by S_{in} the influent substrate concentration. In line with a commonly accepted assumption since Monod introduced it in 1942, we consider that the reaction rate ρ of the above microbial growth is the product of the autocatalyst X and a proportionality coefficient called specific (i.e. per unit biomass) growth rate μ ($\rho = \mu X$).

Then by considering (2.1), the mass balance equations for S and X are written as follows:

$$\frac{dS}{dt} = DS_{in} - DS - \frac{1}{Y}\mu X \qquad (2.16)$$

$$\frac{dX}{dt} = -DX + \mu X \qquad (2.17)$$

Let us now take a closer look at this model and highlight some important features.

The conversion rate of the biomass, ϕ_X, is simply the reaction rate (i.e. the growth rate)($\phi_X = \rho = \mu X$). Indeed this corresponds to a standardisation of the yield coefficients with respect to one component (here, the biomass). This standardisation is obviously a priori arbitrary (it could as well have been performed with respect to the substrate, in the above example), although the use in biochemical engineering consists of standardising the yield coefficients of growth reactions with respect to the related biomass.

It is also usual in biochemical engineering to define yield coefficients in growth reactions as ratios of biomass production over substrate consumption (as Y above), or of product synthesis over biomass production (see section 2.2.4 below): this implies that in the conversion terms of the (dynamical) mass balance equations, the yield coefficient divides the reaction rate when the component is a substrate of the reaction, and it multiplies the reaction rate when it is a reaction product. Hence, in equation (2.16), the conversion term ϕ_S is the product of the yield coefficient $\frac{1}{Y}$ and of the reaction rate μX ($\phi_S = \frac{1}{Y}\rho = \frac{1}{Y}\mu X$).

Strictly speaking, if we refer to the reaction network formalism considered in (2.15), the yield coefficient Y should be a stoichiometric coefficient, i.e. here Y is the quantity of biomass X (in appropriate dimensions, e.g. in grammes) which is produced when one (mass) unit of substrate S is consumed *by the reaction (2.15)*. However, this would only be true if the conversion process is completely defined by the reaction (2.15), i.e. if

1. this is the only reaction (i.e. in particular there are no side reactions);
2. the only components are S and X;
3. and if the elemental content (in carbon, nitrogen, oxygen, hydrogen) of S and X are exactly known.

Most often, at least one of the above conditions are not fulfilled in biochemical processes. The underlying biological process is most often much more complex than the process (2.15), and it is also quite difficult in most instances to precisely characterise the elemental content of the biomass. One important consequence is that, instead of being constant (like they should if they were stoichiometric coefficients), the yield coefficients in biochemical processes may be in practice changing with time (within some bounds) e.g. when the wastewater composition or operating conditions are (significantly) changing, e.g. because side reactions which were previously neglected are no longer negligible.

2.2.3 Biomass Death and Substrate Maintenance

Biomass death and/or substrate maintenance terms are sometimes added in the simple microbial growth models. These reactions can be formalised as follows (see also [14]):

$$\text{biomass death:} \qquad X \longrightarrow X_d \qquad\qquad (2.18)$$
$$\text{maintenance:} \qquad S + X \longrightarrow X \qquad\qquad (2.19)$$

Note that in the above reactions, X no longer plays the role of an autocatalyst: it is a simple reactant (transformed in dead biomass X_d) in reaction (2.18), and a catalyst in reaction (2.19): then it is neither consumed or produced, that's why we have put X on both sides of the reaction scheme (2.19). The dynamical equations (2.16)(2.17) are then modified as follows:

$$\frac{dS}{dt} = DS_{in} - DS - \frac{1}{Y}\mu X - m_S X \qquad\qquad (2.20)$$

$$\frac{dX}{dt} = -DX + \mu X - bX \tag{2.21}$$

where m_S and b are maintenance and death coefficients, respectively.

2.2.4 (Liquid and Gaseous) Product Dynamics

The growth of microorganisms in bioreactors is often accompanied by the formation of products which are either soluble in the culture or which are given off in gaseous form. The mass balance relative to the product in the bioreactor is given by (assuming no product in the inlet):

$$\frac{dP}{dt} = -DP - Q + \nu X \tag{2.22}$$

with P being the product concentration (in the liquid phase), Q the rate of mass outflow of the product from the reactor in gaseous form, and ν the specific production rate.

The term νX in (2.22) represents the rate of product formation: it expresses the fact that the production is, in some way, "catalysed" by the biomass X.

In some instances (e.g. methane CH_4), the liquid concentration can be assumed to be negligible ($P \approx 0$). The gaseous outflow rate is then usually considered as being equal to the production rate and is written as follows:

$$Q = \nu X \tag{2.23}$$

We shall later (Section 2.9) introduce mathematical tools that give a sound basis to "reduce" the mass balance equation (2.22) to quation (2.23).

An important special case arises when the product formation is *growth-associated*, i.e. the specific production rate is assumed to be proportional to the specific growth rate:

$$\nu = Y_P \mu \tag{2.24}$$

with Y_P a yield coefficient. However, the specific production rate may also be completely or partially independent of the specific growth rate. A classical example is the lactic fermentation for which Luedeking and Piret (1959) [164] have proposed the following expression:

$$\nu = Y_P \mu + \eta \tag{2.25}$$

where η is the non-growth associated specific production rate.

2.2.5 Oxygen Dynamics

Aerobic bioprocesses are processes in which the microorganisms need oxygen in order to develop properly. A typical example is the activated sludge process. In such cases dissolved oxygen in the mixed liquor can be considered as an additional substrate.

$$S + S_O \longrightarrow X \tag{2.26}$$

The dissolved oxygen (DO) mass balance in the bioreactor is described as follows:

$$\frac{dS_O}{dt} = DS_{O,in} - DS_O + OTR - OUR \tag{2.27}$$

where S_O is the DO concentration in the reactor, OTR is the oxygen transfer rate and OUR is the oxygen uptake rate. Note the presence of the liquid inlet term $DS_{O,in}$ in the mass balance equation of oxygen. This term (which is often neglected) is introduced here to emphasise the possible presence of dissolved oxygen in the influent e.g. as a consequence of the oxygen transfer to the wastewater in the sewer system before it is fed to the tank.

The oxygen uptake rate OUR obviously depends on the growth of the biomass. This is often expressed as follows:

$$OUR = \frac{1}{Y_{O_2}} \mu X \tag{2.28}$$

with Y_{O_2} being a yield coefficient. Note that the usual formalism in activated sludge processes consists of considering that if Y is the biomass growth yield, Y biomass units (in COD (Chemical Oxygen Demand) units) are produced from one unit (in COD units) of the substrate, i.e. S, and the rest of the available COD, i.e. $1 - Y$ (in COD units) is oxidised to produce carbon dioxide CO_2. This implies by considering the above line of reasoning that the yield coefficient Y_{O_2} is equal to:

$$Y_{O_2} = \frac{Y}{1 - Y} \tag{2.29}$$

A term proportional to the biomass concentration ($b_{O_2} X$) is often included in equation (2.28) to account for the oxygen consumption for endogenous metabolism (to maintain viability of the cell). Hence, the total oxygen uptake rate OUR is composed of two terms, a growth related "exogenous" respiration rate OUR_{ex} and a cell maintenance related "endogenous" respiration rate OUR_{end}:

$$OUR = OUR_{ex} + OUR_{end} = \frac{1 - Y}{Y} \mu X + b_{O_2} X \tag{2.30}$$

By using a line of reasoning based on Henry's law to model the liquid-gas transfer dynamics, the oxygen transfer rate, OTR, is expressed as follows:

$$OTR = k_L a(S_O^* - S_O) \tag{2.31}$$

where $k_L a$ is the mass transfer coefficient and S_O^* is the oxygen saturation concentration.

However expression (2.31) may have to be used with precaution because the coefficients S_O^* and $k_L a$ may be unknown and may vary greatly with time: it is well known that the oxygen saturation concentration S_O^* depends on variables such as the oxygen partial pressure in the surrounding atmosphere, temperature, salinity and concentration of chemicals in the liquid, and that factors such as the type and geometry of the aerator, the air flow rate, the presence of surfactants or the biomass concentration determine the value of $k_L a$. In some applications, however, the input and output gaseous oxygen mass flow rates are measured on-line using off-gas analysis equipment [116]. Hence, provided the liquid-gas transfer dynamics are negligible, the OTR can simply be expressed from the gaseous oxygen balance:

$$OTR = Q_{O_2,in} - Q_{O_2,out} \qquad (2.32)$$

where $Q_{O_2,in}$ and $Q_{O_2,out}$ are respectively the input and output oxygen flow rates (per reactor volume unit).

2.2.6   Models of Reaction Rates

It can clearly be seen from equations (2.16), (2.17), (2.24), (2.25) and (2.28) that the reaction rate ρ, and more particularly the specific growth rate μ is a key parameter for the description of biomass growth, substrate consumption and product formation.

Let us briefly introduce some basic notions of kinetics, before introducing some typical models of the specific growth rate (see also e.g. [88], [14], [91] for surveys on the subject; note also that more than 60 kinetic models for bioprocesses are listed in Bastin and Dochain [14]).

Basic kinetic law.   The basic kinetic model for (bio)chemical reactions takes the following form:

$$\rho = k_0 \prod_i C_i^{\alpha_i} \qquad (2.33)$$

where C_i are the concentrations of the reactants involved in the reaction, k_0 is the kinetic constant, and α_i are defined as the orders of the reaction with respect to the ith reactant. The total *reaction order* is equal to the sum of the α_i ($\sum_i \alpha_i$). When the reaction orders α_i are equal to the stoichiometric coefficients, the reaction is referred to as an *elementary* reaction.

Note that, for instance, for the biomass death reaction (2.18), if the death coefficient b is constant, then (2.18) is a first order reaction.

One of the essential consequences of the above kinetic law (2.33) is that the reaction rate ρ will be equal to zero if the concentration of one of the reactants involved in the reaction is equal to zero:

$$\rho = 0 \quad if \quad C_i = 0 \qquad (2.34)$$

It is obvious that this is a minimal requirement for any reaction rate model, i.e. also for specific growth rate models.

Models of the specific growth rate.   Biochemical experiments carried out over more than half a century on pure cultures as well as on open cultures (with non-sterile substrates) have clearly indicated that the parameter μ varies with time and is influenced by many physico-chemical and biological environmental factors among which the most important ones are: substrate concentration, biomass concentration, product concentration, pH, temperature, dissolved oxygen concentration, light intensity and various inhibitors of microbial growth.

The specific growth rate is then commonly (but not systematically, see e.g. the Contois model here below) expressed by the multiplication of individual terms, each of them referring to one of the influencing factors:

$$\mu(t) = \mu(S)\mu(X)\mu(P)\mu(S_O)\mu(pH)\mu(T)... \tag{2.35}$$

where X, S, P, S_O have been defined above while T refers to temperature.

We shall present, in the following paragraphs, some of the most commonly used kinetic models for the different terms of equation (2.35).

It is important to draw the attention that any specific growth rate must follow some basic modelling rules in order to be a correct representation of the phenomenon that it is supposed to describe. First of all, recall that the specific growth rate is part of the reaction rate ($\rho = \mu X$) of a reaction which transforms substrate(s) into biomass (maybe some additional products). This means that the specific growth rate has to be positive (otherwise it would represent the reverse reaction of transformation of biomass into substrate(s)!):

$$\mu \geq 0, \qquad \text{for all time} \tag{2.36}$$

Secondly it must fulfill the basic kinetic law condition (2.34), i.e. the specific growth rate must equal to zero when the concentration of one of the substrates S_i (i = 1, 2,...) of the reaction is equal to zero.

$$\mu(S_i) = 0 \qquad \text{if} \qquad S_i = 0 \tag{2.37}$$

Influence of the substrate concentration S.   The most widespread analytical specific growth rate model is certainly the "Michaelis-Menten law", also often called the "Monod law", which expresses the dependence of μ on the substrate concentration S as follows (Figure 2.4):

$$\mu = \frac{\mu_{max} S}{K_S + S} \tag{2.38}$$

where μ_{max} is the maximum growth rate and K_S is the "Michaelis-Menten" or saturation constant. The subscript S in the parameter K_S refers to the involved substrate. Note that the value of K_S corresponds to the substrate concentration for which the specific growth rate μ is half its maximum value μ_{max} (see Figure 2.4):

$$\mu(S = K_S) = \frac{\mu_{max}}{2} \tag{2.39}$$

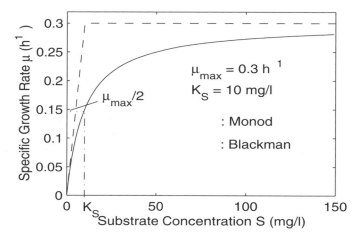

FIG. 2.4. The Monod law and the Blackman law.

In fact, this expression was initially proposed by Michaelis and Menten in 1913 [173] and physically justified by Briggs and Haldane [43] later, in 1925, to express the reaction rate of enzyme-catalysed reactions with a single substrate. It was extended by Monod in 1942 [176] to the case of microorganism growth, but without any specific physical justification.

Expression (2.38) was adopted by Monod because it fits experimental data well. But, as Monod himself recognised, "toute courbe d'allure sigmoïde pourrait être ajustée aux données expérimentales" (any sigmoidal curve could be fitted to the experimental data)[176]. Besides, another expression was suggested by Tessier [247] in the same year 1942:

$$\mu = \mu_{max}(1 - e^{\frac{-S}{K_S}}) \tag{2.40}$$

This equation could fit the Monod data equally well. Many different (and more or less esoteric) formulas have been proposed since then (see [14]).

Another simple and often used model for the specific growth rate dependence with respect to the substrate concentration is the Blackman law (Figure 2.4):

$$\mu = \frac{\mu_{max}}{K_S} S \qquad \text{if } S \leq K_S \tag{2.41}$$

$$= \mu_{max} \qquad \text{if } S > K_S \tag{2.42}$$

which is a combination of first order kinetics with respect to S for $S \leq K_S$, and zero order kinetics for $S > K_S$.

A drawback of the Monod, Tessier or Blackman laws is that they do not allow any account to be taken of possible substrate inhibitory effects at high concentrations (overloading). Typical examples in biological wastewater treatment are

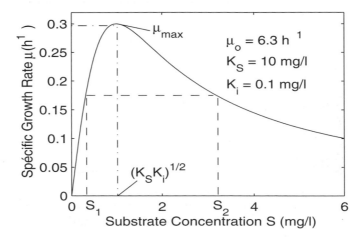

FIG. 2.5. The Haldane law.

NH_4-inhibition of nitrification or self-inhibition of phenol degradation. Andrews [8] suggested that substrate inhibition be treated by the "Haldane law" which was initially derived by Haldane [107] to describe inhibition in enzyme-substrate reactions (Figure 2.5):

$$\mu = \frac{\mu_0 S}{K_S + S + \frac{S^2}{K_i}} \tag{2.43}$$

where K_i is the "inhibition parameter" and:

$$\mu_0 = \mu_{max}\left(1 + 2\sqrt{\frac{K_S}{K_i}}\right) \tag{2.44}$$

If the substrate inhibition is neglected ($K_i \to \infty$), the Haldane law reduces to the Monod law.

One important feature of the Haldane model is shown in Figure 2.5: for each value of the specific growth rate, there exist two possible substrate concentrations (denoted by S_1 and S_2 in Figure 2.5), which are distributed on both sides of the maximum specific growth rate (which corresponds to the value of S equal to $\sqrt{K_S K_i}$). This feature has the following consequences in CSTRs with a simple microbial growth reaction (2.15) (i.e. described by equations (2.16)(2.17)). First, it allows the emphasis of the possible existence of three equilibrium points: beside the wash-out ($X = 0$, $S = S_{in}$), there are two possible steady states, each one corresponding to one of the two values S_1 and S_2. Secondly, the Haldane relation allows the emphasis of the possible existence of instability in bioprocesses: indeed, it will be shown in Section 2.7 that in a CSTR (2.16)(2.17), the steady state corresponding to S_1 is a stable equilibrium point, and the one corresponding to S_2

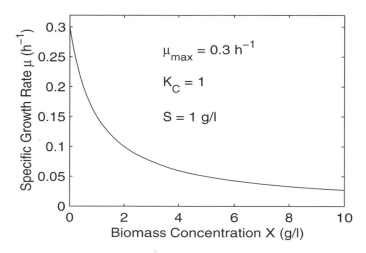

FIG. 2.6. The Contois law.

is an unstable one. The latter unstable state corresponds to the experimentally observed situation when there is accumulation of inhibiting substrate which results in a wash-out of the biomass. A typical example is the accumulation of volatile fatty acids in anaerobic digestion processes (e.g. [29], [92]).

Influence of the biomass concentration X. The biomass growth is often observed to be slowed down at high biomass concentration (and this has been experimentally observed in particular instances). A simple model that accomodates for this situation assumes that the specific growth rate decreases linearly with increasing biomass concentration:

$$\mu(X) = \mu_{max}(1 - aX) \tag{2.45}$$

where μ_{max} is the maximum growth rate and a $(= \frac{1}{X_{max}})$ the inhibition constant. It is often called the "logistic model" and was proposed by Verhulst in 1838 [271]. Another model which is a function of both S and X is the following:

$$\mu = \frac{\mu_{max} S}{K_c X + S} \tag{2.46}$$

with K_c constant. This model was proposed by Contois in 1959 [61], and is illustrated in Figure 2.6, which shows the inhibition dependence of μ with respect to X (for constant S).

Influence of the product concentration P. It is a well known fact that, in particular bioreactions, the product of the reaction can also inhibit the biomass growth. A typical example is anaerobic digestion when acetate/propionate and/or hydrogen

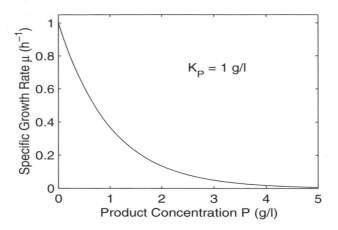

FIG. 2.7. Product inhibition model $\mu(P) = e^{-K_P P}$.

are inhibitory to acidogenesis reactions. Typical models for product inhibition are e.g. (Figure 2.7) [4]:

$$\mu(P) = \frac{K_p}{K_p + P} \qquad (2.47)$$

$$\mu(P) = e^{-K_p P} \qquad (2.48)$$

with K_p constant.

These inhibition models are also used to emphasise for instance the inhibition of denitrification by oxygen and are often termed "switching functions".

Influence of pH. As we have indicated here above, the biomass growth can actually takes place only if pH and temperature lie within (usually small) ranges of admissible values.

The pH often inhibits the biological activity due to non dissociated acids or bases (ammonium, fatty acids, nitrite) in the mixed liquor. Up to now, few good models for this pH-dependency are presented. One possibility is a bell-shaped function:

$$K_{pH} = K_{\max} \cdot \frac{1}{1 + 10^{pK_1 - pH} + 10^{pH - pK_2}}$$

where pK_1 and pK_2 are the low and the high pH with half of the maximum activity (Figure 2.8).

In anaerobic digestion, for instance, the process is known to operate correctly only for nearly neutral pH ($= 7$). For this process, Rozzi [222] proposes treating the influence of pH by a parabolic law derived from experimental data:

$$\mu(pH) = a\, pH^2 + b\, pH + c \qquad \text{if} \quad pH_{min} \le pH \le pH_{max} \qquad (2.49)$$

$$= 0 \qquad\qquad\qquad \text{otherwise} \qquad (2.50)$$

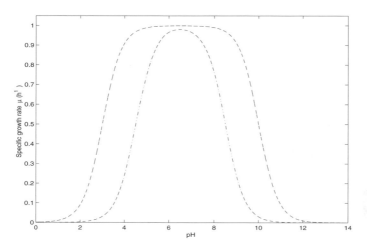

FIG. 2.8. Effect of pH on biological activity (.-: pK$_1$ = 4.5, pK$_2$ = 8.5; - -: pK$_1$ = 3, pK$_2$ = 10).

with a, b, c constants, and pH_{min} and pH_{max} are the pH bounds within which growth is possible.

Influence of temperature. On the other hand, the influence of temperature is most often modelled by an Arrhenius-type law, as has been done, for instance, by Topiwala and Sinclair [250]:

$$\mu(T) = a_1 e^{-\frac{E_1}{RT}} - a_2 e^{-\frac{E_2}{RT}} - b \qquad (2.51)$$

where E_1, E_2 are activation energies, R is the gas constant (8.314 J.g.mol/K), and a_1, a_2, b are constants. This expression shows that the specific growth rate increases continuously with temperature up to a maximum value T_{max} (at which the cells die) (see Figure 2.9).

This is often simplified to:

$$\mu(T) = K_{T1} \cdot 10^{-\alpha(T_1 - T_2)} \text{ or } K_{T1} \cdot \theta^{-(T_1 - T_2)} \qquad (2.52)$$

where θ is around 1.03 to 1.05 for most processes (corresponding to a doubling of the reaction rate with a temperature increase of 10°). On the other hand, nitrification is strongly dependent on temperature ($\theta = 1.15$). More detailed models for description of the effects of temperature can be found in the overview by Zwietering *et al.* [291]. See also [115], [147], [210].

2.3 Examples of Biological Wastewater Treatment Process Models

In this section, we introduce some typical examples of biological wastewater processes and their mass balance dynamical models:

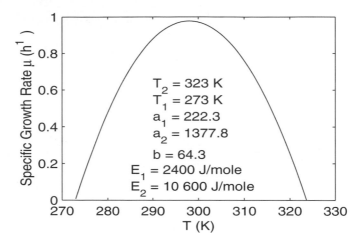

FIG. 2.9. Temperature model $\mu(T)$.

- Anaerobic digestion: The 4 population model
- Activated sludge process: The basic model
- Activated sludge process: The IWA Activated Sludge Model No. 1
- Two step nitrification
- Two step denitrification

The section will be used to extend the ideas developed in Section 2.2 for simple examples to more complex ones. As it has been suggested in the preceding section, the model building basically considers the following two steps:

1. building of an appropriate reaction network
2. derivation of the dynamical mass balances

We shall use the first example (anaerobic digestion) to illustrate the model building procedure with mass balances for processes which involve a large number of components (10) and reactions (4).

2.3.1 *Anaerobic Digestion: The 4 Population Model*

Anaerobic digestion is a biological wastewater treatment process which produces methane. Four metabolic paths [178] can be identified in this process: two for acidogenesis and two for methanisation (Figure 2.10).

In the first acidogenic path (Path 1), glucose (or another complex substrate) is decomposed into volatile fatty acids (VFAs)(acetate, propionate), hydrogen and inorganic carbon by acidogenic bacteria. In the second acidogenic path (Path 2), Obligate Hydrogen Producing Acetogens (OHPA) decompose propionate into acetate, hydrogen and inorganic carbon. In a first methanisation path (Path 3), acetate is transformed into methane and inorganic carbon by acetoclastic methanogenic

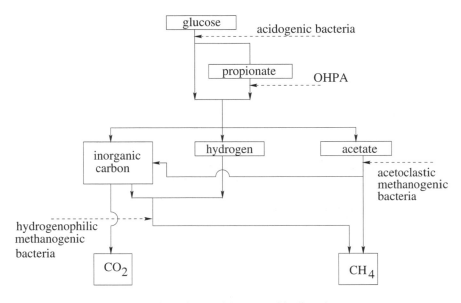

FIG. 2.10. Scheme of the anaerobic digestion.

bacteria, while in the second methanisation path (Path 4), hydrogen combines with inorganic carbon to produce methane under the action of hydrogenophilic methanogenic bacteria. The process can then be described by the following reaction network:

$$S_1 \longrightarrow X_1 + S_2 + S_3 + S_4 + S_5 \qquad (2.53)$$
$$S_2 \longrightarrow X_2 + S_3 + S_4 + S_5 \qquad (2.54)$$
$$S_3 \longrightarrow X_3 + S_5 + P_1 \qquad (2.55)$$
$$S_4 + S_5 \longrightarrow X_4 + P_1 \qquad (2.56)$$

where S_1, S_2, S_3, S_4, S_5, X_1, X_2, X_3, X_4 and P_1 are respectively glucose, propionate, acetate, hydrogen, inorganic carbon, acidogenic bacteria, OHPA, acetoclastic methanogenic bacteria, hydrogenophilic methanogenic bacteria and methane.

The conversion terms for the above reactions (2.53) to (2.56) are characterised by:

1. a reaction rate, denoted ρ_i, the index i = 1, 2, 3, 4 being related to reaction (2.53), (2.54), (2.55), (2.56), respectively. Each reaction rate is a growth rate, and each growth reaction involves a different microorganism population: therefore in line with the usual biochemical engineering notations, each reaction rate may be written as the product of a specific growth rate by the concentration of the biomass involved in the reaction:

$$\rho_i = \mu_i X_i, \ i = 1 \ to \ 4 \qquad (2.57)$$

2. stoichiometry (yield coefficients) associated to each component of each reaction. Here we have considered to use the symbols Y_{ji} for the yield coefficients, where the index j corresponds to the component (ranging in the
following order: X_1, S_1, X_2, S_2, X_3, S_3, X_4, S_4, S_5 and P_1), and the index
i to the reaction (ranged as above). This choice looks a priori arbitrary, but,
as we shall see in Section 2.4, corresponds to the position of elements in a
matrix.

Beside the conversion terms, the mass balance model also contains transport terms:
let us consider that the anaerobic digestion process is operated in a CSTR, and that
the only inlet substrate is the organic matter S_1.

We now have all the necessary information to derive the dynamical mass balance model of the process. The dynamics of the anaerobic digestion process are
then described by the following equations:

$$\frac{dX_1}{dt} = \mu_1 X_1 - DX_1 \tag{2.58}$$

$$\frac{dS_1}{dt} = -\frac{1}{Y_{21}}\mu_1 X_1 + DS_{in} - DS_1 \tag{2.59}$$

$$\frac{dX_2}{dt} = \mu_2 X_2 - DX_2 \tag{2.60}$$

$$\frac{dS_2}{dt} = Y_{41}\mu_1 X_1 - \frac{1}{Y_{42}}\mu_2 X_2 - DS_2 \tag{2.61}$$

$$\frac{dX_3}{dt} = \mu_3 X_3 - DX_3 \tag{2.62}$$

$$\frac{dS_3}{dt} = Y_{61}\mu_1 X_1 + Y_{62}\mu_2 X_2 - \frac{1}{Y_{63}}\mu_3 X_3 - DS_3 \tag{2.63}$$

$$\frac{dX_4}{dt} = \mu_4 X_4 - DX_4 \tag{2.64}$$

$$\frac{dS_4}{dt} = Y_{81}\mu_1 X_1 + Y_{82}\mu_2 X_2 - \frac{1}{Y_{84}}\mu_4 X_4 - DS_4 - Q_1 \tag{2.65}$$

$$\frac{dS_5}{dt} = Y_{91}\mu_1 X_1 + Y_{92}\mu_2 X_2 + Y_{93}\mu_3 X_3 - \frac{1}{Y_{94}}\mu_4 X_4 - DS_4 - Q_2 \tag{2.66}$$

$$\frac{dP_1}{dt} = Y_{03}\mu_3 X_3 + Y_{04}\mu_4 X_4 - DP_1 - Q_3 \tag{2.67}$$

where μ_1, μ_2, μ_3, μ_4 are the specific growth rates (h^{-1}) of reactions (2.53), (2.54),
(2.55), (2.56), respectively, and S_{in}, Q_1, Q_2 and Q_3 represent respectively the
influent glucose concentration (g/L) and the gaseous outflow rates of H_2, CO_2 and
CH_4 (g/L.h).

Note that if we intend to use the model for (dynamic) simulation, we still need
expressions for the specific growth rates. This can be done by choosing e.g. one

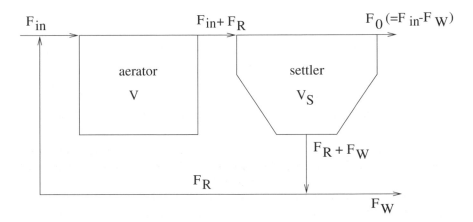

FIG. 2.11. Schematic view of an activated sludge process.

(or a combination) of the expressions presented in Section 2.2. The problem of choosing appropriate expressions for specific growth rates and calibrating parameter values will be discussed in Chapters 3, 4 and 5.

2.3.2   Activated Sludge Process: The Basic Model

The activated sludge process is one other classical biological (but aerobic) wastewater treatment process. It is usually operated in (at least) two sequential tanks (see Figure 2.11): an aerator (in which the degradation of the pollutants S takes place) and a settler (which is used to separate biomass X from the treated wastewater and recycle it to the aerator).

The reaction in the aerator may be described by a simple microbial growth reaction (see e.g. [108], [169]).

$$S + S_O \longrightarrow X \tag{2.68}$$

It is often assumed that the settler works perfectly, i.e. there is no biomass in the overflow of the settler (characterised by the flow rate F_o ($= F_{in} - F_W$)). The dynamics of the plant (aerator + settler) are then described by the following mass balance equations:

$$\frac{dS}{dt} = \frac{-1}{Y_S}\mu X - \frac{F_R + F_{in}}{V}S + \frac{F_{in}}{V}S_{in} \tag{2.69}$$

$$\frac{dS_O}{dt} = \frac{-1}{Y_0}\mu X + \frac{F_{in}}{V}S_{O,in} - \frac{F_R + F_{in}}{V}S_O + k_La(S_O^* - S_O) \tag{2.70}$$

$$\frac{dX}{dt} = \mu X + \frac{F_R}{V}X_R - \frac{F_R + F_{in}}{V}X \tag{2.71}$$

$$\frac{dX_R}{dt} = \frac{F_{in} + F_R}{V_S}X - \frac{F_R + F_W}{V_S}X_R \tag{2.72}$$

where X_R is the concentration of the recycled biomass (g/L), F_{in}, F_R and F_W are the influent, recycle and waste flow rates (g/L.h), respectively, and V and V_S the aerator and settler volumes (L), respectively.

2.3.3 Activated Sludge Process: The IWA Activated Sludge Model No. 1

Generally speaking, the simple aerobic microbial growth reaction (2.68) is not the only one which takes place in an activated sludge process. The simple reaction results in a rather simple dynamical model which may be convenient in many instances, e.g. for control design. However, in other instances, e.g. for an exhaustive process description, the dynamical model should include the influences of the different reactions that are taking place in activated sludge processes. One of the objectives of the Task Group that eventually produced the IWA Activated Sludge Model No. 1 [120] was to gather these reactions (or at least the most important ones) into one model framework. There are 13 important components and 8 important reactions, which are carried out by two types of microorganisms: heterotrophs (their carbon comes from an organic compound) and autotrophs (they require only CO_2 to supply their carbon needs). Next to their growth (for the heterotrophs both under aerobic and anoxic conditions) and decay reactions (reactions 1-5), 3 other reactions take place: ammonification of soluble organic nitrogen (6), hydrolysis of entrapped organics (7) and of entrapped organic nitrogen (8). These can be formalised into the following reaction network:

1. Aerobic growth of heterotrophs: $S_S + S_O + S_{NH} \longrightarrow X_{B,H}$
2. Anoxic growth of heterotrophs: $S_S + S_{NO} + S_{NH} \longrightarrow X_{B,H}$
3. Aerobic growth of autotrophs: $S_O + S_{NH} \longrightarrow X_{B,A} + S_{NO}$
4. Decay of heterotrophs: $X_{B,H} \longrightarrow X_P + X_S + X_{ND}$
5. Decay of autotrophs: $X_{B,A} \longrightarrow X_P + X_S + X_{ND}$
6. Ammonification of soluble organic nitrogen: $S_{ND} \longrightarrow S_{NH}$
7. Hydrolysis of entrapped organics: $X_S \longrightarrow S_S$
8. Hydrolysis of entrapped organic nitrogen: $X_{ND} \longrightarrow S_{ND}$

where S_S, S_O, S_{NH}, S_{NO}, S_{ND}, X_{ND}, $X_{B,H}$, $X_{B,A}$, X_P, X_S are the readily biodegradable substrate, the dissolved oxygen concentration, the soluble ammonia nitrogen, the nitrate nitrogen, the soluble organic nitrogen, the particulate biodegradable organic nitrogen, the heterotrophic bacteria, the autotrophic bacteria, the particulate products arising from biomass decay, and the slowly biodegradable substrate, respectively.

The dynamics in a stirred tank reactor are given by the following mass balance equations (with the yield coefficient nomenclature of the IWA Activated Sludge Model No. 1):

$$\frac{dS_S}{dt} = -\frac{1}{Y_H}\mu_1 X_{B,H} - \frac{1}{Y_H}\mu_2 X_{BH} + \rho_7 + DS_{S,in} - DS_S \qquad (2.73)$$

$$\frac{dX_S}{dt} = (1 - f_P)b_H X_{B,H} + (1 - f_P)b_A X_{B,A} - \rho_7 + DX_{S,in} - DX_S \tag{2.74}$$

$$\frac{dX_{B,H}}{dt} = \mu_1 X_{B,H} + \mu_2 X_{B,H} - b_H X_{B,H} + DX_{B,H,in} - DX_{B,H} \tag{2.75}$$

$$\frac{dX_{B,A}}{dt} = \mu_3 X_{B,A} - b_A X_{B,A} + DX_{B,A,in} - DX_{B,A} \tag{2.76}$$

$$\frac{dX_P}{dt} = f_P b_H X_{B,H} + f_P b_A X_{B,A} + DX_{P,in} - DX_P \tag{2.77}$$

$$\frac{dS_O}{dt} = -\frac{1 - Y_H}{Y_H}\mu_1 X_{B,H} - \frac{4.57 - Y_A}{Y_A}\mu_3 X_{B,A}$$
$$+ k_L a(S_O^* - S_O) + DS_{O,in} - DS_O \tag{2.78}$$

$$\frac{dS_{NO}}{dt} = -\frac{1 - Y_H}{2.86 Y_H}\mu_2 X_{B,H} + \frac{1}{Y_A}\mu_3 X_{B,A} + DS_{NO,in} - DS_{NO} \tag{2.79}$$

$$\frac{dS_{NH}}{dt} = -i_{XB}\mu_1 X_{B,H} - i_{XB}\mu_2 X_{B,H} - (i_{XB} + \frac{1}{Y_A})\mu_3 X_{B,A} + DS_{NH,in}$$
$$- DS_{NH} \tag{2.80}$$

$$\frac{dS_{ND}}{dt} = \rho_8 - k_a S_{ND} X_{B,H} + DS_{ND,in} - DS_{ND} \tag{2.81}$$

$$\frac{dX_{ND}}{dt} = (i_{XB} - f_P i_{XP})b_H X_{B,H} + (i_{XB} - f_P i_{XP})b_A X_{B,A} - \rho_8$$
$$+ DX_{ND,in} - DX_{ND} \tag{2.82}$$

$$\rho_7 = k_h \frac{X_S/X_{B,H}}{K_X + X_S/X_{B,H}} X_{B,H} \tag{2.83}$$

$$\rho_8 = \rho_7 \frac{X_{ND}}{X_S} \tag{2.84}$$

In the above equations, ρ_7 and ρ_8 hold for the reaction rates of reaction 7(hydrolysis of entrapped organics) and 8 (hydrolysis of entrapped organic nitrogen). In the IWA Activated Sludge Model No. 1, the model equations are summarised in a table (see Table 2.1[1]) which is very convenient in the sense that it clearly emphasises

---

[1]"S_I and X_I are not involved in any conversion process. Nevertheless they are included because they are important to the performance of the process"; "Inclusion of alkalinity (S_{ALK}) is not essential, but its inclusion is desirable because it provides information whereby undue changes in pH can be predicted" [120].

Table 2.1. Tabular format of the stoichiometry of the IWA Activated Sludge Model No. 1

Components → i	1	2	3	4	5	6	7	8	9	10
j Process ↓	S_S	X_S	$X_{B,H}$	$X_{B,A}$	X_P	S_O	S_{NO}	S_{NH}	S_{ND}	X_{ND}
1 Aerobic growth of heterotrophs	$-\frac{1}{Y_H}$		1			$-\frac{1-Y_H}{Y_H}$		$-i_{XB}$		
2 Anoxic growth of heterotrophs	$-\frac{1}{Y_H}$		1				$-\frac{1-Y_H}{2.86Y_H}$	$-i_{XB}$		
3 Aerobic growth of autotrophs				1		$-\frac{4.57-Y_A}{Y_A}$	$\frac{1}{Y_A}$	$-i_{XB}-\frac{1}{Y_A}$		
4 'Decay' of heterotrophs		$1-f_P$	-1		f_P					$i_{XB}-f_Pi_{XP}$
5 'Decay' of autotrophs		$1-f_P$		-1	f_P					$i_{XB}-f_Pi_{XP}$
6 Ammonification of soluble organic nitrogen								1	-1	
7 'Hydrolysis' of entrapped organics	1	-1								
8 'Hydrolysis' of entrapped organic nitrogen									1	-1

the interaction and interconnections of the different process components within the different process reactions. The adoption of a Petersen matrix [196] format in which components, reaction kinetics and stoichiometry are clearly separated has shown to be very fruitful. In the General Dynamical Model section (Section 2.4), we shall introduce another, yet very similar, compact form of mass balance models (i.e. in a mathematical matrix form) which will be very convenient for mathematical manipulations (e.g. for model reduction and monitoring design).

2.3.4   Two Step Nitrification

Nitrification is the biological oxidation of ammonium-nitrogen to nitrate-nitrogen. Yet nitrification is known to be a two-step reaction (e.g. [5], [11], [286]):

$$S_{NH} \longrightarrow X_1 + S_{NO_2} \tag{2.85}$$

$$S_{NO_2} \longrightarrow X_2 + S_{NO_3} \tag{2.86}$$

where S_{NH}, S_{NO_2}, S_{NO_3}, X_1 and X_2 hold respectively for ammonium-nitrogen (NH_4-N), nitrite-nitrogen (NO_2-N), nitrate-nitrogen (NO_3-N), Nitrosomonas bacteria and Nitrobacter bacteria, respectively.

In the IWA Activated Sludge Model No. 1, nitrification is represented by only one reaction: reaction 3 (aerobic growth of autotrophs, see previous section). Indeed, the two sequential reactions (2.85)(2.86) of nitrification can be reduced to the single reaction 3 if the reaction (2.85) is assumed to be slower than reaction (2.86). In the model order reduction section (Section 2.9) we shall introduce a systematic approach for reducing reaction systems with slow and fast reactions.

The dynamics of the two-step nitrification in a stirred tank reactor fed with only ammonium containing wastewater are given by the following mass balance equations:

$$\frac{dS_{NH}}{dt} = -\frac{1}{Y_1}\mu_1 X_1 + DS_{NH,in} - DS_{NH} \tag{2.87}$$

$$\frac{dS_{NO_2}}{dt} = -\frac{1}{Y_2}\mu_2 X_2 + Y_3\mu_1 X_1 - DS_{NO_2} \tag{2.88}$$

$$\frac{dS_{NO_3}}{dt} = Y_4\mu_2 X_2 - DS_{NO_3} \tag{2.89}$$

$$\frac{dX_1}{dt} = \mu_1 X_1 - DX_1 \tag{2.90}$$

$$\frac{dX_2}{dt} = \mu_2 X_2 - DX_2 \tag{2.91}$$

where μ_1 and μ_2 are the specific growth rates of reaction (2.85) and (2.86), respectively, Y_1 to Y_4 are yield coefficients, and $S_{NH,in}$ the influent concentration of S_{NH}.

2.3.5 Two Step Denitrification

Denitrification plays an important role not only in wastewater treatment (see reaction 2 in the IWA Activated Sludge Model No. 1) but also in the production of drinking water. As this is a multi-step process and some toxic intermediates are formed, nitrite in particular, a process description involving nitrite is needed as one desires its complete absence from drinking water. In wastewater treatment, important cost savings can be obtained if one adheres to what has been termed "Nitrit Verfahren" for nitrogen removal, i.e. partial nitrification to nitrite and denitrification from nitrite to nitrogen gas [1],[5], [117].

For such processes the denitrification reaction scheme is also composed of two sequential reactions:

$$S_{NO_3} + S \longrightarrow X + S_{NO_2} \tag{2.92}$$

$$S_{NO_2} + S \longrightarrow X + N_2 \tag{2.93}$$

where S_{NO_3}, S_{NO_2}, S, N_2 and X are nitrate (NO_3^-), nitrite (NO_2^-), organic carbon, nitrogen, and denitrifying bacteria, respectively. In the above reaction scheme, the denitrifying bacteria are of the same type and are growing in a similar environment: that's why only one biomass X is considered in both reactions.

The dynamical mass balances in a stirred tank reactor are then equal to:

$$\frac{dS_{NO_3}}{dt} = -\frac{1}{Y_1}\mu_1 X + D S_{NO_3,in} - D S_{NO_3} \tag{2.94}$$

$$\frac{dS_{NO_2}}{dt} = -\frac{1}{Y_2}\mu_2 X + Y_3 \mu_1 X - D S_{NO_2} \tag{2.95}$$

$$\frac{dS}{dt} = -\frac{1}{Y_4}\mu_1 X - \frac{1}{Y_5}\mu_2 X + D S_{in} - D S \tag{2.96}$$

$$\frac{dX}{dt} = \mu_1 X + \mu_2 X - D X \tag{2.97}$$

$$\frac{dN_2}{dt} = Y_6 \mu_2 X - D N_2 - Q_{N_2} \tag{2.98}$$

where μ_1 and μ_2 are the specific growth rates of reaction (2.92) and (2.93), respectively, Y_1 to Y_6 are yield coefficients, $S_{NO_3,in}$ and S_{in} the influent concentration of nitrate S_{NO_3} and organic carbon S, and Q the gaseous nitrogen outflow rate.

2.4 General Dynamical Model of Stirred Tank Reactors

So far, we have formalised each biological wastewater treatment process example by considering reaction networks. The notion of reaction scheme (as it is suggested in the above examples) is a useful tool to derive the dynamical mass balance model of the process: indeed, the dynamical model can be derived on the basis of the reaction network and put in a matrix form that is now introduced in this section. But before, let us draw the attention on the following points.

- As suggested in the above examples, a reaction system (and in particular a biological wastewater treatment process) can be viewed in the context of mass balance modelling as a set of M (bio)chemical reactions involving N components. The reactions most often encountered in bioprocesses are microbial growth (in which the biomass plays the role of an autocatalyst) and enzyme catalysed reactions (in which the biomass can be viewed as a simple catalyst); but many other reactions can also take place, like microorganism death, maintenance,... The main components are microorganism populations, enzymes, substrates and products.
- The reaction schemes as introduced here do not represent a stoichiometric relation between the process components, in contrast to the common practice in chemical kinetics. They are sometimes qualitative (in line with the comments on yield coefficients of Section 2.2.2). Yet this approach is not exclusive, and the user has the flexibility to explicitly include yield or stoichiometric coefficients in the reaction scheme, if he feels this more convenient.
- The reaction network is basically a tool for process description and model derivation, and therefore does not need to be an exhaustive (i.e. with *all* the reactions and *all* the components) description of the process. For instance, side reactions and by-products may not appear in the reaction network if it appears that their role is negligible in the process description, or even more precisely, in the context of the considered study.

2.4.1 *The General Dynamical Model*

Let us now go back, for instance, to the denitrification example (2.94)-(2.98). The model can be rewritten in the following matrix form:

$$
\frac{d}{dt}\begin{pmatrix} S_{NO_3} \\ S_{NO_2} \\ S \\ X \\ N_2 \end{pmatrix} = -D\begin{pmatrix} S_{NO_3} \\ S_{NO_2} \\ S \\ X \\ N_2 \end{pmatrix} + \begin{pmatrix} -\frac{1}{Y_1} & 0 \\ Y_3 & -\frac{1}{Y_2} \\ -\frac{1}{Y_4} & -\frac{1}{Y_5} \\ 1 & 1 \\ 0 & Y_6 \end{pmatrix}\begin{pmatrix} \mu_1 X \\ \mu_2 X \end{pmatrix} + \begin{pmatrix} DS_{NO_3,in} \\ 0 \\ DS_{in} \\ 0 \\ 0 \end{pmatrix}
$$

$$
- \begin{pmatrix} 0 \\ 0 \\ 0 \\ 0 \\ Q_{N_2} \end{pmatrix} \tag{2.99}
$$

This suggests to rewrite the dynamical mass balance model in the following compact matrix form:

$$
\frac{d\xi}{dt} = -D\xi + Y\rho(\xi) + F - Q(\xi) \tag{2.100}
$$

where ξ is the vector of the bioprocess component concentrations $(\dim(\xi) = N)$, Y is the yield coefficient matrix $(\dim(Y) = N \times M)$, $\rho(\xi)$ is the vector of reaction rates (that may be function of the component concentrations ξ $(\dim(\rho) = M)$, F is the feed rate vector and Q the vector of gaseous outflow rates that also may be influenced by the component concentrations ξ $(\dim(F) = \dim(Q) = N)$.

The model (2.100) has been called the *General Dynamical Model* for stirred tank bioreactors (see [14]). The derivation of the dynamical mass balance model from a reaction network is straightforward by noting that each component Y_{ij} of the yield coefficient matrix

$$Y = [Y_{ij}] \qquad i = 1 \ to \ N, \ j = 1 \ to \ M \qquad (2.101)$$

is representative of the ith component: it is negative if the component is a reactant, it is positive if it is a product and it is equal to zero if the component does not intervene in the reaction.

Note that the mass balance general dynamical model is composed of the two terms that we have mentioned in the introduction and in section 2.2.1:

- transport dynamics: $-D\xi + F - Q(\xi)$,
- conversion: $Y\rho(\xi)$

As it has been pointed out in section 2.2.1, note finally that the General Dynamical Model remains valid for processes with variable volumes via the addition of a dynamical equation describing the volume dynamics.

2.4.2   Examples (Continued)

Let us see how the different process examples that we have previously presented fit into the General Dynamical Model (2.100) framework.

Anaerobic digestion: The 4 population model.   Equations (2.58) to (2.67) can be rewritten in the matrix form (2.100) by considering the following vectors and matrices (N = 4, M = 10):

$$\xi = \begin{pmatrix} X_1 \\ S_1 \\ X_2 \\ S_2 \\ X_3 \\ S_3 \\ X_4 \\ S_4 \\ S_5 \\ P_1 \end{pmatrix}, \ Y = \begin{pmatrix} 1 & 0 & 0 & 0 \\ -\frac{1}{Y_{21}} & 0 & 0 & 0 \\ 0 & 1 & 0 & 0 \\ Y_{41} & -\frac{1}{Y_{42}} & 0 & 0 \\ 0 & 0 & 1 & 0 \\ Y_{61} & Y_{62} & 0 & -\frac{1}{Y_{63}} \\ 0 & 0 & 0 & 1 \\ Y_{81} & Y_{82} & 0 & -\frac{1}{Y_{84}} \\ Y_{91} & Y_{92} & Y_{93} & -\frac{1}{Y_{94}} \\ 0 & 0 & Y_{03} & Y_{04} \end{pmatrix} \qquad (2.102)$$

$$
F = \begin{pmatrix} 0 \\ DS_{in} \\ 0 \\ 0 \\ 0 \\ 0 \\ 0 \\ 0 \\ 0 \\ 0 \end{pmatrix}, \quad Q = \begin{pmatrix} 0 \\ 0 \\ 0 \\ 0 \\ 0 \\ 0 \\ 0 \\ Q_1 \\ Q_2 \\ Q_3 \end{pmatrix}, \quad \rho = \begin{pmatrix} \rho_1 \\ \rho_2 \\ \rho_3 \\ \rho_4 \end{pmatrix} = \begin{pmatrix} \mu_1 X_1 \\ \mu_2 X_2 \\ \mu_3 X_3 \\ \mu_4 X_4 \end{pmatrix} \quad (2.103)
$$

Activated sludge process: The basic model. An interesting feature of the basic model of activated sludge process is that due to its multi-tank characteristics (aerator + settler), the dilution rate becomes a matrix. In the General Dynamical Model (2.100) framework, the mass balance equations (2.69) to (2.72) become ($N = 4$, $M = 1$):

$$
\xi = \begin{pmatrix} S \\ S_O \\ X \\ X_R \end{pmatrix}, \quad Y = \begin{pmatrix} -\frac{1}{Y_S} \\ -\frac{1}{Y_O} \\ 1 \\ 0 \end{pmatrix}, \quad F = \begin{pmatrix} D_{in} S_{in} \\ D_{in} S_{O,in} + k_L a(S_O^* - S_O) \\ 0 \\ 0 \end{pmatrix} \quad (2.104)
$$

$$
\rho = \mu X, \quad Q = 0, \quad D = \begin{pmatrix} D_1 & 0 & 0 & 0 \\ 0 & D_1 & 0 & 0 \\ 0 & 0 & D_1 & -D_2 \\ 0 & 0 & -D_3 & D_4 \end{pmatrix} \quad (2.105)
$$

with the following definitions for D_{in}, D_1, D_2, D_3, and D_4:

$$
D_{in} = \frac{F_{in}}{V}, \quad D_2 = \frac{F_R}{V}, \quad D_1 = D_{in} + D_2, \quad D_3 = \frac{F_{in} + F_R}{V_S}, \quad D_4 = \frac{F_R + F_W}{V_S}
$$

$$(2.106)$$

Activated sludge process: The IWA Activated Sludge Model No. 1. For the IWA Activated Sludge Model No. 1, the model equations are written in the General Dynamical Model (2.100) framework with the following definitions ($N = 10$, $M = 8$):

$$\xi = \begin{pmatrix} S_S \\ X_S \\ X_{B,H} \\ X_{B,A} \\ X_P \\ S_O \\ S_{NO} \\ S_{NH} \\ S_{ND} \\ X_{ND} \end{pmatrix}, \quad F = \begin{pmatrix} DS_{S,in} \\ 0 \\ 0 \\ 0 \\ 0 \\ DS_{O,in} + k_L a(S_O^* - S_O) \\ 0 \\ DS_{NH,in} \\ 0 \\ 0 \end{pmatrix}, \tag{2.107}$$

$$Y = \begin{pmatrix} -\frac{1}{Y_H} & -\frac{1}{Y_H} & 0 & 0 & 0 & 0 & 1 & 0 \\ 0 & 0 & 0 & 1 - f_P & 1 - f_P & 0 & -1 & 0 \\ 1 & 1 & 0 & -1 & 0 & 0 & 0 & 0 \\ 0 & 0 & 1 & 0 & -1 & 0 & 0 & 0 \\ 0 & 0 & 0 & f_P & f_P & 0 & 0 & 0 \\ -\frac{1-Y_H}{Y_H} & 0 & -\frac{4.57-Y_A}{Y_A} & 0 & 0 & 0 & 0 & 0 \\ 0 & -\frac{1-Y_H}{2.86 Y_H} & \frac{1}{Y_A} & 0 & 0 & 0 & 0 & 0 \\ -i_{XB} & -i_{XB} & -i_{XA} & 0 & 0 & 1 & 0 & 0 \\ 0 & 0 & 0 & 0 & 0 & -1 & 0 & 1 \\ 0 & 0 & 0 & Y_B & Y_B & 0 & 0 & -1 \end{pmatrix} \tag{2.108}$$

$$\rho = \begin{pmatrix} \rho_1 \\ \rho_2 \\ \rho_3 \\ \rho_4 \\ \rho_5 \\ \rho_6 \\ \rho_7 \\ \rho_8 \end{pmatrix} = \begin{pmatrix} \mu_1 X_{B,H} \\ \mu_2 X_{B,H} \\ \mu_3 X_{B,A} \\ b_H X_{B,H} \\ b_A X_{B,A} \\ k_a S_{ND} X_{B,H} \\ \rho_{X7} X_{B,H} \\ \rho_{X8} X_{B,H} \end{pmatrix}, \quad Q = 0 \tag{2.109}$$

with ρ_{X7} and ρ_{X8} the specific (i.e. per $X_{B,H}$ units) hydrolysis rates of reactions (6) and (7), $Y_B = i_{XB} - f_P i_{XP}$, and $i_{XA} = i_{XB} + \frac{1}{Y_A}$.

Note the great similarity between the Petersen matrix of the IWA Activated Sludge Model No. 1 and the matrix Y. The matrix is simply the transpose[2] of the Petersen matrix where columns 1, 3 and 13 have been deleted (see footnote of Section 2.3.3).

Two step nitrification. In the nitrification example, the different terms of (2.100) are defined as follows (N = 5, M = 2):

[2]Indeed the rows correspond to the reactions in the IWA Activated Sludge Model No. 1, and to the process components in the matrix Y of the General Dynamical Model (2.100). Conversely, the columns correspond to the process components in the matrix Y of the General Dynamical Model (2.100), and to the reactions in the IWA Activated Sludge Model No. 1.

$$\xi = \begin{pmatrix} S_{NH} \\ S_{NO_2} \\ S_{NO_3} \\ X_1 \\ X_2 \end{pmatrix}, \quad Y = \begin{pmatrix} -\frac{1}{Y_1} & 0 \\ Y_3 & -\frac{1}{Y_2} \\ 0 & Y_4 \\ 1 & 0 \\ 0 & 1 \end{pmatrix}, \quad F = \begin{pmatrix} DS_{NH,in} \\ 0 \\ 0 \\ 0 \\ 0 \end{pmatrix} \qquad (2.110)$$

$$\rho = \begin{pmatrix} \rho_1 \\ \rho_2 \end{pmatrix} = \begin{pmatrix} \mu_1 X_1 \\ \mu_2 X_2 \end{pmatrix}, \quad Q = 0 \qquad (2.111)$$

Two step denitrification. In the denitrification example, the different terms of (2.100) are defined as follows:

$$\xi = \begin{pmatrix} S_{NO_3} \\ S_{NO_2} \\ S \\ X \\ N_2 \end{pmatrix}, \quad Y = \begin{pmatrix} -\frac{1}{Y_1} & 0 \\ Y_3 & -\frac{1}{Y_2} \\ -\frac{1}{Y_4} & -\frac{1}{Y_5} \\ 1 & 1 \\ 0 & Y_6 \end{pmatrix}, \quad F = \begin{pmatrix} DS_{NO_3,in} \\ 0 \\ DS_{in} \\ 0 \\ 0 \end{pmatrix}, \quad Q = \begin{pmatrix} 0 \\ 0 \\ 0 \\ 0 \\ Q_{N_2} \end{pmatrix}$$

$$\rho = \begin{pmatrix} \rho_1 \\ \rho_2 \end{pmatrix} = \begin{pmatrix} \mu_1 X \\ \mu_2 X \end{pmatrix} \qquad (2.112)$$

2.5   Multi Tank and Non Completely Mixed Reactors

In the preceding sections, we have considered the mass balances in tanks where the medium is homogeneous. However in many practical situations, this is not the case due to a lack of sufficient mixing: this is typically the situation of large-scale reactors, for which it is almost impossible to provide proper stirring to homogenise the reactor medium. However, the non complete mixing may also be deliberate; there is even a trend for increased use of processes which intrinsically are characterised by non homogeneous conditions. Typical examples are fixed bed and fluidised bed reactors, in which the (auto)catalysts are either fixed on some support, or maintained in suspension in the tank. Another obvious example is the settler which is used to separate biomass from the treated wastewater. Finally, plug flow reactors are also characterised by non-perfect mixing conditions resulting in (deliberate) concentration profiles along the length of the reactor.

Well-known from engineering text books is the simplest way to model non homogeneous tanks by considering them as a sequence of reactors. More fundamentally, the dynamics of fixed bed or fluidised bed reactors and settlers can be derived by considering mass balances in small ("infinitesimal") volumes: this will result in partial differential equation (PDE) models, i.e. models in which the state variables (the process components) do not only depend on time but also on the position in the tank; these models also most often include partial derivatives of the state variables with respect to time and space. The finite difference of fixed bed reactor or settler PDE models will result in sequential reactor models: this clearly emphasises the relation between both types of models. In the following, we shall first consider

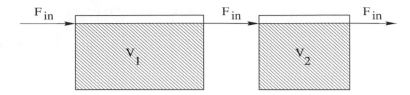

FIG. 2.12. Two sequential tanks.

the sequential reactor model, then the PDE models for fixed bed reactors (with and without axial and radial dispersion), fluidised bed reactors, and settlers.

2.5.1 Sequential Reactors

Let us first consider two sequential stirred tank reactors (Figure 2.12) of volumes V_1 and V_2, respectively, and in which a simple microbial growth reaction $S \longrightarrow X$ takes place. Let us denote each tank by an index i: i = 1 for the first tank, and 2 for the second. The mass balance for the substrate concentration S in both tanks is written as follows:

$$\frac{dS_1}{dt} = \frac{F_{in}}{V_1}S_{in} - \frac{F_{in}}{V_1}S_1 - \frac{1}{Y}\mu_1 X_1 \tag{2.113}$$

$$\frac{dS_2}{dt} = \frac{F_{in}}{V_2}S_1 - \frac{F_{in}}{V_2}S_2 - \frac{1}{Y}\mu_2 X_2 \tag{2.114}$$

Similarly for the biomass, the mass balance in both tanks gives rise to the following set of differential equations:

$$\frac{dX_1}{dt} = -\frac{F_{in}}{V_1}X_1 + \mu_1 X_1 \tag{2.115}$$

$$\frac{dX_2}{dt} = \frac{F_{in}}{V_2}X_1 - \frac{F_{in}}{V_2}X_2 + \mu_2 X_2 \tag{2.116}$$

in which μ_i is a function of the variables in tank i (i = 1 or 2), e.g. of S_1 and X_1 for μ_1, and of S_2 and X_2 for μ_2. An important feature of the substrate concentration equations is that the influent substrate concentration in tank 2 is the effluent substrate concentration from tank 1. The same remark holds for the biomass.

The generalisation to N (> 2) sequential reactors is then straigthforward. For instance, the mass balance of the substrate concentration in N STRs is written as follows:

$$\frac{dS_1}{dt} = \frac{F_{in}}{V_1}S_{in} - \frac{F_{in}}{V_1}S_1 - \frac{1}{Y}\mu_1 X_1 \tag{2.117}$$

$$\frac{dS_2}{dt} = \frac{F_{in}}{V_2}S_1 - \frac{F_{in}}{V_2}S_2 - \frac{1}{Y}\mu_2 X_2 \tag{2.118}$$

$$\vdots$$

$$\frac{dS_N}{dt} = \frac{F_{in}}{V_N}S_{N-1} - \frac{F_{in}}{V_N}S_N - \frac{1}{Y}\mu_N X_N \tag{2.119}$$

For the biomass, it is equal to:

$$\frac{dX_1}{dt} = -\frac{F_{in}}{V_1}X_1 + \mu_1 X_1 \tag{2.120}$$

$$\frac{dX_2}{dt} = \frac{F_{in}}{V_2}X_1 - \frac{F_{in}}{V_2}X_2 + \mu_2 X_2$$

$$\vdots \tag{2.121}$$

$$\frac{dX_N}{dt} = \frac{F_{in}}{V_N}X_{N-1} - \frac{F_{in}}{V_N}X_N + \mu_N X_N \tag{2.122}$$

The General Dynamical Model (2.100) formalism is still valid. Indeed, if for instance we choose to group the components by tank, i.e. if we define the component concentration vector ξ as follows:

$$\xi^T = \begin{pmatrix} S_1 & X_1 & S_2 & X_2 & \cdots & S_N & X_N \end{pmatrix} \tag{2.123}$$

the above mass balance equations can be written in the General Dynamical Model (2.100) framework by considering the following definitions:

$$Y = \begin{pmatrix} -\frac{1}{Y} & 0 & \cdots & 0 \\ 1 & 0 & \cdots & 0 \\ 0 & -\frac{1}{Y} & \cdots & 0 \\ 0 & 1 & \cdots & 0 \\ \vdots & \vdots & \vdots & \vdots \\ 0 & \cdots & 0 & -\frac{1}{Y} \\ 0 & \cdots & 0 & 1 \end{pmatrix} \tag{2.124}$$

$$D = \begin{pmatrix} \frac{F_{in}}{V_1} & 0 & 0 & 0 & \cdots & 0 & 0 & 0 \\ 0 & \frac{F_{in}}{V_1} & 0 & 0 & \cdots & 0 & 0 & 0 \\ -\frac{F_{in}}{V_2} & 0 & \frac{F_{in}}{V_2} & 0 & \cdots & 0 & 0 & 0 \\ 0 & -\frac{F_{in}}{V_2} & 0 & \frac{F_{in}}{V_2} & \cdots & 0 & 0 & 0 \\ \vdots & \vdots & \vdots & \vdots & \vdots & \vdots & \vdots & \vdots \\ 0 & 0 & 0 & \cdots & -\frac{F_{in}}{V_N} & 0 & \frac{F_{in}}{V_N} & 0 \\ 0 & 0 & 0 & \cdots & 0 & -\frac{F_{in}}{V_N} & 0 & \frac{F_{in}}{V_N} \end{pmatrix} \tag{2.125}$$

$$\rho = \begin{pmatrix} \mu_1 X_1 \\ \mu_2 X_2 \\ \vdots \\ \mu_N X_N \end{pmatrix}, \quad F = \begin{pmatrix} D S_{in} \\ 0 \\ \vdots \\ 0 \end{pmatrix}, \quad Q = 0 \tag{2.126}$$

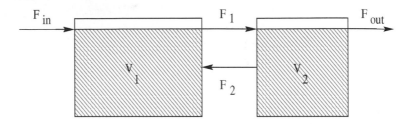

FIG. 2.13. Two interconnected tanks.

A first difference between the single tank model and the multi-tank one is, as it was already mentioned in the basic activated sludge model, that now the dilution rate is a matrix. A second difference is the increase of the dimension of the reaction rate ρ, and in consequence, of the yield coefficient matrix Y. Note finally the above matrix presentation is arbitrary: we could also have chosen to write the vector ξ by grouping the terms by component, i.e. the first rows for S and the remaining ones for X (yet the General Dynamical Model (2.100) formalism works!)(see also [56] for further details on multi-tank bioreactor models).

Let us end this section by the model of two interconnected CSTRs (Figure 2.13). Note that if the volumes V_1 and V_2 are constant, then $F_{out} = F_{in}$ and $F_2 = F_1 - F_{in}$. If we consider as before a simple microbial growth reaction in both tanks, the mass balance of the substrate concentration is then equal to:

$$\frac{dS_1}{dt} = \frac{F_{in}}{V_1}S_{in} + \frac{F_1 - F_{in}}{V_1}S_2 - \frac{F_1}{V_1}S_1 - \frac{1}{Y}\mu_1 X_1 \qquad (2.127)$$

$$\frac{dS_2}{dt} = \frac{F_1}{V_2}S_1 - \frac{F_1}{V_2}S_2 - \frac{1}{Y}\mu_2 X_2 \qquad (2.128)$$

The use of interconnected CSTR equations like (2.127)(2.128) is often considered to simulate the lack of medium homogeneity, e.g. due to insufficient stirring in reactors.

2.5.2 Fixed Bed Reactor: The Basic Mass Balance Model

Let us consider an example of a non completely mixed reactor: the fixed bed reactor, i.e. a tank in which the (auto)catalysts are *fixed* on some (solid) support and in which the reactants (in the liquid phase) are flowing through and are transformed when in contact with the (fixed) (auto)catalysts. Recently, a number of such reactor systems have been developed as they are capable of maintaining high autocatalyst concentrations and, hence, can sustain a high volumetric loading rate. Examples are submerged trickling filters for denitrification.

There is no specific stirring mechanism, and generally speaking, the reactor is not in completely mixed conditions anymore. Therefore in terms of (dynamic) modelling, the approach has to be modified: the mass balance has to be computed

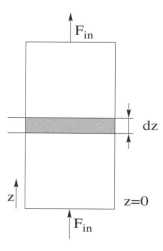

FIG. 2.14. Schematic view of a fixed bed reactor.

on a thin section dz of the reactor (and not anymore on the whole volume of the reactor) (see Figure 2.14).

Let us first concentrate on the dynamical model derivation for fixed bed reactors with axial dispersion, and for which radial dispersion is negligible. Assume that the dispersion phenomenon obeys Fick's diffusion law which expresses that the perpendicular flux of particles j through a unit surface is proportional to the gradient of concentration at the surface according to the following equation:

$$j = -D\frac{\partial C}{\partial z} \qquad (2.129)$$

where D is a diffusion coefficient and z is the direction of the flux j. Indeed the "microscopic" phenomenon of diffusion has been by analogy extended to the "macroscopic" reactor situation, and the terminology "dispersion" is more often used than "diffusion" in fixed bed or fluidised bed reactor modelling. Let us note D_{ma} $(m^2.s^{-1})$ the axial mass dispersion coefficient.

For simplicity, we also consider that the reactor is tubular; this implies in particular that the cross-section of the reactor is constant (equal to A (m^2)). Furthermore this section is the sum of the section occupied by the (auto)catalysts (i.e. a "solid" section, A_S (m^2)) and of the section left free for the flow of (liquid) reactants and products, A_L (m^2) $(A = A_S + A_L)$.

The reactions are assumed to be autocatalytic and to take place on the *solid* catalytic support. Let us further assume that the mass transfer dynamics between the solid and liquid phases are negligible (This assumption may be formally justified by considering a *quasi steady state approximation*, e.g. [85], see also Section 2.9)

In order to explicit the model derivation, let us begin with the mass balance on the thin section dz of a component of concentration C ($kg.m^{-3}$) which intervenes in one reaction characterised by a reaction rate ρ ($kg.m^{-3}.s^{-1}$) (since the reaction takes place on the *solid* support, the dimension of the reaction is mass per time unit per *solid* volume (i.e. $A_S dz$):

$$\frac{\partial}{\partial t}[A_L dz C] \qquad = \qquad F_{in} C \qquad \qquad -F_{in}(C + \frac{\partial C}{\partial z} dz)$$

$$\text{time variation} \qquad\qquad \text{inflow} \qquad\qquad\qquad \text{outflow}$$
$$\text{of the mass of C} \qquad\qquad \text{at position z} \qquad\qquad \text{at position z + dz}$$

$$-D_{ma} A_L \frac{\partial C}{\partial z} \qquad\qquad -(-D_{ma} A_L \frac{\partial}{\partial z}[C + \frac{\partial C}{\partial z} dz])$$
$$\text{diffusion} \qquad\qquad\qquad \text{diffusion}$$
$$\text{at position z} \qquad\qquad\qquad \text{at position z + dz}$$

$$+A_S dz\, \rho(z,t)$$
$$\text{reaction rate}$$
$$\text{between z and z + dz}$$

where F_{in} is the influent flow rate ($m^3.s^{-1}$). If one divides both sides by $A_L dz$ and defines the void fraction ϵ and the fluid superficial velocity u ($m.s^{-1}$):

$$\epsilon = \frac{A_L}{A}, \ u = \frac{F_{in}}{A_L} \qquad\qquad (2.130)$$

the above equation can be reduced to the following expression:

$$\frac{\partial C}{\partial t} = -u\frac{\partial C}{\partial z} + D_{ma}\frac{\partial^2 C}{\partial z^2} + \frac{1-\epsilon}{\epsilon} r(z,t) \qquad\qquad (2.131)$$

Note that the above development assumes a constant cross-sectional area A and constant individual terms A_L and A_S.

2.5.3   *General Dynamical Model of Fixed and Fluidised Bed Reactors*

Fixed Bed Reactors with Axial Dispersion. Let us now generalise the model for fixed bed reactors with N components involved in M reactions [73]. Let us first assume that among the N process components, N_{fi} are microorganisms entrapped or fixed on some support and, hence, remain within the reactor, and N_{fl} other reactants (essentially substrates and products) flow through the reactor. In addition, let us define two associated component vectors, x_{fi} and x_{fl}, respectively.

From mass balance considerations on a section dz, we can deduce the following *General Dynamical Model for fixed bed reactors with axial dispersion*:

$$\frac{\partial x_{fi}}{\partial t} = Y_{fi}\, \rho(x_{fi}, x_{fl}) \qquad\qquad (2.132)$$

$$\frac{\partial x_{fl}}{\partial t} = -u\frac{\partial x_{fl}}{\partial z} + D_{ma}\frac{\partial^2 x_{fl}}{\partial z^2} + \frac{1-\epsilon}{\epsilon}Y_{fl}\,\rho(x_{fi}, x_{fl}) \qquad (2.133)$$

where $\rho(x_{fi}, x_{fl})$ is the reaction rate vector $(kg.m^{-3}.s^{-1})$:

$$\rho^T(x_{fi}, x_{fl}) = [\rho_1(x_{fi}, x_{fl}), \rho_2(x_{fi}, x_{fl}), ..., \rho_M(x_{fi}, x_{fl})] \qquad (2.134)$$

and Y_{fi} and Y_{fl} are the yield coefficient matrices.

Remark: Note that, as in the completely mixed reactors, the above matrix formulation emphasises the presence of two terms in the process dynamics:

- *conversion:* $Y_{fi}\,\rho(x_{fi}, x_{fl})$ and $\frac{1-\epsilon}{\epsilon}Y_{fl}\,\rho(x_{fi}, x_{fl})$;
- *transport dynamics:* $-u\frac{\partial x_{fl}}{\partial z} + D_{ma}\frac{\partial^2 x_{fl}}{\partial z^2}$.

Boundary conditions have still to be added to complete the model. Because of the presence of the second order derivative with respect to z, two boundary conditions are required: typically, one at the reactor inlet ($z = 0$) and one at the outlet ($z = L$). Although they are the object of criticism in some instances (e.g. [85]), the most largely used boundary conditions are those of Danckwerts [63]:

$$D_{ma}\frac{\partial x_{fl}}{\partial z} = -u(x_{in} - x_{fl}) \text{ at } z = 0 \qquad (2.135)$$

$$\frac{\partial x_{fl}}{\partial z} = \quad 0 \quad \text{ at } z = L \qquad (2.136)$$

with x_{in} the influent value of x_{fl}.

Example: biochemical reactor with a growth reaction and a death/detachment reaction
Let us now consider a fixed bed reactor with a growth reaction and a death/detachment reaction:

$$\text{growth} : S \longrightarrow X + P \qquad (2.137)$$

$$\text{death} : X \longrightarrow X_d \qquad (2.138)$$

If we assume that the non-active microorganisms detach and leave the bioreactor, the dynamics of the bioprocess (2.137)(2.138) will be described by the following equations:

$$\frac{\partial X}{\partial t} = \mu X - bX \qquad (2.139)$$

$$\frac{\partial S}{\partial t} = -u\frac{\partial S}{\partial z} + D_{ma}\frac{\partial^2 S}{\partial z^2} - \frac{1-\epsilon}{\epsilon}\frac{1}{Y_S}\mu X \qquad (2.140)$$

$$\frac{\partial P}{\partial t} = -u\frac{\partial P}{\partial z} + D_{ma}\frac{\partial^2 P}{\partial z^2} + \frac{1-\epsilon}{\epsilon}Y_P\mu X \tag{2.141}$$

$$\frac{\partial X_d}{\partial t} = -u\frac{\partial X_d}{\partial z} + D_{ma}\frac{\partial^2 X_d}{\partial z^2} + \frac{1-\epsilon}{\epsilon}bX \tag{2.142}$$

$$z = 0 \;:\; D_{ma}\frac{\partial S(0,t)}{\partial z} = -u(S_{in}(t) - S(0,t))$$

$$D_{ma}\frac{\partial P(0,t)}{\partial z} = uP(0,t)$$

$$D_{ma}\frac{\partial X_d(0,t)}{\partial z} = uX_d(0,t) \tag{2.143}$$

$$z = L \;:\; \frac{\partial S(L,t)}{\partial z} = \frac{\partial P(L,t)}{\partial z} = \frac{\partial X_d(L,t)}{\partial z} = 0 \tag{2.144}$$

The above equations can be rewritten in the formalism of equations (2.132)(2.133) (2.135)(2.136) by considering the following definitions:

$$x_{fi} = X, \; x_{fl} = \begin{pmatrix} S \\ P \\ X_d \end{pmatrix} \tag{2.145}$$

$$x_{in} = \begin{pmatrix} S_{in} \\ 0 \\ 0 \end{pmatrix}, \; \rho = \begin{pmatrix} \rho_1 \\ \rho_2 \end{pmatrix} = \begin{pmatrix} \mu X \\ bX \end{pmatrix} \tag{2.146}$$

$$Y_{fi} = \begin{pmatrix} 1 & -1 \end{pmatrix}, \; Y_{fl} = \begin{pmatrix} -\frac{1}{Y_S} & 0 \\ Y_P & 0 \\ 0 & 1 \end{pmatrix} \tag{2.147}$$

Extension 1: Plug Flow Reactor. The dynamical model of the fixed bed reactor in absence of dispersion, i.e. of the plug flow reactor, is readily obtained from the equations (2.132)(2.133)(2.135) by simply setting the dispersion coefficient D_{ma} to zero ($D_{ma} = 0$). Since there is then only a first order derivative of the state variable x_{fl} with respect to the space variable z, only the boundary condition at the reactor input ($z = 0$) (2.135) is kept, and the boundary condition at the reactor output (2.136) is dropped. The model equations are then:

$$\frac{\partial x_{fi}}{\partial t} = Y_{fi}\,\rho(x_{fi}, x_{fl}) \tag{2.148}$$

$$\frac{\partial x_{fl}}{\partial t} = -u\frac{\partial x_{fl}}{\partial z} + \frac{1-\epsilon}{\epsilon}Y_{fl}\,\rho(x_{fi}, x_{fl}) \tag{2.149}$$

$$x_{fl} = x_{in} \qquad\qquad at\; z = 0 \tag{2.150}$$

The frequent description of an activated sludge plant as a plug flow system is just a special case of the above in which no fixed biomass is present and hence equation(2.148) vanishes.

Extension 2: Gas phase. Assume that one product gives off in the gaseous phase, as in high rate fixed film anaerobic digesters (CO_2, CH_4) [67] or denitrification reactors (N_2) [102], [131]; or that the fixed bed reactor is aerated for aerobic carbon removal [172] or nitrification [283]. Note that the very widespread trickling filter also has to be described as a three phase fixed bed reactor [224]. The equations for fixed bed systems (2.132)-(2.136) will be modified by introducing the gaseous flow rate vector Q ($kg.s^{-1}$). By introducing similar arguments as in equation (2.131), a new term, the derivative of Q with respect to z, is introduced in the model formulation:

$$\frac{\partial x_{fi}}{\partial t} = Y_{fi}\rho(x_{fi}, x_{fl}) \tag{2.151}$$

$$\frac{\partial x_{fl}}{\partial t} = -u\frac{\partial x_{fl}}{\partial z} + D_{ma}\frac{\partial^2 x_{fl}}{\partial z^2} + \frac{1-\epsilon}{\epsilon}Y_{fl}\,\rho(x_{fi}, x_{fl}) - \frac{1}{A_L}\frac{\partial Q}{\partial z} \tag{2.152}$$

with the same boundary conditions (2.135)(2.136) as above. Note that the section A is now the sum of three sections, the solid, liquid and gaseous sections: $A = A_S + A_L + A_G$. In the above model formulation, it has been implicitly assumed that the section A_G is negligible with respect to A_S and A_L. If this assumption is not correct, as in trickling filters, then the term $(1 - \epsilon)/\epsilon$ should be replaced by A_S/A_L.

Note that the above model extension to gas production is only valid if the dynamics between the liquid and gas phases are fast enough to be considered as being negligible, i.e. the gas and liquid phases are assumed to be in equilibrium.

As a matter of illustration, assume that in the bioreactor example (2.137) (2.138), the product P gives off in the gaseous phase. Then the model is completed with the following gaseous flow rate vector:

$$Q = \begin{pmatrix} 0 \\ Q_P \\ 0 \end{pmatrix} \tag{2.153}$$

where Q_P is the gaseous flow rate of the product P.

Extension 3: Radial Dispersion. Here again the extension (e.g. [85], [145]) is straightforward if one introduces a new spatial variable, the radial coordinate r in case of a tubular reactor. The model is readily obtained by using arguments similar to those used for the derivation of equations (2.132)(2.133)(2.135)(2.136): it results in an addition of a radial dispersion term and of two boundary conditions (at $r = 0$ and $r = R$, with R the radius of the reactor, see Figure 2.15). In the presence of radial dispersion, the two-phase model (2.132)(2.133)(2.135)(2.136) (i.e. without the gas phase) becomes:

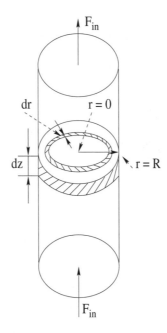

FIG. 2.15. Schematic view of a fixed bed reactor with radial dispersion.

$$\frac{\partial x_{fi}}{\partial t} = Y_{fi}\ \rho(x_{fi}, x_{fl}) \tag{2.154}$$

$$\frac{\partial x_{fl}}{\partial t} = -u\frac{\partial x_{fl}}{\partial z} + D_{ma}\frac{\partial^2 x_{fl}}{\partial z^2} + D_{mr}\frac{1}{r}\frac{\partial}{\partial r}(r\frac{\partial x_{fl}}{\partial r})$$

$$+\frac{1-\epsilon}{\epsilon}Y_{fl}\ \rho(x_{fi}, x_{fl}) \tag{2.155}$$

$$D_{ma}\frac{\partial x_{fl}}{\partial z} = -u(x_{in} - x_{fl}) \qquad\qquad at\ z = 0 \tag{2.156}$$

$$\frac{\partial x_{fl}}{\partial z} = 0 \qquad\qquad at\ z = L \tag{2.157}$$

$$\frac{\partial x_{fl}}{\partial r} = 0 \qquad\qquad at\ r = 0 \tag{2.158}$$

$$\frac{\partial x_{fl}}{\partial r} = 0 \qquad\qquad at\ r = R \tag{2.159}$$

where D_{mr} $(m^2.s^{-1})$ is the radial mass dispersion coefficient.

Extension 4: Fluidised Bed Reactors.   Fluidised bed technology as many others stems from chemical engineering technology, and has its roots in solid bed reactors

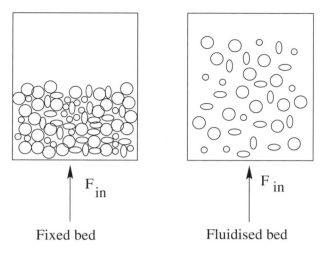

FIG. 2.16. Fixed bed versus fluidised bed.

[177]. In fluidised bed reactors, the solid (catalytic) particles are "fluidised" by the stream of liquid (most common) or gas [54] [249] from below the reactor as the upflow velocity increases. Figure 2.16 illustrates the difference between a fixed (or "packed") bed reactor and a fluidised bed. If the fluid velocity is low, the particles are packed at the bottom of the reactor: the reactor behaves like a fixed bed reactor since the solid particles do not move upwards. For larger values of the fluid or gas velocity, the solid particles will be "pushed" upwards and may be maintained in suspension for appropriate hydrodynamical conditions, i.e. an expanded bed is formed.

The oldest application of fluidisation [227] in biotechnology is in the area of nitrification and denitrification of wastewater and drinking water treatment [243] [84]. Another important application area of biofluidisation is in the field of anaerobic digestion with the very widespread use of the UASB (Upflow Anaerobic Sludge Bed) reactors [67] [82].

The dynamical modelling of fluidised bed reactors requires even more attention and care. In terms of dynamical modelling, the basic difference between fixed bed and fluidised bed reactors is that in the latter, the solid phase (basically the catalyst on its support) is not fixed but is composed of particles in suspension in the reactor. A major implication is that now, generally speaking, the void fraction ϵ cannot be considered as being constant, especially during expansion and contraction of the bed.

At this point there are two possible options.

1. Either to consider that the transient phase during bed expansion and contraction is very fast and to neglect its dynamics. Then the void fraction ϵ in steady state can be deduced by considering the (heuristic but largely used)

Richardson and Zaki's law [216]:

$$\epsilon = (\frac{u_0}{U_T})^{\frac{1}{n}} \qquad (2.160)$$

where u_0 is the fluid superficial velocity in the absence of solid particles, U_T is the terminal settling velocity of the particles and n is the expansion index (which typically depends on the type of particles and medium).

2. Or to consider the dynamics of the bed expansion and contraction. Then extra equations are required. Typically the mass balances equations (also sometimes called the *continuity equations*)(for instance, (2.132)(2.133) or (2.148)(2.149)) are completed with momentum equations [89]. In the plug flow case, these are written as follows for the liquid and particle phases, respectively:

$$\rho_L \frac{\partial}{\partial t}[\epsilon u] = -\frac{\partial}{\partial z}[\epsilon(p_L + \frac{1}{2}\rho_L u^2)] - \epsilon\, \rho_L\, g - F_i$$

$$\rho_S \frac{\partial}{\partial t}[(1-\epsilon)u_S] = -\frac{\partial}{\partial z}[(1-\epsilon)(p_S + \frac{1}{2}\rho_L u^2)] - (1-\epsilon)\, \rho_S\, g + F_i$$

where ρ_L, ρ_S, p_L, p_S, g, F_i and u_S are the liquid phase and particle phase densities $(kg.m^{-3})$, the liquid and particle pressures $(N.m^{-2})$, the gravity constant $(m.s^{-2})$, the liquid-particle interaction force per unit volume $(N.m^{-3})$ and the particles' velocity in the reactor $(m.s^{-1})$, respectively.

Moreover the Richardson and Zaki's equation has to be modified to account for the transient of the solid particles (e.g. [199]:

$$u - u_S\epsilon = U_T\epsilon^{n-1} \qquad (2.161)$$

The choice of appropriate expressions for F_i, p_L and p_S is still a matter of discussion in the scientific literature. As a matter of illustration, the following expressions are recommended e.g. in Foscolo and Gibilaro [89], and Jean and Fan [132] for F_i, p_L and p_S:

$$F_i = (1-\epsilon)\rho_L g + (1-\epsilon)(\rho_S - \rho_L)g(\frac{u_0 - u_S}{U_T})^{\frac{4.8}{n}}\epsilon^{-4.8} \quad (2.162)$$

$$p_S = 0, \quad \epsilon p_L = -(\epsilon\rho_L + (1-\epsilon)\rho_S)gz \qquad (2.163)$$

Another important equation is the volume balance at any position in the reactor:

$$\epsilon u + (1-\epsilon)u_S = u_0 \qquad (2.164)$$

Because of the much lower complexity of the model, the first option is very often considered in many applications.

An example of dynamical model of an anaerobic digestion process in a fluidised bed reactor is given in Bonnet *et al.* [30].

2.5.4 *Settlers*

Solid flux theory basic model.   In Section 2.5.2, we have considered that the dynamics of the settler (Figure 2.17) are often described by those of a CSTR. Largely because the model neglects sedimentation effects, the CSTR assumption is more and more an object of criticism in the scientific community, and alternative models are presently recommended. Let us first recall that the main functions of secondary settlers in wastewater treatment plants are double:

1. **clarification**, i.e. separation of the biomass from the treated wastewater in order to produce a solid-free effluent;
2. **thickening** of the biomass at the bottom of the settler to be recycled back into the aerator.

In some cases the settler is also considered for more advanced use, i.e.

1. **sludge storage**, i.e. in the bottom part of the settler sludge is stored for subsequent use under high waste load conditions;
2. **the settler as a reactor** where additional aerobic conversions can occur or where denitrification may take place [230] (with the danger of formation of nitrogen gas bubbles and rising sludge problems [119]).

Hence, to be able to evaluate the functioning of the settler under dynamic conditions, models must be able to describe these different aspects in more or less detail [83].

The alternative models are usually based on (or at least refer to) mass balances (here again, the terminology "continuity equation" is largely used) and on the Solid Flux Theory [69]. The mass balance is computed not anymore on the whole tank but on a small section dz (see Figure 2.17). If we recall the mass balance considerations of Section 2.5.2, the dynamical model in its simplest form (e.g. [170]) is written as follows:

$$\frac{\partial X}{\partial t} = -\frac{\partial F_l}{\partial z} \tag{2.165}$$

$$X(z = H, t) = X_R(t) \tag{2.166}$$

where F_l is the biomass solid flux $(kg.m^{-2}.h^{-1}])$ at depth z of the settler ($z \in [0, H]$, where 0 and H correspond to the top and bottom of the settler, respectively). The flux is basically due to two phenomena: sedimentation (*settling*) due to gravity, and "bulk" flow due to sludge withdrawal at the bottom. Note that, in the above equation (2.165), only convection ($\partial F_l / \partial z$) is considered.

The total flux $F_l(t, z)$ can be expressed as follows:

$$F_l(z, t) = vX + uX \tag{2.167}$$

where v is the settling velocity, and u is the bulk velocity, i.e. the velocity of the recycled biomass mixed liquor due to underflow pumping. With u simply the ratio of the flow rate $F_R + F_w$ over the settler section A_{se}:

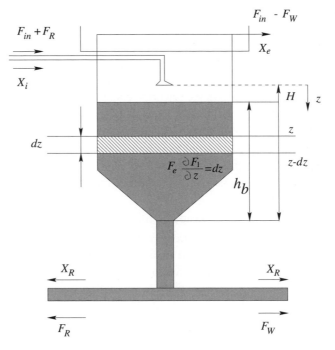

FIG. 2.17. Schematic view of a secondary settler.

$$u = \frac{F_R + F_w}{A_{se}} \qquad (2.168)$$

and if we denote:

$$vX = f(X) \qquad (2.169)$$

then the mass balance equations (2.165)(2.166) are rewritten as follows:

$$\frac{\partial X}{\partial t} = -(\frac{F_R + F_w}{A_{se}} + \frac{\partial f(X)}{\partial X})\frac{\partial X}{\partial z} + X\frac{F_R + F_w}{A_{se}^2}\frac{dA_{se}}{dz} \qquad (2.170)$$

$$X(H, t) = X_R(t) \qquad (2.171)$$

There exists in the literature a number of different models for the settling velocity v, e.g.:

- Vesilind exponential model:

$$v = v_0 e^{-aX}, \qquad v_0, a > 0 \qquad (2.172)$$

- Power law:

$$v = nX^{-a}, \qquad n, a > 0 \qquad (2.173)$$

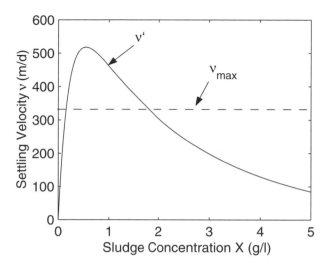

FIG. 2.18. Settling velocity: Takacs model.

- Inverse power law:

$$v = \frac{a}{b + X^n}, \qquad n, a, b > 0 \qquad (2.174)$$

- Takacs model ("generalised Vesilind model") [245]:

$$v = v' \qquad if\, v' \leq v_{max} \qquad (2.175)$$
$$ = v_{max} \qquad if\, v' > v_{max} \qquad (2.176)$$
$$with \quad v' = v_0(e^{-r_h(X - X_{min})} - e^{-r_p(X - X_{min})}),$$
$$v_0, X_{min} > 0, \ r_p > r_h > 0 \qquad (2.177)$$

with v_{max} the maximum attainable settling velocity, X_{min} the minimum attainable suspended solids concentration, and r_h and r_p the settling characteristics of the hindered settling zone and of the low solids concentration, respectively. The model is graphically represented in Figure 2.18. Note that with that model, the maximum attainable settling velocity v_{max} is generally speaking different from (lower than) the maximum of v' which is reached for a value of X equal to:

$$X = X_{min} + \frac{1}{r_p - r_h} \ln\left(\frac{r_p}{r_h}\right) \qquad (2.178)$$

- Cho model [59]

$$v = v_0 \frac{e^{-aX}}{X}, \qquad v_0, a > 0 \qquad (2.179)$$

Remark: the model (2.170)(2.171) only considers one spatial coordinate: this means that we have implicitly assumed that the settler dynamics are uniform at each section dz. More complicated models can also be derived when this assumption is not fulfilled in practice, with 2 or 3 (for non-circular section settlers) space coordinates (see for instance [244]): this type of model also includes momentum conservation equations.

Steady-state curves of the settler model. Before going further, let us see what is the relative importance of the settling term and of the "bulk" flow term, at least in steady state. Figure 2.19 exhibits some curves, in steady state, of solid fluxes F_l as a function of the biomass X for the Vesilind model (2.172) and with the bulk velocity u as an operational parameter. It shows that for a given value of u, the recycled biomass concentration X_R is uniquely determined by the solid flux established in the settler. Indeed the minimum solid flux F^* represents the maximum rate of solids transferred through the liquid/solid interface for a given settler geometry and sludge settling behaviour. Excess solids which cannot be transferred to the bottom accumulate above the interface. On the other hand, in steady state, the underflow solids concentration is not a dynamic variable but depends on the limiting flux and available mass above the interface. If the settler is critically loaded, the total limiting flux through the thickening zone, i.e. $A_{se}F^*$ is equal to the underflow withdrawal, i.e. $(F_R + F_W)X_R$:

$$A_{se}F^* = (F_R + F_W)X_R \qquad (2.180)$$

This gives the value of the recycled biomass:

$$X_R = \frac{A_{se}F^*}{F_R + F_W} \qquad (2.181)$$

For the Vesilind model, v tends to zero if X tends to infinity:

$$\lim_{X \to \infty} v = 0 \qquad (2.182)$$

This implies that the solid flux F_l tends to uX for large values of X. Therefore the value of X_R can be graphically (Figure 2.19) determined from the intersection of F^* and uX (which is the asymptote of F_l for X tending to infinity).

Limitations of the solid flux theory basic model. The major drawback of the Solid Flux Theory basic model (2.165)(2.166) is its inability to emphasise concentration gradients at least in steady state. Indeed in steady state ($\frac{\partial X}{\partial t} = 0$) for settlers with constant section, we have:

$$\frac{dF}{dz} = 0 \qquad (2.183)$$

i.e., since all the proposed models for the settling velocity v are only functions of the sludge concentration X:

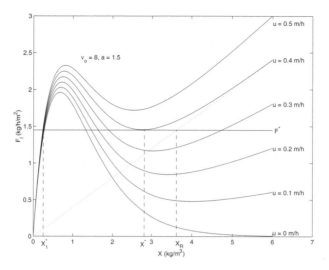

FIG. 2.19. Solid flux vs biomass concentration.

$$(u + v + X\frac{\partial v}{\partial X})\frac{dX}{dz} = 0 \qquad (2.184)$$

There are therefore a priori (mathematically) two possible steady state solutions of (2.183) [75], [205]:

$$1)\ u + v + X\frac{\partial v}{\partial X} = 0 \qquad (2.185)$$

$$2)\qquad \frac{dX}{dz} = 0 \qquad (2.186)$$

Generally speaking, the first steady state has to be rejected, because it implies that:

$$v = -u + K\frac{1}{X}, \qquad K \text{ constant} \qquad (2.187)$$

e.g. that v is a function of u in steady state. This also imposes one model structure for the dependence of the settling velocity v with respect to the sludge concentration X and the superficial fluid velocity u. v will even be negative for sufficiently large values of X ($> u/K$).

Therefore we end up with one possible steady state,

$$\frac{dX}{dz} = 0 \qquad (2.188)$$

which means that there is no spatial profile in steady state, in major contradiction with the experimental evidence (as illustrated in Figure 2.21 here below).

To circumvent the difficulty, different modifications have been proposed to the mass balance model. The most popular and most successful presently is the model proposed by e.g. [278],[245]. The model is based on two important assumptions: the first one is to consider a layered settler model (Figure 2.20), which can be viewed as a finite difference approximation of the space derivative in the mass balance model; the second one is a limitation of flux from one layer to another ("The mass flux into a layer cannot exceed the mass flux the volume is capable of passing, nor can it exceed the mass flux which the volume immediately below it is capable of passing").

The gap between the layered model and its assumed original model, the Solid Flux Theory based mass balance equations (2.165)(2.166)(2.167), with only a convection term ($\frac{\partial F_l}{\partial z}$), is indeed quite large. It is well known (e.g. [93], [73] and also Section 2.9.5) that a large number of "layers" (typically larger than 100) are necessary to obtain a satisfactory approximation of hyperbolic systems like (2.165)(2.166)(2.167) or plug flow reactor models (2.149)(2.150) that basically behave like delay systems (see also section 2.10), while a lower number of "layers" are necessary for approximating correctly parabolic systems with second order derivatives like the modified settler model (2.190) here below or the axial dispersion model (2.133)(2.135)(2.136). In practice, a limited number of layers are usually considered in the application of the layered model. This, combined with the flux limitation, contributes to make the layered model closer to a model with the second order derivatives (as proposed in the modified version of the settler model). Strangely, the bad level of approximation of the original model (i.e. a selection of not enough layers for a good approximation) results to a simulation model that tends to mimic and therefore behaves closer to a model that is able to emphasize the spatial gradient in settlers. This probably explains at least partially the popularity of the layered models in numerical simulation applications. We see therefore that the original model has been approximated in order to obtain a model that behaves more closely to the physical reality, but the connection of the resulting model with the original model is difficult to see. This renders even more difficult the interpretation of the resulting model with respect to the physical phenomenon that it is supposed to describe. It further motivates to go back to the basic physical phenomena to build physically sound models.

As we have just said, although the model proves quite successful in applications, the physical basis of the model is questionable. As clearly shown by Jeppsson ([133], there is no rigourous connection between the layered settler model and the "convection-type" Kynch theory derived model. The main problem is that the flux limitation should be inherently connected to the mass balance, and not introduced in the lumped model after the approximation procedure. In other words, the question could be formulated as follows: is it possible to derive the mass balance model such that a limitation of flux is implicitly included (i.e. a consequence of the structure of the physical model)?

A *possible* solution to the above question and to that of the presence of spatial gradient in settlers is indeed to introduce a second order derivative term with re-

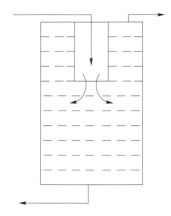

FIG. 2.20. Layered settler scheme.

spect to z in the mass balance model. A possible interpretation of such a term is that it represents diffusion following Fick's law. This type of term is largely used e.g. in chemical engineering to emphasise non perfect plug flow hydrodynamics and the presence of "back-mixing" effects in reactors like fixed-bed reactors. In such a case, the flux F is written as follows:

$$F = uX + vX - D_a \frac{\partial X}{\partial z} \tag{2.189}$$

and the mass balance model becomes:

$$\frac{\partial X}{\partial t} = -u \frac{\partial X}{\partial z} - \frac{\partial (vX)}{\partial z} + D_a \frac{\partial^2 X}{\partial z^2} \tag{2.190}$$

Let us now explain why this is a possible solution to introduce a spatial gradient in the settler model. In order to keep the line of reasoning as simple as possible, let us assume that the settling velocity v is constant (this assumption is only needed for this purpose, and this does not mean that we impose v to be constant in the settler model).

Let us then calculate the steady state solution (i.e. when $\frac{\partial X}{\partial t} = 0$) of the PDE equation with the second order derivative term. The above assumption (constant v) allows us to have an analytical solution of the steady state (it is indeed its main merit). In steady state, the mass balance equation then becomes:

$$D_a \frac{d^2 X}{dz^2} - (u + v) \frac{dX}{dz} = 0 \tag{2.191}$$

Let us consider here the following boundary conditions [3]:

[3] These have been chosen in accordance with the physical reality, but will not be very useful in practice for simulating the model, because it requires the values of the sludge concentrations at the settler's input and output.

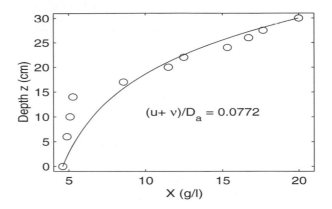

FIG. 2.21. Identification of model (2.193) with experimental data.

$$X(z = H) = X_R, \ X(z = 0) = X_i \qquad (2.192)$$

where X_R and X_i are the sludge concentrations at the output and at the input of the settler, respectively. Then the analytical solution has the following form:

$$X(z) = \frac{1}{1 - e^{\frac{u+v}{D_a}H}}[(X_i(e^{\frac{u+v}{D_a}z} - e^{\frac{u+v}{D_a}H}) - X_R(e^{\frac{u+v}{D_a}z} - 1)] \qquad (2.193)$$

Note that the above steady state expression of X depends explicitly on the spatial z ($e^{\frac{u+v}{D_a}z}$) in accordance with the experimental evidence (accumulation of sludge from the top to the bottom of the settler). This is illustrated on one typical steady state data set (from a cylindrical laboratory settler from the CEIT (San Sebastian, Spain) [253]): Figure 2.21 presents the results of the identification of the model equation (2.193) via a Levenberg-Marquardt identification routine compared to the data. The identification procedure gives the following value to the parameter:

$$\frac{u + v}{D_a} = 0.0772 \qquad (2.194)$$

It is worth noting that the above suggested approach is in line with a number of recent works in the area. A similar model with a dispersion term is also considered in [109]. The report of Vanrolleghem *et al.* [260] is dealing with the parameter identification of such a model. And another approach which results in a model which has some similarities is considered by Cacossa and Vaccari [46]. In their paper, the authors consider the following settling velocity expression:

$$v = v_m(1 - \frac{1}{K}\frac{\partial X}{\partial z}) \qquad (2.195)$$

where v_m and K are the maximum settling velocity and the compressibility function, respectively. Then the dynamical model of the settler is written as follows:

$$\frac{\partial X}{\partial t} = -u\frac{\partial X}{\partial z} - v_m\frac{\partial X}{\partial z} + \frac{v_m}{K}(\frac{\partial X}{\partial z})^2 + \frac{v_m}{K}X\frac{\partial^2 X}{\partial z^2} \qquad (2.196)$$

Here again a second order derivative term with respect to z has been introduced in the model.

The settler model equation (2.190) has only been considered to draw attention to the fact that with a second order derivative term with respect to z, a gradient in the sludge concentration can be emphasised. This is its prime advantage, but we do not suggest modelling settler dynamics using this format for, because it has at least two drawbacks to be feasible in more general practical situations.

1. The settling dynamics with respect to X are linear (v is constant): all the models proposed in the literature are nonlinear.

2. The boundary conditions are not very convenient, at least for numerical simulation (indeed the values of the concentrations at both inputs and outputs of the settler are needed to solve the equations, while the objective should be to be able to predict one (and also the distribution inside the settler) from the other). That's why we suggest to use, in line with those used for non completely mixed reactors (see Section 1.5.2), the following boundary conditions:

$$X(z = 0) = X_i, \quad \frac{\partial X}{\partial z}(z = H) = 0 \qquad (2.197)$$

A clarifier/sedimentation model. By using the arguments considered above, we can derive a model that combines the clarification and sedimentation aspects in the settler. The model derivation is based on Figure 2.22. It is also a one-dimensional model with respect to space. An important question is that of the boundary conditions at the inlet. Because the model considers only one dimension (vertical), the interface corresponding to the inlet can be considered as homogeneous. If we denote by X_1 and X_2 the sludge concentration in the settler and in the clarifier, respectively, the mass balance equations for clarifiers/settlers with constant cross sections are written as follows:

- Settler

$$\frac{\partial X_1}{\partial t} = -\frac{F_R + F_W}{A_{se}}\frac{\partial X_1}{\partial z} - \frac{\partial(v_1 X_1)}{\partial z} + D_a\frac{\partial^2 X_1}{\partial z^2} \qquad (2.198)$$

$$\frac{\partial X_1}{\partial z} = 0 \qquad\qquad\qquad\qquad\qquad \text{for } z = H_1 \quad (2.199)$$

$$D_a\frac{\partial X_1}{\partial z} = \frac{F_R + F_W}{A_{se}}(X_1 - X_i) - v_i X_i + v_1 X_1 \qquad \text{for } z = 0 \quad (2.200)$$

- Clarifier

$$\frac{\partial X_2}{\partial t} = -\frac{F_{in} - F_W}{A_{se}}\frac{\partial X_2}{\partial z} - \frac{\partial(v_2 X_2)}{\partial z} - D_a'\frac{\partial^2 X_2}{\partial z^2} \qquad (2.201)$$

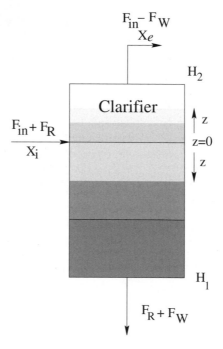

FIG. 2.22. Clarifier/settler scheme.

$$\frac{\partial X_2}{\partial z} = 0 \qquad\qquad\qquad \text{for } z = H_2 \text{ (2.202)}$$

$$D'_a \frac{\partial X_2}{\partial z} = \frac{F_{in} - F_W}{A_{se}}(X_2 - X_i) - v_i X_i + v_2 X_2 \quad \text{for } z = 0 \quad \text{(2.203)}$$

In line with other works [154], [281], we have introduced the flexibility of the potential existence of different dispersion coefficients for the clarifier and for the settler.

Determination of the sludge blanket height. There is still an important variable: the sludge blanket height h_b. The easiest way is to calculate it as the location where the concentration X becomes larger than a certain threshold value when starting from the top of the settler. One possible alternative to determine it has been proposed by Stehfest [241]. It is based on the following line of reasoning (the technical mathematical details are extensively discussed in [241]). Let us consider the mass balance equation (2.165) around the interface liquid/solid, and integrate it with respect to the spatial coordinate z (with z_a and z_b (unknown) on each side of h_b):

$$\frac{d}{dt} \int_{z_a}^{z_b} X dz = F(X(z_b)) - F(X(z_a)) \qquad\qquad (2.204)$$

If the concentration X_- and X_+ are the values of X for $z \to h_b$ from below and above, then the above equation becomes:

$$\frac{dh_b}{dt} = \frac{F(X_+) - F(X_-)}{X_- - X_+} \quad (2.205)$$

The computation of h_b also requires the knowledge of X_- and X_+. These are given by mass balance equations for the quantities of biomass above and below h_b. In order to have a simple operational model, the biomass concentrations are assumed to be homogeneous in both zones. This means that the quantities of biomass in the zones above and below h_b are equal to $(H - h_b)X_a$ and $h_b X_b$. After some mathematical manipulations, the dynamical model for determining h_b proposed by [241] has the following form:

$$\frac{dX_a}{dt} = \frac{1}{H - h_b}(F(X_i) - F(X_a)) \quad (2.206)$$

$$\frac{dX_b}{dt} = \frac{1}{h_b}(F(X_a) - F(X_0)) + (X_b - X_a)\frac{F(X_a) - F(X_-)}{X_a - X_-} \quad (2.207)$$

$$\frac{dh_b}{dt} = \frac{F(X_a) - F(X_-)}{X_- - X_a} \quad (2.208)$$

with:

$$X_- = min(X_b, X_1^*), \quad X_0 = max(X_b, X^*) \quad (2.209)$$

where X_1^* and X^* are the steady-state values defined in Figure 2.19 and Section 2.5.4 (subsection "steady-state curves of the settler model").

2.6   Linear vs Nonlinear Models

Up to now we have introduced a number of models, and we have formalised all the models based on mass balance in one general dynamical model framework, for stirred tank reactors on one hand, and for non completely mixed reactors on the other hand. In Section 1.2, we have introduced a brief classification of models. In particular, we have made a distinction between linear and nonlinear models. Strictly speaking, what we had considered was the notion of systems linear (or nonlinear) *in the state x and the input u*, and also linear (or nonlinear) *in the parameters*. Let us illustrate further these notions by an example: let us consider the simple microbial growth model (2.16) (2.17) repeated below as (2.210) (2.211):

$$\frac{dS}{dt} = DS_{in} - DS - \frac{1}{Y}\mu X \quad (2.210)$$

$$\frac{dX}{dt} = -DX + \mu X \quad (2.211)$$

With Monod kinetics, the above equations are written as follows:

$$\frac{dS}{dt} = DS_{in} - DS - \frac{1}{Y}\frac{\mu_{max}S}{K_S + S}X \quad (2.212)$$

$$\frac{dX}{dt} = -DX + \frac{\mu_{max}S}{K_S + S}X \tag{2.213}$$

These equations are clearly nonlinear in the state (via the conversion term which includes the multiplication of two "states" S and X, nd the division by a function of S, i.e. $K_S + S$). It is also nonlinear in the parameters Y, μ_{max} and K_S. And if the dilution rate D is considered as an input, then there is a bilinear term appearing in both equations: DS and DX, respectively.

Let us consider another viewpoint. For instance, let us consider zero order kinetics with respect to S for the growth rate ρ ($= \mu X$ in the above equations (2.210) (2.211)) (because for instance in the considered application, the substrate concentration S is large compared to the substrate affinity constant K_S). Hence, the specific growth rate μ can be assumed to be almost constant and equal to the maximum specific growth rate μ_{max}. This means that the growth rate can be written as follows:

$$\rho = \mu_{max} \tag{2.214}$$

and the mass balance equations take the following form:

$$\frac{dS}{dt} = DS_{in} - DS - \frac{1}{Y}\mu_{max}X \tag{2.215}$$

$$\frac{dX}{dt} = -DX + \mu_{max}X \tag{2.216}$$

Now the model is linear in the states S and X. However, it is nonlinear in the parameters Y and μ_{max} (division by Y and "multiplication" $\frac{1}{Y}\mu_{max}$. Yet, by considering the following (one-to-one) transformation:

$$\theta_1 = \frac{1}{Y}\mu_{max}, \quad \theta_2 = \mu_{max} \tag{2.217}$$

the model can be rewritten as follows:

$$\frac{dS}{dt} = DS_{in} - DS - \theta_1 X \tag{2.218}$$

$$\frac{dX}{dt} = -DX + \theta_2 X \tag{2.219}$$

which is linear in the parameters θ_1 and θ_2. (Note that we could have used the same argument with the Monod kinetics mass balance model if the problem had been to estimate the parameters $\frac{1}{Y}$ and μ_{max} under the assumption that K_S is known, albeit that this model remains nonlinear in the states.)

Finally, assume that the operating conditions are such that the dilution rate D is fixed and that the influent substrate concentration S_{in} can be manipulated, then S_{in} is the system input and D can be considered as a parameter: this implies that the model is also linear in the input. Then the model (2.218)(2.219) can be written in the linear model framework:

$$\frac{d}{dt}\begin{pmatrix} S \\ X \end{pmatrix} = \begin{pmatrix} -D - \theta_1 & 0 \\ \theta_2 & -D \end{pmatrix}\begin{pmatrix} S \\ X \end{pmatrix} + \begin{pmatrix} D \\ 0 \end{pmatrix} S_{in} \qquad (2.220)$$

The objective of the above discussion is motivated by the following arguments:

- on one hand, as suggested above, the models used in biological wastewater treatment are, generally speaking, nonlinear;
- on the other hand, the tools of system analysis are, generally speaking, much simpler, easier to apply and/or of larger use for linear systems than for non-linear ones.

That's why we shall prefer the use of linear system methods whenever possible, either because the model is linear in the context of the study or because by a proper manipulation we have been able to transform the nonlinear model into a linear one. We have already considered a transformation for the parameters. In other instances, we might have to use another transformation, the most largely used one consists of linearising the nonlinear model around some equilibrium point (or steady state).

Finally, it is important to point out that all the analysis results which can be obtained from a "linearised" model are usually not global (i.e. not valid for all the values of the variables and parameters) but local: for instance, analysis results based on a model linearised around some equilibrium points will only be valid close to these points. Note that even the above proposed parameter transformation exhibits a singular point: the combination ($\theta_1 = 0$, $\theta_2 \neq 0$) gives $Y = \infty$! And therefore strictly speaking the transformation is one-to-one except at that singular point.

2.7   Equilibrium Points, Linearisation and Stability Analysis

An equilibrium point (or steady state) of a dynamical model is, by definition, a constant state, i.e. its time derivative is equal to zero. If we consider the general dynamical model (2.100), the equilibrium points (that we shall denote $\bar{\xi}$) are such that:

$$\frac{d\bar{\xi}}{dt} = 0 \qquad (2.221)$$

This implies that equilibrium points are solutions of the algebraic equation:

$$-\bar{D}\bar{\xi} + Y\rho(\bar{\xi}) + \bar{F} - Q(\bar{\xi}) = 0 \qquad (2.222)$$

for given constant values of $\bar{D}$ and $\bar{F}$ of the dilution rate D and of the feed rates F. The problem of calculating the equilibrium points $\bar{\xi}$ of a biological wastewater treatment process whose dynamics are described by equations (2.100) is that of solving equation (2.222). The latter has no general analytical solution and can only be solved in specific applications.

Furthermore an important feature of (bio)chemical processes is that they can exhibit multiple steady states, i.e. several solutions to (2.222) exist. Let us illustrate this via the simple microbial growth model with different kinetic models (first order, Monod, and Haldane).

2.7.1 First Order Kinetics

Let us consider the mass balance model (2.215) (2.216) of the preceding section:

$$\frac{dS}{dt} = DS_{in} - DS - \frac{1}{Y}\mu_{max}X \tag{2.223}$$

$$\frac{dX}{dt} = -DX + \mu_{max}X \tag{2.224}$$

which, as we already said, is linear. As a consequence, there will be no multiple equilibrium points, i.e. to one set of values of $(\bar{D}, \bar{S}_{in})$ corresponds one and only one set of values of $\bar{S}$ and $\bar{X}$. Indeed the equilibrium points are here the solution of the equations:

$$0 = \bar{D}\bar{S}_{in} - \bar{D}\bar{S} - \frac{1}{Y}\mu_{max}\bar{X} \tag{2.225}$$

$$0 = -\bar{D}\bar{X} + \mu_{max}\bar{X} \tag{2.226}$$

The equilibrium point is then given by the following expressions:

$$\bar{S} = \frac{\bar{D}}{\bar{D} + \frac{1}{Y}\mu_{max}}\bar{S}_{in} \tag{2.227}$$

$$\bar{X} = \frac{k_0}{\bar{D} + \frac{1}{Y}\mu_{max}}\bar{S}_{in} \tag{2.228}$$

2.7.2 Monod and Haldane Kinetics

Let us recall the dynamical mass balance equations of the simple microbial growth process (2.13) (2.14):

$$\frac{dS}{dt} = DS_{in} - DS - \frac{1}{Y}\mu X \tag{2.229}$$

$$\frac{dX}{dt} = -DX + \mu X \tag{2.230}$$

The equilibrium points are therefore the solutions of the following set of algebraic equations:

$$0 = \bar{D}\bar{S}_{in} - \bar{D}\bar{S} - \frac{1}{Y}\bar{\mu}\bar{X} \tag{2.231}$$

$$0 = -\bar{D}\bar{X} + \bar{\mu}\bar{X} \tag{2.232}$$

From the second equation, we note that there are two possible solutions:

1. $\bar{X} = 0$
2. $\bar{\mu} = \bar{D}$

The first equilibrium point corresponds to the *wash-out* steady state:

$$\bar{X} = 0, \ \bar{S} = \bar{S}_{in} \qquad (2.233)$$

It corresponds to a wash-out of the biomass from the reactor. It can occur for any value of $\bar{D}$ and $\bar{S}_{in}$. It is also the only possible equilibrium point when $\bar{D} > \mu_{max}$. It is obvious that wash-out is undesirable and should be avoided as far as possible in practice.

The second equilibrium point corresponds to operational steady states. Expressions of the specific growth rate are necessary to obtain explicit expressions of the equilibrium points. Let us consider two typical specific growth rate expressions.

Monod kinetics.   The operational steady state equation $\bar{\mu} = \bar{D}$ with Monod kinetics is written as follows:

$$\frac{\mu_{max}\bar{S}}{K_S + \bar{S}} = \bar{D} \qquad (2.234)$$

We can deduce the following equilibrium point from the above equation and the steady state equation (2.231) of $\bar{S}$:

$$\bar{S} = \frac{\bar{D}K_S}{\mu_{max} - \bar{D}} \qquad (2.235)$$

$$\bar{X} = Y(S_{in} - S) = Y\left(S_{in} - \frac{\bar{D}K_S}{\mu_{max} - \bar{D}}\right) \qquad (2.236)$$

Note in particular that the steady state of the substrate concentration $\bar{S}$ is independent of the influent substrate concentration $\bar{S}_{in}$. This property is indeed generic for any model of the specific growth rate which *only* depends on the substrate concentration. (Exercise: calculate the steady states of the simple microbial growth process with Tessier and with Contois specific growth rate models, and compare the results.)

Haldane kinetics.   For the Monod kinetics we have obtained two possible equilibrium points: one wash-out steady state, and one operational steady state. Let us now consider the same microbial growth process but with Haldane kinetics (2.43). There is still obviously a wash-out equilibrium point. As suggested in Figure 2.5, there are now two operational equilibrium points (instead of one with the Monod model). Indeed the equation $\bar{\mu} = \bar{D}$ with the Haldane model is written as follows:

$$\frac{\mu_{max}\bar{S}}{K_S + \bar{S} + \frac{\bar{S}^2}{K_i}} = \bar{X} \qquad (2.237)$$

Therefore the steady state value $\bar{S}$ is the solution of the following equation:

$$\frac{\bar{D}}{K_i}\bar{S}^2 + (\bar{D} - \mu_{max})\bar{S} + \bar{K}_S = 0 \tag{2.238}$$

and the two solutions S_1 and S_2 in Figure 2.5 are equal in steady state to:

$$\bar{S}_1 = \frac{(\mu_{max} - \bar{D})K_i}{2\bar{D}} - \frac{K_i}{2\bar{D}}\sqrt{(\mu_{max} - \bar{D})^2 - 4\bar{D}^2\frac{K_S}{K_i}} \tag{2.239}$$

$$\bar{S}_2 = \frac{(\mu_{max} - \bar{D})K_i}{2\bar{D}} + \frac{K_i}{2\bar{D}}\sqrt{(\mu_{max} - \bar{D})^2 - 4\bar{D}^2\frac{K_S}{K_i}} \tag{2.240}$$

The above two operational equilibrium points have fundamentally different dynamical properties in terms of stability. This is what we shall briefly analyse in Section 2.7.4. However since the analysis is based on a linearised model of the process, we shall first introduce the linearised tangent model of nonlinear models.

2.7.3 Linearised Tangent Model of Nonlinear Models

Let us consider a nonlinear dynamical model:

$$\frac{dx}{dt} = f(x, u) \tag{2.241}$$

where x is the state vector and u the input vector. The equilibrium point(s) are the values of $\bar{x}$, the solution of:

$$0 = f(\bar{x}, \bar{u}) \tag{2.242}$$

for constant values of $\bar{u}$. The derivation of the linearised tangent model around the steady state $\bar{x}$ is based on the Taylor series' expansion of the function f around this steady state for x and u.

Let us define the deviation variables $\tilde{x}$ and $\tilde{u}$:

$$\tilde{x} = x - \bar{x}, \quad \tilde{u} = u - \bar{u} \tag{2.243}$$

The Taylor series expansion of f around $x = \bar{x}$ and $u = \bar{u}$ is equal to:

$$f(x, u) = f(\bar{x}, \bar{u}) + \frac{\partial f(\bar{x}, \bar{u})}{\partial x}\tilde{x} + \frac{\partial f(\bar{x}, \bar{u})}{\partial u}\tilde{u} + \frac{1}{2!}\frac{\partial^2 f(\bar{x}, \bar{u})}{\partial x^2}\tilde{x}^2 + \frac{1}{2!}\frac{\partial^2 f(\bar{x}, \bar{u})}{\partial u^2}\tilde{u}^2 + \dots \tag{2.244}$$

If we stop the expansion at the first order derivative and we consider the deviation state variable $\tilde{x}$, the nonlinear dynamical model (2.241) can be rewritten under the following linear form:

$$\frac{d\tilde{x}}{dt} = A\tilde{x} + B\tilde{u} \tag{2.245}$$

with:

$$A = \frac{\partial f(\bar{x}, \bar{u})}{\partial x}, \quad B = \frac{\partial f(\bar{x}, \bar{u})}{\partial u} \tag{2.246}$$

The above model (2.245)(2.246) is the linearised tangent model of the nonlinear model (2.241). (It is indeed tangent of the nonlinear model at the equilibrium point $\bar{x}$).

As a matter of illustration, let us apply the above linearisation procedure to the simple microbial growth process equations (2.16) (2.17). Assume that the specific growth rate is, generally speaking, a function of S and X ($\mu(S, X)$), and D and S_{in} are both inputs. Then the linearised tangent model around operational equilibrium points $S = \bar{S}$ and $X = \bar{X}$ ($\bar{\mu} = \bar{D}$!) is equal to:

$$\frac{d}{dt} \begin{pmatrix} \tilde{s} \\ \tilde{x} \end{pmatrix} = \begin{pmatrix} -\bar{D} - \frac{1}{Y}\bar{\mu}_S \bar{X} & -\frac{1}{Y}(\bar{D} + \bar{\mu}_X \bar{X}) \\ \bar{\mu}_S \bar{X} & 0 \end{pmatrix} \begin{pmatrix} \tilde{s} \\ \tilde{x} \end{pmatrix}$$

$$+ \begin{pmatrix} \bar{D} & \bar{S}_{in} - \bar{S} \\ 0 & -\bar{X} \end{pmatrix} \begin{pmatrix} \tilde{s}_{in} \\ \tilde{d} \end{pmatrix} \tag{2.247}$$

with:

$$\tilde{s} = S - \bar{S}, \quad \tilde{x} = X - \bar{X}, \quad \tilde{s}_{in} = S_{in} - \bar{S}_{in}, \quad \tilde{d} = D - \bar{D} \tag{2.248}$$

$$\bar{\mu}_S = \frac{\partial \mu(\bar{S}, \bar{X})}{\partial S}, \quad \bar{\mu}_X = \frac{\partial \mu(\bar{S}, \bar{X})}{\partial X} \tag{2.249}$$

For the Haldane model for instance:

$$\bar{\mu}_X = 0, \quad \bar{\mu}_S = \frac{\mu_{max}(K_S - \frac{\bar{S}^2}{K_i})}{(K_S + \bar{S} + \frac{\bar{S}^2}{K_i})^2} \tag{2.250}$$

and the state matrix A is then equal to:

$$A = \begin{pmatrix} -\bar{D} - \frac{1}{Y}\bar{\mu}_S \bar{X} & -\frac{1}{Y}\bar{D} \\ \bar{\mu}_S \bar{X} & 0 \end{pmatrix} \tag{2.251}$$

2.7.4 Stability of Equilibrium Points

Stability theory of dynamical systems is a very old subject. Yet, significant advances in the stability analysis are due to the Russian mathematician Lyapunov (1892), whose results were largely confined to Eastern European countries until about 1960. Lyapunov's first method [284] utilises the eigenvalues of the state matrix A of the linear model (2.245) to check the stability of the equilibrium state. Recall that the eigenvalues λ of a square matrix A are the roots of the characteristic polynomial:

$$det(\lambda I - A) = 0, \quad I : \text{identity matrix} \tag{2.252}$$

If the real parts of all the eigenvalues are negative, the equilibrium point is stable. If any of the real parts of the eigenvalues are positive, the equilibrium is unstable. No conclusion can be drawn in case of eigenvalues having zero real parts.

Let us apply the analysis to the simple microbial growth model with Haldane kinetics. Let us calculate the characteristic polynomial of the linearised tangent model state matrix (2.251):

$$det(\lambda I - A) = \lambda^2 + (\bar{D} + \frac{1}{Y}\mu_S\bar{X})\lambda + \frac{1}{Y}\mu_S\bar{X}\bar{D} \qquad (2.253)$$

A necessary and sufficient condition for a second order polynomial to have roots with all negative real parts is that all the coefficients of the polynomial have the same sign. This means that here in order to have stability, $\bar{D} + \frac{1}{Y}\bar{\mu}_S\bar{X}$ and $\frac{1}{Y}\bar{\mu}_S\bar{X}\bar{D}$ must be positive. If we recall the value of $\bar{\mu}_S$ (2.250), this condition will be fulfilled if and only if $\bar{S} < \sqrt{K_S K_i}$, i.e. for S_1 in Figure 2.5. Conversely, the equilibrium point $\bar{S}_2$ will be unstable since $\bar{S} > \sqrt{K_S K_i}$, and (at least) one coefficient in the polynomial will be negative.

This is a quite interesting feature of the Haldane model that it is able to emphasise the possible presence of unstable steady states. These states might be of practical interest in industrial applications but are not reachable without any appropriate external control. It is clearly a major issue in automatic control to regulate and stabilise processes around equilibrium points which are open-loop (i.e. without external action) unstable (see e.g. [14]).

2.8   A Key State Transformation

The key result of this section is the introduction of a state transformation by which part of the dynamical model (2.100) becomes independent of the reaction kinetics ρ (see [14], [56]). This transformation will play a very important role in the design of asymptotic observers (Chapter 7). The proposed transformation readily derives from the notion of invariants in reaction systems (see e.g. [95], [86]).

2.8.1   Definition of the State Transformation

The transformation is defined as follows. Let us denote rank(Y) = R and consider a state partition (in which the state vector is decomposed in two sub-vectors ξ_a and ξ_b):

$$\xi = \begin{bmatrix} \xi_a \\ \xi_b \end{bmatrix} \qquad (2.254)$$

where ξ_a contains R (arbitrarily chosen) process variables and ξ_b the others, but such that the corresponding submatrix Y_a is full rank (rank(Y_a) = R). Let us define the state transformation into the auxiliary variable ζ (dim(ζ) = N-R):

$$\zeta = C_a\xi_a + C_b\xi_b \qquad (2.255)$$

where C_a and C_b are solutions of the matrix equation:

$$C_aY_a + C_bY_b = 0 \qquad (2.256)$$

In the particular (but quite general) situation of M independent irreversible reactions, then R = M and C_b may be chosen to be a full rank square matrix (the simplest choice being obviously the identity matrix I). Then C_a is found from

$$C_a = -C_b Y_b Y_a^{-1} \tag{2.257}$$

Let us illustrate the dynamics of ζ from the General Dynamical Model (2.100) or (2.132)(2.133) and the definition (2.255). Note that the vectors F, Q, D are partitioned according to the partitioning of the state vector ξ.

1) Single reactor: D = scalar

$$\frac{d\zeta}{dt} = -D\zeta + C_a(F_a - Q_a) + C_b(F_b - Q_b) \tag{2.258}$$

2) Multi-reactor: D = matrix

$$\frac{d\zeta}{dt} = -(C_b D_{bb} + C_a D_{ab})C_b^{-1}\zeta + C_a(F_a - Q_a) + C_b(F_b - Q_b)$$

$$+[(C_b D_{bb} + C_a D_{ab})C_b^{-1} - C_b D_{ba} - C_a D_{aa}]\xi_a \tag{2.259}$$

with:

$$D = \begin{pmatrix} D_{aa} & D_{ab} \\ D_{ba} & D_{bb} \end{pmatrix} \tag{2.260}$$

3) Fixed bed reactor

For simplicity reasons, let us consider here $C_b = I$ and let us put the vector ξ_{fi} of the fixed components in ξ_b:

$$\xi_b = \begin{pmatrix} \xi_{bf} \\ \xi_{fi} \end{pmatrix} \tag{2.261}$$

Then we can rewrite the auxiliary variable ζ as follows:

$$\zeta = \begin{pmatrix} \zeta_{fl} \\ \zeta_{fi} \end{pmatrix} = \begin{pmatrix} \xi_{bf} \\ \xi_{fi} \end{pmatrix} + \begin{pmatrix} C_{af} \\ C_{ae} \end{pmatrix} \xi_a \tag{2.262}$$

The dynamics of ζ can then be written as follows:

$$\frac{\partial \zeta_{fl}}{\partial t} = -\frac{F_{in}}{A}\frac{\partial \zeta_{fl}}{\partial z} + D_{am}\frac{\partial^2 \zeta_{fl}}{\partial z^2} \tag{2.263}$$

$$\frac{\partial \zeta_{fi}}{\partial t} = -\frac{F_{in}}{A}C_{ae}\frac{\partial \xi_a}{\partial z} + D_{am}C_{ae}\frac{\partial^2 \xi_a}{\partial z^2} \tag{2.264}$$

Note that the dynamical equations of ζ (2.258), (2.259) and (2.263)(2.264) are independent of the reaction kinetics $\rho(\xi)$.

2.8.2 *Example 1: Two Step Denitrification*

Let us consider as a first example the denitrification process described by equations (2.99). One possible choice for the state partition is the following:

$$\xi_a = \begin{pmatrix} X \\ N_2 \end{pmatrix}, \ \xi_b = \begin{pmatrix} S_{NO_3} \\ S_{NO_2} \\ S \end{pmatrix} \tag{2.265}$$

Indeed, the submatrices Y_a and Y_b are then equal to:

$$Y_a = \begin{pmatrix} 1 & 1 \\ 0 & Y_6 \end{pmatrix}, \ Y_b = \begin{pmatrix} -\frac{1}{Y_1} & 0 \\ Y_3 & -\frac{1}{Y_2} \\ -\frac{1}{Y_4} & -\frac{1}{Y_5} \end{pmatrix} \tag{2.266}$$

and Y_a is full rank (since Y_6 is by definition strictly positive). If we consider the simplest possible choice for C_b (= I), then C_a is equal to:

$$C_a = -Y_b Y_a^{-1} = \begin{pmatrix} \frac{1}{Y_1} & -\frac{1}{Y_1 Y_6} \\ -Y_3 & \frac{Y_2 Y_3 + 1}{Y_2 Y_6} \\ \frac{1}{Y_4} & \frac{Y_5 - Y_4}{Y_4 Y_5 Y_6} \end{pmatrix} \tag{2.267}$$

Therefore the auxiliary variable ζ is equal to:

$$\zeta = \begin{pmatrix} \zeta_1 \\ \zeta_2 \\ \zeta_3 \end{pmatrix} = \begin{pmatrix} S_{NO_3} + \frac{1}{Y_1} X - \frac{1}{Y_1 Y_6} N_2 \\ S_{NO_2} - Y_3 X + \frac{Y_2 Y_3 + 1}{Y_2 Y_6} N_2 \\ S + \frac{1}{Y_4} X + \frac{Y_5 - Y_4}{Y_4 Y_5 Y_6} N_2 \end{pmatrix} \tag{2.268}$$

and its dynamics are written as follows:

$$\frac{d}{dt} \begin{pmatrix} \zeta_1 \\ \zeta_2 \\ \zeta_3 \end{pmatrix} = -D \begin{pmatrix} \zeta_1 \\ \zeta_2 \\ \zeta_3 \end{pmatrix} + \begin{pmatrix} DS_{NO_3,in} + \frac{1}{Y_1 Y_6} Q_{N_2} \\ -\frac{Y_2 Y_3 + 1}{Y_2 Y_6} Q_{N_2} \\ DS_{in} - \frac{Y_5 - Y_4}{Y_4 Y_5 Y_6} Q_{N_2} \end{pmatrix} \tag{2.269}$$

Remark: all the above calculations are based on an arbitrary choice of the state partition ξ_a, ξ_b. Many other choices would have been appropriate. For instance we could have considered the first two entries of the state vector, i.e. S_{NO_3} and S_{NO_2} for ξ_a and the other for ξ_b, and this would have resulted in the following state transformation:

$$\zeta = \begin{pmatrix} \zeta_1 \\ \zeta_2 \\ \zeta_3 \end{pmatrix} = \begin{pmatrix} S - \frac{Y_1(Y_5 + Y_2 Y_3 Y_4)}{Y_4 Y_5} S_{NO_3} - \frac{Y_2}{Y_5} S_{NO_2} \\ X + Y_1(1 + Y_2 Y_3) S_{NO_3} + Y_2 S_{NO_2} \\ N_2 + Y_1 Y_2 Y_3 Y_6 S_{NO_2} + Y_2 Y_6 S_{NO_3} \end{pmatrix} \tag{2.270}$$

It is important to remember that the time evolution of these auxiliary variables can be calculated without knowledge of the reaction kinetics, i.e. without any requirement concerning the knowledge about its model structure and/or its parameters.

2.8.3   Example 2: Activated Sludge Process: The Basic Model

As already mentioned, the basic activated sludge process model is a two-tank re-actor model, with a dilution rate matrix D.

There is one reaction: ξ_a will be a scalar. Let us for instance consider X for ξ_a (note that we could also have taken S or S_O but not X_R). Then we have the following state partition:

$$\xi_a = X, \ \xi_b = \begin{pmatrix} S \\ S_O \\ X_R \end{pmatrix} \tag{2.271}$$

This means that the different entries D_{aa}, D_{ab}, D_{ba} and D_{bb} of the matrix D are equal to:

$$D_{aa} = D_1, \ D_{ab} = \begin{pmatrix} 0 & 0 & -D_2 \end{pmatrix} \tag{2.272}$$

$$D_{ba} = \begin{pmatrix} 0 \\ 0 \\ -D_3 \end{pmatrix}, \ D_{bb} = \begin{pmatrix} D_1 & 0 & 0 \\ 0 & D_1 & 0 \\ 0 & 0 & D_4 \end{pmatrix} \tag{2.273}$$

The matrices Y_a, Y_b and C_a are equal to:

$$Y_a = 1, \ Y_b = \begin{pmatrix} -\frac{1}{Y_S} \\ -\frac{1}{Y_O} \\ 0 \end{pmatrix}, \ C_a = -Y_b Y_a^{-1} = \begin{pmatrix} \frac{1}{Y_S} \\ \frac{1}{Y_O} \\ 0 \end{pmatrix} \ (C_B = I) \tag{2.274}$$

i.e. ζ is equal to:

$$\zeta = \begin{pmatrix} S + \frac{1}{Y_S} X \\ S_O + \frac{1}{Y_O} X \\ X_R \end{pmatrix} \tag{2.275}$$

2.8.4   Example 3: Activated Sludge Process: The IWA Activated Sludge Model No. 1

A careful look at the reaction network of the IWA Activated Sludge Model No. 1 reveals that there is a loop with reactions 1, 4 and 7. Indeed the readily biodegrad-able substrate S_S is transformed in reaction 1 in heterotrophic bacteria X_{BH}, which in turn is transformed in the decay reaction 4 in slowly biodegradable substrate X_S. And in reaction 7 X_S is hydrolysed in S_S.

This reaction loop has an important consequence on the above state transforma-tion. Indeed the process components and reactions are not completely independent from each other. This means that not any choice of state partition will give a full rank Y_a. For instance, if one takes the first 8 components for ξ_a, the related Y_a will be singular, since row 7 will be a linear combination of rows 1 to 4 and row 6, or in other words, in terms of reaction kinetics, S_{NO} is a linear combination of S_S, X_S, X_{BH}, X_{BA} and S_O.

A possible choice for the state partition is to take the first 6 components in ξ (2.107) plus S_{NH} and S_{ND} for the subvector ξ_a:

$$\xi_a = \begin{pmatrix} S_S \\ X_S \\ X_{B,H} \\ X_{B,A} \\ X_P \\ S_O \\ S_{NH} \\ S_{ND} \end{pmatrix}, \quad \xi_b = \begin{pmatrix} S_{NO} \\ X_{ND} \end{pmatrix} \tag{2.276}$$

$$Y_a = \begin{pmatrix} -\frac{1}{Y_H} & -\frac{1}{Y_H} & 0 & 0 & 0 & 0 & 1 & 0 \\ 0 & 0 & 0 & 1-f_P & 1-f_P & 0 & -1 & 0 \\ 1 & 1 & 0 & -1 & 0 & 0 & 0 & 0 \\ 0 & 0 & 1 & 0 & -1 & 0 & 0 & 0 \\ 0 & 0 & 0 & f_P & f_P & 0 & 0 & 0 \\ -\frac{1-Y_H}{Y_H} & 0 & -\frac{4.57-Y_A}{Y_A} & 0 & 0 & 0 & 0 & 0 \\ -i_{XB} & -i_{XB} & -i_{XB} - \frac{1}{Y_A} & 0 & 0 & 1 & 0 & 0 \\ 0 & 0 & 0 & 0 & 0 & -1 & 0 & 1 \end{pmatrix} \tag{2.277}$$

$$Y_b = \begin{pmatrix} 0 & -\frac{1-Y_H}{2.86Y_H} & \frac{1}{Y_A} & 0 & 0 & 0 & 0 & 0 \\ 0 & 0 & 0 & i_{XB} - f_P i_{XP} & i_{XB} - f_P i_{XP} & 0 & 0 & -1 \end{pmatrix} \tag{2.278}$$

Then the matrix C_a is equal to:

$$C_a = \begin{pmatrix} c_1 & c_1 & c_2 & c_2 & c_3 & \frac{1}{2.86} & 0 & 0 \\ c_4 & c_4 & c_5 & c_5 & c_6 & 0 & 1 & 1 \end{pmatrix} \tag{2.279}$$

with

$$c_1 = \frac{Y_H}{2.86}\left(-\frac{1-Y_H}{Y_H} + \frac{1.71-Y_A}{Y_A}\right) \tag{2.280}$$

$$c_2 = \frac{1.71-Y_A}{2.86Y_A} \tag{2.281}$$

$$c_3 = \frac{(1-f_P)(Y_A - 1.71Y_H) + 1.71Y_A}{2.86f_P Y_A} \tag{2.282}$$

$$c_4 = \frac{Y_H}{Y_A} \tag{2.283}$$

$$c_5 = i_{XB} + \frac{1}{Y_A} \tag{2.284}$$

$$c_6 = \frac{1 - Y_H + f_P(Y_H - Y_A i x_P)}{f_P Y_A} \tag{2.285}$$

and the auxiliary variables ζ are written as follows:

$$\zeta = \begin{pmatrix} S_{NO} + c_1(S_S + X_S) + c_2(X_{B,H} + X_{B,A}) + c_3 X_P + \frac{1}{2.86} S_O \\ X_{ND} + c_4 S_S + c_4 X_S + c_5 X_{B,H} + c_5 X_{B,A} + c_6 X_P + S_{NH} + S_{ND} \end{pmatrix} \tag{2.286}$$

2.8.5 *Example 4: Fixed Bed Reactor Model With a Growth Reaction and a Death/Detachment Reaction*

Let us consider for instance the following state partition:

$$x_a = \begin{pmatrix} S \\ X_d \end{pmatrix}, \quad x_b = \begin{pmatrix} P \\ X \end{pmatrix} \tag{2.287}$$

The matrices Y_a and Y_b are equal to:

$$Y_a = \begin{pmatrix} -\frac{1}{Y_S} & 0 \\ 0 & 1 \end{pmatrix}, \quad Y_b = \begin{pmatrix} Y_P & 0 \\ \frac{\epsilon}{1-\epsilon} & -\frac{\epsilon}{1-\epsilon} \end{pmatrix} \tag{2.288}$$

Therefore the matrix C_a is equal to:

$$C_a = -Y_b Y_a^{-1} = \begin{pmatrix} Y_S Y_P & 0 \\ \frac{Y_S \epsilon}{(1-\epsilon)} & \frac{\epsilon}{1-\epsilon} \end{pmatrix} \tag{2.289}$$

ζ is then equal to:

$$\zeta = \begin{pmatrix} P + Y_P Y_S S \\ X + \frac{Y_S \epsilon}{(1-\epsilon)} S + \frac{\epsilon}{1-\epsilon} X_d \end{pmatrix} \tag{2.290}$$

2.9 Model Order Reduction

The examples of bioprocesses presented in the preceding sections have shown that a bioreactor dynamical model may be fairly complex in some instances and involve a large number of differential equations. But there are many practical applications where a simplified reduced order model is sufficient from an engineering viewpoint. One possible systematic approach to achieve model simplification is to use the singular perturbation method, which is a technique that allows to transform a set of $n + m$ differential equations into a set of n differential equations and a set of m algebraic equations. It is based on the partition of the state equations into two sets of dynamical equations characterised by the variables x_1 and x_2, of dimension n-m and m, respectively, i.e.:

$$\frac{dx_1}{dt} = f_1(x_1, x_2, u) \tag{2.291}$$

$$\delta \frac{dx_2}{dt} = f_2(x_1, x_2, u, \delta) \qquad (2.292)$$

i.e. the time derivative of some of the state variables are multiplied by a small parameter δ (also sometimes called the perturbation parameter). For sufficiently small parameter δ, the parameter δ can considered as being negligible, and the singular perturbation consists of setting δ to zero, and to replace the $n + m$ differential equations (2.292) by algebraic equations:

$$f_2(x_1, x_2, u, \delta) = 0 \qquad (2.293)$$

More precisely, if $\bar{x}_2$ is the solution of the above algebraic equation, then the model (2.291)(2.292) is replaced by the following set of equations:

$$\frac{dx_1}{dt} = f_1(x_1, \bar{x}_2, u) \qquad (2.294)$$

$$\bar{x}_2 = g(x_1, u) \qquad (2.295)$$

The model (2.294) is sometimes called a *quasi-steady state* model, which is often considered for instance in (bio)chemical engineering, e.g. [37], [114], [255].

This technique is suitable when neglecting the dynamics of products with low solubility in the liquid phase or of substrates in fast reactions. The method will be illustrated with two specific examples (low solubility product, and substrates in a fast reaction) before stating the general rule for order reduction.

2.9.1   Singular Perturbation Technique for Low Solubility Products

Let us consider a reaction described by the following reaction scheme:

$$S \longrightarrow P \qquad (2.296)$$

where P is a volatile product which can be given off in gaseous form and has low solubility in the liquid phase (e.g. H_2 in anaerobic digestion). The dynamical model is as follows:

$$\frac{dS}{dt} = -\rho - DS + DS_{in} \qquad (2.297)$$

$$\frac{dP}{dt} = Y\rho - DP - Q \qquad (2.298)$$

For the consistency of this model, the product concentration P can be expressed with respect to the saturation concentration representative of the product solubility as follows:

$$P = \Pi P_{sat}, \qquad\qquad 0 \le \Pi(t) \qquad (2.299)$$

where P_{sat} is the saturation concentration which is constant in a stable physico-chemical environment. The model (2.297)(2.298) is rewritten in the standard singular perturbation form, with $\delta = P_{sat}$:

$$\frac{dS}{dt} = -\rho - DS + DS_{in} \qquad (2.300)$$

$$\delta \frac{d\Pi}{dt} = Y\rho - \delta D\Pi - Q \tag{2.301}$$

If the solubility is very low, we obtain a reduced order model by setting $\delta = 0$ and replacing the differential equation (2.301) by the algebraic one:

$$Q = Y\rho \tag{2.302}$$

2.9.2 Singular Perturbation Technique for Substrates of Fast Reactions

Singular perturbation can also be applied to reduce the order of the model dynamics for reactors with fast and slow reactions. For simplicity we shall concentrate on an example, i.e. the nitrification process which is characterised by the following two sequential reactions (2.85) (2.86) repeated below as (2.303)(2.304):

$$S_{NH} \longrightarrow X_1 + S_{NO_2} \tag{2.303}$$

$$S_{NO_2} \longrightarrow X_2 + S_{NO_3} \tag{2.304}$$

The dynamics in a CSTR are given by the following equations:

$$\frac{dS_{NH}}{dt} = -\frac{1}{Y_1}\rho_1 + DS_{NH,in} - DS_{NH} \tag{2.305}$$

$$\frac{dS_{NO_2}}{dt} = -\frac{1}{Y_2}\rho_2 + Y_3\rho_1 - DS_{NO_2} \tag{2.306}$$

$$\frac{dS_{NO_3}}{dt} = Y_4\rho_2 - DS_{NO_3} \tag{2.307}$$

$$\frac{dX_1}{dt} = \rho_1 - DX_1 \tag{2.308}$$

$$\frac{dX_2}{dt} = \rho_2 - DX_2 \tag{2.309}$$

In line with the basic kinetics rules (Section 2.2.6), the reaction rates ρ_1 and ρ_2 can be written as follows:

$$\rho_1 = k_1 S_{NH}^{\alpha_1} \phi_1(S_{NH}, X_1, S_{NO_2}) \tag{2.310}$$

$$\rho_2 = k_2 S_{NO_2}^{\alpha_2} \phi_2(S_{NO_2}, X_2, S_{NO_3}) \tag{2.311}$$

where k_i (i = 1, 2) are the maximum reaction rates, α_i (i = 1, 2) are coefficients (which might be the reaction orders), and ϕ_i (i = 1, 2) positive functions of the state ($\phi_i > 0$). Assume now that the first reaction is slow and the second one is fast. This can be formalised by considering that the maximum reaction rate of the second reaction k_2 is much larger than the one of the first reaction k_1:

$$k_2 \gg k_1 \tag{2.312}$$

Let us now consider the following state transformation (the general formulation is given in [255]):

$$\zeta_1 = Y_4 Y_2 S_{NO_2} + S_{NO_3} \tag{2.313}$$

$$\zeta_2 = Y_2 S_{NO_2} + X_2 \tag{2.314}$$

Then we can rewrite the process dynamics as follows:

$$\frac{dS_{NH}}{dt} = -\frac{1}{Y_1}\rho_1 + DS_{NH,in} - DS_{NH} \tag{2.315}$$

$$\frac{dS_{NO_2}}{dt} = -\frac{1}{Y_2}\rho_2 + Y_3\rho_1 - DS_{NO_2} \tag{2.316}$$

$$\frac{d\zeta_1}{dt} = Y_2 Y_3 Y_4 \rho_1 - D\zeta_1 \tag{2.317}$$

$$\frac{dX_1}{dt} = \rho_1 - DX_1 \tag{2.318}$$

$$\frac{d\zeta_2}{dt} = Y_2 Y_3 \rho_1 - D\zeta_2 \tag{2.319}$$

The choice of the above transformation can be briefly intuitively motivated as follows. First, the second equation (2.316) is the only one that still contains the kinetics of both reactions, while the other four equations only contain the reaction rate of the slow reaction. Secondly the auxiliary variables ζ_1 and ζ_2 are the algebraic sum of S_{NO_2} (i.e. the "intermediate" component: product of the first reaction, and substrate of the second), and of the products of the second (fast) reaction, S_{NO_3} and X_2 respectively.

Let us define the singular perturbation δ as the inverse of the maximum reaction rate k_2:

$$\delta = \frac{1}{k_2} \tag{2.320}$$

By considering the expression (2.311) and the singular perturbation parameter δ, the dynamical equation of S_{NO_2} can be rewritten as follows:

$$\delta \frac{dS_{NO_2}}{dt} = -\frac{1}{Y_2} S_{NO_2}^{\alpha_2} \phi_2(S_{NO_2}, X_2, S_{NO_3}) + \delta Y_3 \rho_1 - \delta DS_{NO_2} \tag{2.321}$$

If the second reaction is sufficiently fast, then we can apply the singular perturbation and set δ to zero. This implies that the above equation reduces to:

$$\frac{1}{Y_2} S_{NO_2}^{\alpha_2} \phi_2(S_{NO_2}, X_2, S_{NO_3}) = 0 \tag{2.322}$$

Since Y_2 and ϕ_2 are strictly positive, this implies: $S_{NO_2} = 0$. Then by recalling the definition of ζ, the equations (2.315)(2.317)(2.318)(2.319) of the dynamical model can be rewritten as follows:

$$\frac{dS_{NH}}{dt} = -\frac{1}{Y_1}\rho_1 + DS_{NH,in} - DS_{NH} \tag{2.323}$$

$$\frac{dS_{NO_3}}{dt} = Y_2 Y_3 Y_4 \rho_1 - DS_{NO_3} \qquad (2.324)$$

$$\frac{dX_1}{dt} = \rho_1 - DX_1 \qquad (2.325)$$

$$\frac{dX_2}{dt} = Y_2 Y_3 \rho_1 - DX_2 \qquad (2.326)$$

i.e. it is as if the two sequential reactions (2.303)(2.304) have been reduced to one reaction:

$$S_{NH} \longrightarrow X_1 + X_2 + S_{NO_3} \qquad (2.327)$$

2.9.3 A General Rule for Order Reduction

The above examples show that the rule for model simplification is actually very simple and that an explicit singular perturbation analysis is not really needed. Consider that, for some i, the dynamics of the component ξ_i are to be neglected. The dynamics of ξ_i are described by equation (2.100):

$$\frac{d\xi_i}{dt} = -D\xi_i + Y_i \rho + F_i - Q_i \qquad (2.328)$$

where Y_i is the row of Y corresponding to the component ξ_i. The simplification is then achieved by setting ξ_i and $d\xi_i/dt$ to zero i.e. by replacing the differential equation (2.328) by the following algebraic equation:

$$Y_i \rho = -F_i + Q_i \qquad (2.329)$$

It has been shown that the above model order reduction rule is not only valid for low solubility products but also for bioprocesses with fast and slow reactions. Then the above order reduction rule (2.329) applies to substrates of fast reactions (as long as they intervene only in fast reactions) (see [255] for further details).

Note the close connection between the singular perturbation reduction and the quasi steady state (QSS) approximation, which is largely used in (bio)chemical engineering. This suggests the following comment: singular perturbation can be viewed as an efficient mathematical tool to rigorously justify QSS approximations on a systematic basis via an appropriate analysis (including the choice of an appropriate *small* perturbation parameter).

The above considerations also apply to models with other hydrodynamics, like the fixed bed reactor models with or without dispersion, mutatis mutandis, i.e. by setting the different derivative ($\frac{\partial x_i}{\partial t}$, $\frac{\partial x_i}{\partial z}$, $\frac{\partial^2 x_i}{\partial z^2}$) of the low solubility product or of the "fast substrate" to zero in its mass balance equation.

2.9.4 Example: the Anaerobic Digestion

Let us see how to apply the above model order reduction rule (2.329) to a specific example, the anaerobic digestion. First of all, it is well-known that methane is a low solubility product. Therefore the above procedure applies. Furthermore, assume that the second methanisation path (hydrogen consumption) is limiting, i.e. that the first three reactions (2.53)(2.54)(2.55) are fast and the fourth one (2.56) is slow. We can then apply the model order reduction rule (2.329) to the the glucose concentration S_1, the propionate concentration S_2, the acetate concentration S_3 and the dissolved methane concentration P_1. By setting their values and their time derivatives to zero:

$$S_1 = S_2 = S_3 = P_1, \quad \frac{dS_1}{dt} = \frac{dS_2}{dt} = \frac{dS_3}{dt} = \frac{dP_1}{dt} = 0 \qquad (2.330)$$

we reduce their differential equations to the following set of algebraic equations:

$$\begin{pmatrix} -\frac{1}{Y_{21}} & 0 & 0 & 0 \\ Y_{41} & -\frac{1}{Y_{42}} & 0 & 0 \\ Y_{61} & Y_{62} & -\frac{1}{Y_{63}} & 0 \\ 0 & 0 & Y_{03} & Y_{04} \end{pmatrix} \begin{pmatrix} \rho_1 \\ \rho_2 \\ \rho_3 \\ \rho_4 \end{pmatrix} = \begin{pmatrix} -DS_{in} \\ 0 \\ 0 \\ Q_3 \end{pmatrix} \qquad (2.331)$$

By inverting the submatrix of the yield coefficients of the left hand side of (2.331), we can express the reaction rates ρ_1, ρ_2, ρ_3 and ρ_4 as functions of the feedrate DS_{in} and of the gaseous methane outflow rate Q_3:

$$\rho_1 = Y_{21} DS_{in} \qquad (2.332)$$

$$\rho_2 = Y_{21} Y_{41} Y_{42} DS_{in} \qquad (2.333)$$

$$\rho_3 = (Y_{41} Y_{42} Y_{62} + Y_{61}) Y_{21} Y_{63} DS_{in} \qquad (2.334)$$

$$\rho_4 = \frac{1}{Y_{04}} Q_3 - \frac{Y_{03}}{Y_{04}} (Y_{41} Y_{42} Y_{62} + Y_{61}) Y_{21} Y_{63} DS_{in} \qquad (2.335)$$

Let us replace the reaction rates ρ_1, ρ_2 and ρ_4 by their above expressions (2.332), (2.333), (2.335) in the dynamical equation of the hydrogen concentration S_4, which is then rewritten as follows:

$$\frac{dS_4}{dt} = -DS_4 - Q_1 - Y_1 Q_3 + Y_2 DS_{in} \qquad (2.336)$$

where Y_1 and Y_2 are defined as follows:

$$Y_1 = \frac{1}{Y_{84} Y_{04}} \qquad (2.337)$$

$$Y_2 = Y_{81} Y_{21} + Y_{82} Y_{21} Y_{41} Y_{42} + \frac{Y_{03}}{Y_{04} Y_{84}} (Y_{41} Y_{42} Y_{62} + Y_{61}) Y_{21} Y_{63} \qquad (2.338)$$

Note that the coefficients Y_1 and Y_2 are nonlinear combinations of the yield coefficients Y_{ij}.

2.9.5 *Specific Approach for Model Reduction of PDE Models*

The model reduction of PDEs is also a key question in dynamical modelling, estimation and control. In Section 2.5, we have already suggested that the fixed bed reactor is often approximated by a sequence of STRs. Here we shall present two approaches (singular perturbation, and Laplace transform) which also allow the connection to be made between the PDE model of fixed bed reactors to STR models. The two approaches are used to reduce the *infinite-dimensional* Distributed Parameter Systems (DPS) to *finite-dimensional* lumped parameter models.

Application of the Singular Perturbations to the Fixed Bed Reactor Model. Here the perturbation method is used in a somewhat different perspective as before (see also [289], [72], [32]). A small parameter δ is also introduced, but here the application of the perturbation method is based on an asymptotic expansion of the state variables in powers of δ, see [148]:

$$x(z, t) = \sum_{i=0}^{\infty} \delta^i x_i(z, t) \tag{2.339}$$

where $x_i(z, t)$ represent the approximation of the variable $x(z, t)$ at order i (as we shall see here below, the order 0 in the approximation of the axial dispersion model, for instance, corresponds to the STR model). The approximation methods are often referred to (but the terminology is not uniform in the literature!) as *perturbation methods* when the series expansion is convergent (or is expected to converge), and as *asymptotic methods* when the series is divergent but asymptotic (so that the first few terms yield a good approximation for very low values of δ). It is a generally admitted viewpoint that proving that the series is either convergent or asymptotic is not essential. Also, it is a general property of asymptotic series that finding additional terms not necessarily improves the approximation since the series is generally divergent. As the perturbation parameter δ decreases, the approximation provided by the series gets better. Thus a key assumption in the following is that δ is very small (see e.g. [290], [25]).

As explained e.g. in [270], the axial dispersion model (presented in a general form (2.132) (2.133)) is an intermediate model between two "extreme" models: the STR model and the plug flow reactor model. For highly diffusive reactors, the behaviour of the axial dispersion reactor tends to the one of a STR, while for low dispersion coefficients, the axial dispersion model tends to a plug flow model.

In the following, we shall consider fixed bed reactors in which the mass dispersion may be assumed as being very important: D_{ma} is very large (with respect to uL). Let us define the parameter δ as the (dimensionless) mass Peclet number Pe_m:

$$\delta = Pe_m = \frac{uL}{D_{ma}} \tag{2.340}$$

Before going any further, note that the singular perturbation will only apply to the variables x_{fl}: therefore in order to keep the notations as simple as possible, the index fl will be dropped in the following. Moreover, we shall compact the writing of the conversion term by defining $Y = \frac{1-\epsilon}{\epsilon} Y_{fl}$.

Then the equations (2.133)(2.135)(2.136) become:

$$\delta \frac{\partial x}{\partial t} = uL \frac{\partial^2 x}{\partial z^2} + \delta(-u \frac{\partial x}{\partial z} + Y\rho) \tag{2.341}$$

$$\frac{\partial x}{\partial z} = -\frac{\delta}{L}(x_{in} - x) \qquad \text{for } z = 0 \tag{2.342}$$

$$\frac{\partial x}{\partial z} = 0 \qquad \text{for } z = L \tag{2.343}$$

By using singular perturbation techniques, away from the boundary layer, the solution of (2.341) is given by:

$$x(z, t) = \sum_{n=0}^{\infty} \delta^n x_n(z, t) \tag{2.344}$$

$$\rho(x) = \sum_{n=0}^{\infty} \delta^n \rho_n(x_0, x_1, ..., x_n) \tag{2.345}$$

with:

$$\rho_0(x_0) = \rho(x_0), \qquad \rho_1(x_0, x_1) = (\frac{\partial \rho}{\partial x})_{(x_0)} x_1 \tag{2.346}$$

The differential equations for the different terms x_0, x_1 of the expansion (2.344) (2.345) are thus obtained by substituting these expansions in the equations (2.341) (2.342) (2.343) and equating the terms of the same power in δ, so we have:

- **zero order approximation**:

$$\frac{\partial^2 x_0}{\partial z^2} = 0 \qquad \text{for } 0 \le z \le L \tag{2.347}$$

$$\frac{\partial x_0}{\partial z} = 0 \qquad \text{for } z = 0 \text{ and } z = L \tag{2.348}$$

- **first order approximation**:

$$uL \frac{\partial^2 x_1}{\partial z^2} = u \frac{\partial x_0}{\partial z} + \frac{\partial x_0}{\partial t} - Y\rho_0 \qquad \text{for } 0 \le z \le L \tag{2.349}$$

$$\frac{\partial x_1^o}{\partial z} = \frac{1}{L}(x_0 - x_{in}) \qquad \text{for } z = 0 \tag{2.350}$$

$$\frac{\partial x_1}{\partial z} = 0 \qquad \text{for } z = L \tag{2.351}$$

- **approximation of order n ≥ 2:**

$$uL\frac{\partial^2 x_n}{\partial z^2} = u\frac{\partial x_{n-1}}{\partial z} + \frac{\partial x_{n-1}}{\partial t} - Y\rho_{n-1} \quad \text{for } 0 \leq z \leq L \quad (2.352)$$

$$\frac{\partial x_n}{\partial z} = \frac{1}{L}x_{n-1} \qquad\qquad\qquad \text{for } z = 0 \quad (2.353)$$

$$\frac{\partial x_n}{\partial z} = 0 \qquad\qquad\qquad\qquad \text{for } z = L \quad (2.354)$$

By integrating the equation (2.347)-(2.348) with respect to the spatial coordinate z, we obtain:

$$x_0(z, t) = x_0(t) \tag{2.355}$$

$$\frac{dx_0}{dt} = \frac{u}{L}(x_{in} - x_0) + Y\rho(x_0) \tag{2.356}$$

The equations (2.355)-(2.356) show that the first term in the approximation, say x_0, is independent on the spatial coordinate z, and consequently, for the zero-order approximation, the process behaves like a stirred tank reactor. So singular perturbation when applied to PDEs also allows the reduction of the PDE fixed bed reactor model to a STR model and emphasises the connection between both types of models on a systematic basis.

2.10   Connection Between Plug Flow Reactors and CSTRs: A Laplace Transform Approach

Laplace transform can be a very interesting tool for transforming a linear distributed parameter model into a lumped parameter one. One of its advantages is that it is an exact lumping method and does not involve any approximation. However, it may look difficult to generalise the approach, especially for nonlinear systems, since the Laplace transform is basically applicable to linear systems only. Let us just illustrate the use of Laplace transform in a simple example: the dynamics of the substrate's concentration in a plug flow reactor where a first order reaction takes place. As we shall see, the Laplace transform will be helpful in systematising the physically admitted idea that if the input is the influent substrate concentration and the output the effluent substrate, then the process basically behaves like the combination of a stirred tank and a time delay (due to the transport of the substrate from the reactor input to the reactor output) (see also [105]).

The dynamics of the reactant concentration C is described by the following equation:

$$\frac{\partial C}{\partial t} = -u\frac{\partial C}{\partial z} - k_0 C \tag{2.357}$$

(where k_0 is the kinetic constant), with the following boundary condition:

$$C(z = 0, t) = C_{in}(t) \tag{2.358}$$

Let us consider the Laplace transform of C(z,t) with respect to time t:

$$\mathcal{L}[C(z, t)] = \mathcal{C}(z, s) \tag{2.359}$$

Let us apply the Laplace transform to the above equation in which we neglect the initial conditions $C(z, t = 0) = 0$:

$$s\mathcal{C}(z, s) = -u\frac{\partial \mathcal{C}}{\partial z} - k_0 \mathcal{C} \tag{2.360}$$

which can be rewritten as follows:

$$\frac{\partial \mathcal{C}}{\partial z} = -\frac{s + k_0}{u}\mathcal{C} \tag{2.361}$$

The solution of the above differential equation (in the independent variable z) is equal to:

$$\mathcal{C}(z, s) = \mathcal{C}(0, s)e^{-(s+k_0)z/u} \tag{2.362}$$

which can be rewritten as follows:

$$\mathcal{C}(z, s) = e^{-sz/u}[\mathcal{F}(z, s)] \tag{2.363}$$
$$\mathcal{F}(z, s) = \mathcal{C}(0, s)e^{-k_0 z/u} \tag{2.364}$$

If one recalls the shift property of the Laplace transform, we can notice that the above equation is characterised by a time delay z/u. Indeed:

$$\mathcal{L}[f(z, t - \frac{z}{u})] = e^{-sz/u}[\mathcal{F}(z, s)] \tag{2.365}$$

Moreover, if we consider the boundary condition (2.358), the function $\mathcal{F}(z, s)$ is rewritten as follows:

$$\mathcal{F}(z, s) = \mathcal{C}_{in}(s)e^{-k_0 z/u} \tag{2.366}$$

The above equations (2.363)(2.366) can be used e.g.

1. *to calculate time responses*
   For instance, if the initial state is assumed to be zero ($C(t = 0, z) = 0$), the response to a step $C_{in}1_+(t)$ is equal to:

$$C(z, t) = 0 \qquad\qquad \text{for } t < \frac{z}{u} \tag{2.367}$$

$$= C_{in}e^{-k_0 z/u} \qquad\qquad \text{for } t \geq \frac{z}{u} \tag{2.368}$$

2. *to compare the plug flow reactor model with the CSTR one*
Then it is first worth noting that the equation (2.366) is also the expression of the equilibrium state of $C(z, t)$ for C_{in} constant, i.e. the solution of:

$$0 = -u\frac{\partial C}{\partial z} - k_0 C \qquad (2.369)$$

Let us compare this expression with the steady state value of $C(t)$ in a CSTR, which is the solution of the following equation:

$$0 = \frac{F}{V}(C_{in} - C) - k_0 C \qquad (2.370)$$

i.e.

$$\bar{C} = \frac{C_{in}}{1 + \frac{k_0 V}{F}} \qquad (2.371)$$

It is worth noting that by definition:

$$\frac{F}{V} = \frac{u}{L} \qquad (2.372)$$

We note then that there is a great similarity between both expressions $e^{-k_0 L/u}$ and $\frac{1}{1+\frac{k_0 V}{F}}$ at the reactor output $z = L$. Note in particular that both expressions have the same first order term in the Taylor series expansion:

$$1 - \frac{k_0 L}{u} \qquad (2.373)$$

This similarity is illustrated in Figure 2.23, where the expressions $e^{-k_0 L/u}$ and $\frac{1}{1+\frac{k_0 V}{F}}$ have been drawn. This illustrates the analogy, in that specific case (when the input is the influent concentration), between the plug flow reactor and the CSTR + a time delay.

The above simple example illustrates the advantage of the Laplace transform to study the properties of distributed parameter models and to reduce them exactly to lumped parameter models. It also emphasises the possible limitations of the method. For instance how can we extend the above results to more complex dynamics, e.g. for nonlinear kinetics?

2.11 Conclusions

This chapter has been dedicated to mass balance modelling. Starting from simple examples, we have emphasised the two basic items in mass balance models in WWTP: transport dynamics, and conversion. We have also used the notion of reaction schemes as a basis for modelling, and illustrated the concept via a number of

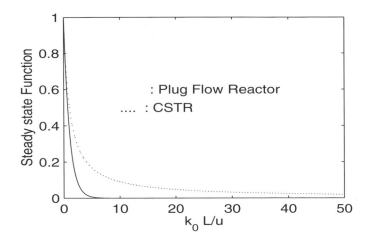

FIG. 2.23. Steady state functions of the CSTR and the plug flow reactor.

WWTP examples. We introduced the General Dynamical Model; this formalism is rather similar to the tabular format (Petersen matrix) used for the IWA model. The General Dynamical Model format is convenient for algebraic manipulations, as has already been illustrated in the present chapter via the state tranformation of Section 2.8 (which will be used later to design software sensors for process monitoring) and the model order reduction approach of Section 2.9. We have illustrated dynamical modelling by using several examples of wastewater treatment processes: anaerobic digestion, activated sludge process (basic model and IWA model), two step nitrification, two step denitrification.

Other examples like waste stabilisation ponds (lagoons) and soil decontamination could also have been considered. A model of detoxification is given in Chapter 7 (Section 7.4.4) in the context of model building of the reaction network independently of the process kinetics, via the use of the asymptotic observers introduced in that chapter. A full study, from model building to parameter identification, of a lagoon is presented in Bonvillain [31] and Grégoire et al. [106]. The model is derived from mass balance considerations by considering three types of microorganisms (micro-algae, aerobic bacteria, anaerobic bacteria) working at different depths in the pond. The resulting model contains 36 parameters, but only 3 variables (temperature, dissolved oxygen and light intensity) were available for measurement. Grégoire et al. [106] summarizes the identification results for the model parameters, which is clearly a challenging issue and has therefore to be very carefully performed. With that respect, the results presented in Grégoire et al. [106] are illustrative and exemplary of a situation largely encountered in parameter identification of WWTP.

We would also like to mention the possibility of including alkalinity in anaerobic digestion models via the introduction of electro-chemical equilibrium equations [223]. A complete study from model building to parameter identification has been performed in the context of an EEC project (AMOCO) for a pilot anaerobic digester [26] [27] [78]. One interesting point of the study is the design of experiments for parameter identification: in order to account for the specific constraints about the operation of the anaerobic digestion process (the process is rather slow and sensitive to disturbances that may easily drive the process to instability, i.e. wash-out), the experiment design for parameter identification has resulted in a sequence of step changes of the influent flow rate and of the influent substrate concentration (or equivalently, of the dilution rate and of the organic load with sufficiently long steady states in between) that basically covers all the operating regions that the process is expected to face. This experiment design can be viewed as an alternative strategy to the one presented in Chapter 5.

So far we had considered single stirred tank reactors. In Section 2.5, we have extended the dynamical models to multi-tank processes, and to non-completely mixed ones, like fixed bed reactors, fluidised bed reactors, and settlers. In the latter case, the dynamical equations are described by PDEs (partial differential equations) instead of ODEs (ordinary differential equations).

Sections 2.6 to 2.10 are dedicated to the dynamical properties of the mass balance models. In Section 2.6, we have drawn the attention to the difference between linear and nonlinear models, and to the fact that the same model can even be either linear or nonlinear depending on the considered input(s) and/or the kinetic models, for instance. In Section 2.7, we have briefly introduced the concept of stability, and the multiple steady states in WWTP: for one set of input values, the process can have different steady states; each of these steady states can even correspond to different stability characteristics, i.e. they can be either stable or unstable. Section 2.8 has introduced a key state transformation that will be very useful to analyse the process model and to design software sensors. Section 2.9 was concerned with a systematic approach for model order reduction (the singular perturbation approach) applied to ODE models as well as to PDE models. Finally, Section 2.10 has considered the use of the Laplace transform as a tool for comparing plug flow reactor models and CSTR models.

3

Structure Characterisation (SC)

3.1 Introduction

As mentioned in the first chapter, Structure Characterisation (SC) is a key step in the model building exercise (see also Figure 1.1 in Chapter 1). After a chapter on mass balance modelling, the present chapter will be concerned with structure characterisation. The objective is to infer the level of model complexity (dimension of the state vector...) and to determine relationships between variables; this can also be viewed, in other words, as a selection of the best model structure among different model structure candidates on the basis of experimental data.

Finding the "true model" $M_T(S_T, P_T)$ with model structure S_T and parameters P_T is utopian. Rather one must aim at finding – from a finite set of N noisy data points – the partial descriptions that are purposeful within the application [159]. Settling for the best possible model $M(S_N, P_N)$, however, induces an error that has two components:

$$M_T(S_T, P_T) - M(S_N, P_N) = [M_T(S_T, P_T) - M(S_N, P^*)]$$
$$+ [M(S_N, P^*) - M(S_N, P_N)] \qquad (3.1)$$

The first term is due to the error between the true model structure S_T and the model structure S_N chosen from the set of candidate models with restricted complexity. This so-called *bias error* reflects the unmodelled dynamics [104]. The second component, the *variance error*, is caused by the particular realisation of the noise in the limited number of data used in the system identification. Each data set will result in different parameter estimates P_N that will only tend to the real P^* (for this structure) when there is no noise or when the number of data points tends to infinity. This variance error also includes the effect of the overparametrisation: the more parameters included in the model, the more uncertain their values will be.

Typically the variance error decreases like $1/N$, but increases like p, with p the number of parameters in the model structure, a measure of its complexity. The bias error, on the other hand, will decrease as p increases, but is independent of N [157], [159]. Hence, as the aim is to obtain the model structure giving the lowest total error, the goal of model structure characterisation will be to find the compromise between bias error and variance error.

This chapter is organised as follows. Section 3.2 starts with the introduction of a model used in respirometry that we shall consider as a case study throughout Chapters 3, 4 and 5. Section 3.3 is dedicated to the presentation of the methods for structure characterisation, basically gathered in a priori and a posteriori methods. Finally, in Section 3.4, we shall deal with optimal experiment design for structure characterisation.

3.2 A Model Case Study

Let us consider the following candidate models that we shall build on the basis of on-line measurements of the oxygen uptake rate via a respirometer ([74]). The considered respirograms are indeed representative of the biodegradation kinetics in the WWTP, and the objective is to identify a kinetic model from the data of the respirometer. Recall that the model considered here expresses the dependence of the exogenous oxygen uptake rate OUR_{ex} on the biodegradation of k substrates S_i present in the mixed liquor:

$$OUR_{ex} = \sum_{i=1}^{k}(1 - Y_i)r_{S_i} \qquad (3.2)$$

In the above expression, Y_i (the yield coefficient) is the fraction of pollutant S_i which is not oxidised but converted into new biocatalyst X, and r_{S_i} is the rate of consumption of S_i. The experiments that we shall consider are performed in batch conditions. In these conditions (and *only* in these conditions! Recall what has been said in Chapter 2, Section 2.2.1), we can write:

$$\frac{dS_i}{dt} = -r_{S_i} \qquad (3.3)$$

The four types of wastewater/sludge interaction that have been included in the set of candidates of model structure are:

Type 1 (Exponential): One pollutant, first order kinetics (k=1)

$$\frac{dS_1}{dt} = -\frac{k_{max1}X}{Y_1}S_1 = -r_{S_1} \tag{3.4}$$

Type 2 (Single Monod): One pollutant, Monod kinetics (k=1)

$$\frac{dS_1}{dt} = -\frac{\mu_{max1}X}{Y_1}\frac{S_1}{K_{S1}+S_1} = -r_{S_1} \tag{3.5}$$

Type 3 (Double Monod): Two pollutants simultaneously degraded without mutual interaction, double Monod kinetics (k=2)

$$\frac{dS_1}{dt} = -\frac{\mu_{max1}X}{Y_1}\frac{S_1}{K_{S1}+S_1} = -r_{S_1} \tag{3.6}$$

$$\frac{dS_2}{dt} = -\frac{\mu_{max2}X}{Y_2}\frac{S_2}{K_{S2}+S_2} = -r_{S_2} \tag{3.7}$$

Type 4 (Modified IWA activated sludge model No. 1) [233]: 3 pollutants, 2 hydrolysed into the first substrate which is used for growth according to the Monod kinetics (k=1)

$$\frac{dS_1}{dt} = -\frac{\mu_{max1}X}{Y_1}\frac{S_1}{K_{S1}+S_1} + k_r X_r + k_s X_s = -r_{S_1} + k_r X_r + k_s X_s \tag{3.8}$$

$$\frac{dX_r}{dt} = -k_r X_r \tag{3.9}$$

$$\frac{dX_s}{dt} = -k_s X_s \tag{3.10}$$

Note that nitrification and its associated oxygen consumption have not been included in this case study. The k_{max1}, μ_{maxi} (i = 1, 2) and k_j (j = r, s) are rate constants, and the K_{Si} (i = 1, 2) are the affinity constants expressing the dependency of the degradation rate on the concentration of pollutant S_i (i = 1, 2). Recall also that the experiments on which the model identification is to be based, are performed in such a way that the change in biomass concentration can be assumed negligible (i.e. $\frac{dX}{dt} = 0$) (which is a fair assumption provided S(t=0) << X(t=0)), that the oxygen uptake rate data are only due to exogenous (= substrate induced) respiration (OUR_{ex}), i.e. endogenous respiration is either assumed negligible or is eliminated from the data, and that the oxygen concentration is always maintained above 2 mg/L so that oxygen is never limiting.

The aim is to select between the above biodegradation models via measurements of the oxygen uptake rate OUR, as these characterise substrate degradation.

3.3 Structure Characterisation Methods

In this section a variety of objective decision tools will be introduced that enable to find the trade-off between bias and variance error. The picture of approaches, methods and results of structure characterisation (SC) is very diverse and scattered over several scientific disciplines. Although a number of studies have evaluated different SC techniques, none of them has resulted in a clear recommendation on a definitely superior method [6], [65], [146], [166], [239]. With the above case study, an analogous exercise was devoted to decide on a SC technique most adapted to the problem at hand. To illustrate the fact that this step of model building is not as mature as, for instance, parameter estimation, we use this case study to also show how new SC methods can be developed. A number of these are specific to the candidate models used in this case study, but others are potentially more generally applicable.

The SC techniques have been classified according to their impact on the total time needed for a model identification (i.e. model selection + parameter estimation).

1. Most existing methods for SC evaluate the quality of the different model structures after fitting each model to the data. Hence, these methods can be termed *a posteriori SC* [262].
2. Methods capable of selecting a model without the need of first estimating the parameters belong to the other class, the *a priori SC* techniques.

In view of the observation that parameter estimation of the nonlinear bioprocess models that are dealt with here takes considerable computing time (see Chapter 6), it is clearly advantageous to apply a priori methods, since they will only require the estimation of the parameters of the a priori selected model.

3.3.1 *A Priori SC*

The approach of a priori SC has also been termed model structure selection based on preliminary data analysis and has been reported to be an underdeveloped field [159].

Two groups of methods can be discerned: one type of method is generally applicable, while the other SC methods take advantage of specific features of the model structures present in the set of candidate models.

General Methods. The pattern recognition capabilities of neural networks [229] have incited a study on their applicability as SC technique. Details on the method developed for the case study can be found elsewhere [272], but the main principles are summarised here (Figure 3.1): A three-layer recurrent neural net architecture receiving preprocessed data was used. Different data preprocessing algorithms were tested, all aiming to perform data reduction (to decrease the number of input nodes to the neural net classifiers) without loss of structure specific information. The learning stage was performed with 750 training-patterns obtained from

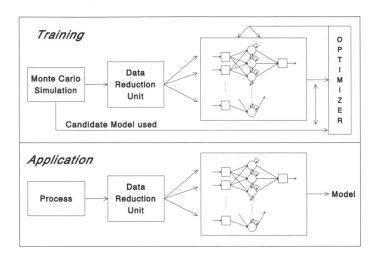

FIG. 3.1. Training and application stages of a neural net used for model selection.

Monte Carlo simulations of the different candidate models with different parameter values and initial conditions. Once the net was trained to recognise the most
appropriate model, it could be used on new (preprocessed) data. The results for the
case study are discussed below.

A next method that can be applied stems from recent developments in identification of linear state space models. Numerical algorithms for Subspace State
Space System Identification (N4SID, read as *enforce it*) combine the estimation of
the order of the state with the identification of the system matrices, but, in contrast
to traditional identification schemes, the order is estimated first [257]. This aspect
of this data-driven approach may be useful for non-linear models as well. Clearly,
for the case study for instance, one cannot expect that N4SID will give a direct
measure of the number of substrates to be included or the degradation mechanism
involved. Rather it can be hoped that the order computed by the algorithm enables
differentiation among the models. The order is determined on the basis of the number of non-zero singular values SV in a matrix H composed of the following Block
Hankel matrices containing "past" (Y_p) and "future" (Y_f) data:

$$Y_p = \begin{pmatrix} OUR_1 & \dots & OUR_j \\ OUR_2 & \dots & OUR_{j+1} \\ \vdots & & \vdots \\ OUR_i & \dots & OUR_{j+i} \end{pmatrix} \tag{3.11}$$

$$Y_f = \begin{pmatrix} OUR_{i+1} & \cdots & OUR_{j+i} \\ OUR_{i+2} & \cdots & OUR_{j+i+1} \\ \vdots & & \vdots \\ OUR_{2i} & \cdots & OUR_{j+2i} \end{pmatrix} \tag{3.12}$$

Different H-matrices can be evaluated:

$$H = Y_f Y_p^T \tag{3.13}$$

$$H = Y_f Y_p^T (Y_p Y_p^T)^{-1} Y_p \tag{3.14}$$

$$H = L_f^{-T} Y_f Y_p^T L_p^{-1} \tag{3.15}$$

with L_f and L_p Cholesky factorisations of $Y_f Y_f^T$ and $Y_p Y_p^T$.

The choices of i and j are important: j must be as large as possible (for the case study j was set $N - 2i$); i must be much smaller than j, but larger than the largest model order expected (i was set 10). For the case study the best results were obtained with the first form of H and singular values considered 0 when below 0.1: type 1 models were selected if $SV \ll 2$; type 2 if $SV = 2$ and type 3 if $SV \gg 2$.

Specific Methods. Structure characterisation on the basis of so-called parameter invariant model features has been advocated, but has found little application due to the difficulty in finding such features from the models under consideration. Indeed, from the data one must find a characteristic that is independent of the parameters of the corresponding model. For instance, we look for features which are scale independent. For the candidate models in the case study, such features can be found [259]: the number of inflection points is 0, 1 and 3 for models of type 1, 2 and 3 respectively (see also Figure 3.2). Though this is analytically correct, the determination of inflection points on noisy data is not straightforward. However, using a moving window regression method in which linear and quadratic regressions are compared with an F-test, the significance of the second derivative can be assessed, leading to a more reliable estimate of the number of inflection points.

Another model-specific approach applicable to the case study consists of fitting an exponential function, a hyperbolic tangent function and a Double hyperbolic tangent function to the data and comparing the resulting sum of square residuals (SSR). In other words, this a priori method for selecting among the biodegradation models is in fact an a posteriori method with respect to the Exp/Tanh/Double Tanh candidate model set. However, these models can be fitted much faster than the biokinetic models. SSR ratios larger than 10 were found necessary before a more complex model was accepted.

A third method developed for the models of the case study is based on the ratio between the area below the OUR_{ex} versus time data (and the square formed by the ($OUR_{ex,max}$, t_{max}) and (0, 0) corners). Some preprocessing of the OUR_{ex} data is performed prior to the determination of this criterion: first, the zero-tail (Figure 3.2) is cut from the data-set using a t-test on the mean value of a data-window and, second, the reduced OUR_{ex}, t-dataset is scaled to the unit square. Type 1 data have the lowest area ($\ll 0.3$) and type 2 the largest area ($\gg 0.5$) under the curve (Figure 3.2). These threshold values for the areas enable the selection among the candidate models type 1, 2 and 3 [259].

These examples of specific methods illustrate that one can devise effective methods for structure characterisation, but this requires special insight in the models to be selected.

3.3.2 A Posteriori SC

Most methods described in this section have found more widespread application. In some applications they seem less appropriate because they may become very time-consuming, especially as the number of candidate models increases. However, it is not inconceivable that a SC strategy is devised in which a priori methods are used to make up a first ranking of the different structures, after which a posteriori methods are called in order to make the final selection among the structures with the highest ranking. In that context, time constraints on the model building process and the selection reliability can be sufficed.

Criteria with Complexity Terms. The criteria on which most of the a posteriori SC methods are based take one of the following two forms [232]:

$$\frac{SSR}{N} [1 + \beta(N, p)] \tag{3.16}$$

$$N log \left(\frac{SSR}{N} \right) + \gamma(N, p) \tag{3.17}$$

with SSR the sum of squared residuals, and N and p as defined before. For $N \gg p$ one can show that both representations are equivalent if [232]:

$$\gamma(N, p) = N\beta(N, p) \tag{3.18}$$

For both cases, the first term decreases with increasing p (increasing complexity) while the second term penalises too complex (overparametrised) models. The model structure with the smallest criterion value is selected. Different authors have proposed several functional forms for the model selection criteria depending on a theoretical starting point. The two best-known are the Final Prediction Error (FPE), with [159]:

$$\beta(N, p) = \frac{2p}{N - p} \tag{3.19}$$

and Akaike's Information Criterion (AIC), with

$$\gamma(N, p) = 2p \tag{3.20}$$

FPE and AIC have been proven not to be consistent (i.e. they do not guarantee that the probability of selecting the wrong model tends to zero as the number of data points tends to infinity) [140]. However, this disadvantage is compensated by the fact that AIC and FPE enjoy certain properties that allow the determination of good prediction models in case the true model does not belong to the set of candidate models [232].

An example of a consistent criterion is the Bayesian Information Criterion (BIC) in which [6], [228]:

$$\gamma(N, p) = p \, log(N) \tag{3.21}$$

If $p \geq 8$, it can easily be seen that BIC will tend to favour models of lower complexity than those chosen by AIC. Another consistent criterion is LILC where [110]:

$$\gamma(N, p) = p \, log \, (log(N)) \tag{3.22}$$

Criteria that Assess Undermodelling.   Goodwin *et al.* [104] decomposed the total error between the true model and a candidate model into three components:

- the effect of the variance of the particular noise realisation
- the effect of the parameter errors due to the noise present in the identification data
- the effect of the undermodelling, corresponding with the bias term in (3.1).

Their General Information Criterion, GIC, is defined as:

$$GIC(N, p) = \hat{\sigma}_v^2 + \frac{p}{N}\hat{\sigma}_v^2 + undermodelling \tag{3.23}$$

In this equation it is essential to have an estimate of the residual error $\sigma_v{}^2$ that is independent of the undermodelling error. Goodwin *et al.* [104] obtained this by fitting a high-dimensional model (assuming it to be the true model) and using the residual variance as an estimate. In the case study of this chapter, such a model is not available, but, fortunately, the zero-tail end of the $OU R_{ex}$-dataset can be used for σ_v^2 estimation since this part of the data is characterised by the absence of biological dynamics [263]. As a result, the variance of this part of the data can be used for an accurate estimate of the residual error. Using the similarities shown by the authors between the expected value of FPE and GIC [104], it is even possible to write an explicit formula for the undermodelling term [259]:

$$undermodelling = \frac{SSR}{N} - \left(\hat{\sigma}_v^2 + \frac{p}{N}\hat{\sigma}_v^2\right)\frac{N - p}{N + p} \tag{3.24}$$

Structure characterisation is then performed by selecting the model with the lowest undermodelling value.

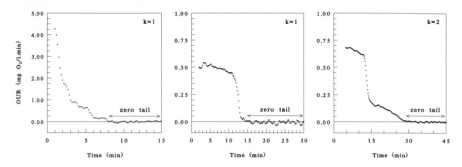

FIG. 3.2. Illustrative OUR_{ex}-data.

Statistical Hypothesis Tests.   The F-test is probably the most frequently applied method to choose among model structures. The statistical test

$$\frac{\left(SSR_i - SSR_j\right)/\left(p_j - p_i\right)}{SSR_j/\left(N - p_j\right)} \tag{3.25}$$

is compared with the $F(p_j - p_i, N - p_j)$ distribution to decide whether the more complex model j is significantly (with a confidence level α) better than model i. The similarities that exist between F- and χ^2-tests and the equivalence between the AIC/FPE criteria and F-tests with a prespecified significance level have been shown [232].

Diagnostic Checking (Analysis of Residuals).   When modelling, some assumptions are made concerning the properties of the noise. In most cases, the prediction errors $\epsilon(t)$ are assumed to be a realisation of independent random variables with zero mean and a defined distribution. Then, the quality of a model can be assessed by analysis of the properties of the calculated residuals. Two approaches can be used to check the independence of the residuals (white noise property): the autocorrelation and the run test. The autocorrelation test [232] is based on the fact that the correlation function for a white noise sequence $\epsilon(t)$:

$$\hat{r}_\epsilon(\tau) = \frac{\displaystyle\sum_{t=\tau}^{N-\tau} \frac{\epsilon(t - \tau)\epsilon(t)}{N - 2\tau}}{\displaystyle\sum_{t=1}^{N} \frac{\epsilon^2(t)}{N}} \tag{3.26}$$

is zero except for $\tau = 0$. Structure characterisation with these tests is performed by selecting the model whose residuals are as white as possible. Two statistical tests are suggested to make objective decisions on the whiteness of the residuals.

- One test compares the correlation for each lag τ (3.26) with the limit value $\frac{N(0,1)}{\sqrt{N}}$, which for $\alpha = 0.05$ means that only 5 % of the autocorrelations may

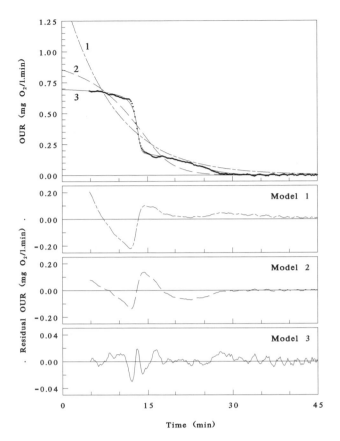

FIG. 3.3. Typical model fitting results for three models to a "Double Monod" style data set.

be larger than $\frac{1.96}{\sqrt{N}}$. In Figure 3.3, an example taken from the case study, the residuals for the three models fitted to a typical type 3 dataset are given. In Figure 3.4 the corresponding autocorrelation function for lags $\tau = 0$ to 20 is depicted. Clearly, the residuals for models 1 and 2 are highly dependent, while for model 3 only the first 4 correlations (20 %) are significantly higher than the prescribed level (indicated by the horizontal lines). Hence, the residuals do not originate from a white noise sequence, indicating some unmodelled dynamics. Looking into some more detail to the residuals of model 3 (Figure 3.3), one distinguishes an oscillatory pattern, probably causing the significant autocorrelation.

- The other autocorrelation test compares a combination of the first m covariances:

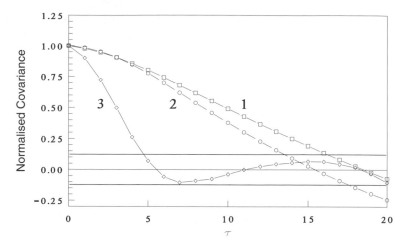

FIG. 3.4. Normalised covariance function of the residuals for the models in figure 3.3.

$$\frac{N}{\hat{r}_\epsilon^2(0)} \sum_{i=1}^{m} \hat{r}_\epsilon^2(i) \qquad (3.27)$$

with the $\chi^2(m)$ distribution giving a significance level for the independence of the residuals [159].

Another residuals test is a so-called non-parametric test in which the number of runs R, calculated easily as the number of sign changes in the sequence of residuals, is evaluated against the expected number of runs, $N/2$ [232]. To assess the significance of a deviation from this number the statistical test

$$\frac{R - \frac{N}{2}}{\sqrt{N/2}} \qquad (3.28)$$

can be compared with $N(0, 1)$. A posteriori SC with this method selects the model with the statistical test closest to zero or the simplest model with a non-significant statistical test.

Results for the Case Study. The different SC methods introduced above were evaluated on the basis of 8 typical real-life $OU\,R_{ex}$-datasets [259]. Only the type 1, 2 and 3 models were included in the candidate model set. In Table 3.1 the selected models are compared with the advice of a human expert. Dataset 4 is difficult to classify since it could be considered very close to both a type 1 and type 2 model. Hence, both were considered correct.

Except for the N4SID method, all a priori SC methods produced very good selection results. Among them, the Tanh and neural net approaches may be preferable considering the noisy data which may cause problems in estimating the number of

Table 3.1 Model selected on the basis of different a priori and a posteriori SC methods. The results are compared with the advice of a human expert (figures in bold indicate "right" choice, underfit and overfit refer to the complexity of the selected model compared to the choice of the expert)

Method of Structure Characterisation	Dataset								Evaluation		
	1	2	3	4	5	6	7	8	underfit	overfit	correct
N4SID	**1**	3	2	**1**	2	2	2	2	3	1	4
Neural Net	3	**2**	**3**	**1**	**3**	**2**	**3**	**2**	0	1	7
Inflection points	**1**	**2**	**3**	2	**3**	**2**	**3**	**2**	0	0	8
Tanh	**1**	**2**	**3**	**1**	**3**	**2**	**3**	**2**	0	0	8
OUR_{ex}-Area	**1**	**2**	**3**	3	**3**	**2**	**3**	**2**	0	1	7
AIC	3	3	**3**	3	**3**	3	**3**	3	0	5	3
FPE	3	3	**3**	3	**3**	3	**3**	3	0	5	3
BIC	3	3	**3**	3	**3**	3	**3**	3	0	5	3
LILC	3	3	**3**	3	**3**	3	**3**	3	0	5	3
Undermodelling-GIC	**1**	**2**	**3**	**1**	**3**	**2**	**3**	**2**	0	0	8
F-test	3	3	**3**	3	**3**	3	**3**	3	0	5	3
Autocorrelation	**1**	**2**	**3**	2	**3**	**2**	**3**	3	0	1	7
Run-test	**1**	**2**	**3**	3	**3**	**2**	**3**	3	0	2	6
Human Expert	**1**	**2**	**3**	**2(1)**	**3**	**2**	**3**	**2**	-	-	-

inflection points. Neural nets have an additional advantage as a potentially general tool for a priori SC.

For the a posteriori methods, an important finding was that all "information" criteria, i.e. AIC, FPE, BIC and LILC, with the notable exception of GIC, result in overfitting of the model compared to the "expert advice". This is probably due to the oscillations that can be observed in the OUR_{ex}-data (see Figure 3.2 and the residuals in Figure 3.3). Since the more complex models possess sufficient flexibility, some of these oscillations can be modelled. This reduces the residual error to such an extent that any penalty for model complexity is compensated for. As the observed oscillations cannot be explained by any biological process and are probably due to some hardware dependent process which is not of interest to the user, this "parasite process" should be eliminated from the data before model identification is initiated.

The F-test also suffers from the modelling of the parasite processes superimposed on the biological response. For the GIC-based method, these overfitting problems were nonexistent because the effect of the oscillations is included in the estimate of the variance (from the zero-tail). In this way only the undermodelling of the biological phenomena is retained leading to model selections congruent with the observation of the human expert.

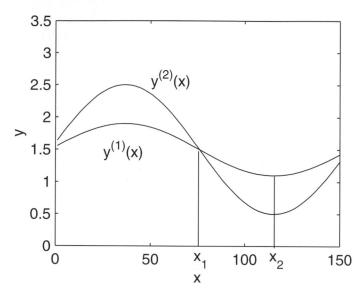

FIG. 3.5. Hunter-Reiner approach for an optimal experiment design to discriminate between two rival models.

3.4　Optimal Experiment Design for Structure Characterisation

3.4.1　*Theoretical Background of OED/SC*

To start with, the OED problem must be clearly stated and translated into an objective function. The goal is, for instance,

To design an experiment to discriminate between rival models y_1 and y_2.

A very useful and intuitive translation into an objective function is given by the Hunter-Reiner approach [130] that chooses the experimental conditions Ψ_i such that the difference between the predictions by both models

$$\left(\hat{y}_1(\Psi_i) - \hat{y}_2(\Psi_i)\right)^2 \qquad (3.29)$$

is maximised.

The approach is illustrated in the examples of Figure 3.5. In both cases it is obvious that no selection can be made between the two models in competition at condition 1 while maximal discriminative power is obtained under experimental condition 2.

While this approach is very appealing (and widely applied), a major drawback is that no consideration is given to the fact that uncertainty exists on the model predictions $\hat{y}_i$. Figure 3.6 illustrates the problem of uncertainty effects for two simple regression models that are in competition:

$$y_1 = bx + a \qquad (3.30)$$

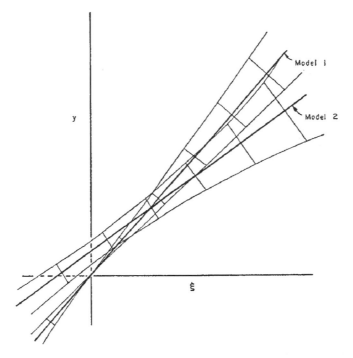

FIG. 3.6. Uncertainty in model predictions of two rival regression models.

$$y_2 = bx \qquad\qquad (3.31)$$

It is straightforward to compute the confidence regions for both models. From the figure it is evident that the confidence regions are important for the assessment of the discriminative power of an experiment. For this particular case, it is clear that model selection will be most reliable under experimental conditions Ψ when equal or lower than zero. Note that one would have preferred high positive values of the experimental conditions Ψ when the confidence regions were not considered.

Box and Hill [39] developed a quantitative description that allows the computation of the divergence between rival models under model prediction uncertainty. They extended the method to the case of discrimination among m rival models and the use of the combined data of n previous experiments so as to design the $(n+1)$th. It would lead us too far to include these complex objective functions for OED/SC here. They can be found in the mentioned reference together with some illustrative examples.

The OED/SC methods mentioned so far were mainly developed for static models (e.g. regression type empirical models). For dynamic models Munack [181] proposed two approaches, one based on the maximisation of the difference in model predictions, i.e. similar to the Hunter-Reiner method, the other being based

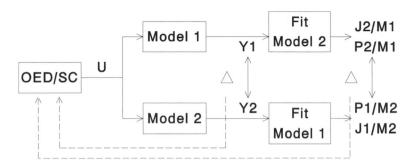

FIG. 3.7. OED/SC: Munack's first approach [181].

on the maximization of the change in parameter estimates when a model is fitted to data obtained under quite different experimental conditions. The latter in fact corresponds to a kind of adequacy evaluation of the rival models since adequate models are characterised by the fact that the parameters do not change with the experimental conditions.

For the first approach the optimisation loop of OED/SC is the following [181] (see Figure 3.7):

1. propose experimental conditions and perform hypothetical experiments through simulation with the 2 models, giving rise to two sets of "raw" data;

2. (a) consider that Model 1 is correct, then fit Model 2 to the Model 1 generated data (evidently, Model 1 must not be fitted since the data originate from this model);

    (b) consider that Model 2 is correct, then fit Model 1 to the data generated from Model 2 (here Model 2 must not be fitted);

3. for both cases calculate the difference between the trajectories simulated with the "best" parameter estimates;

4. maximise the smallest of the two calculated differences (Model 1 - Model 2 fitted), (Model 2 - Model 1 fitted).

It is important to note that the two parameter estimations are included in the procedure, because it is reasonable to expect that a dataset generated by a particular model can be described best by the rival model after a new estimation of its parameters. However, these parameter estimations mean that this OED/SC approach may ask for lengthy computations. These may not be desired for certain applications.

In the second method proposed by Munack [181], the change in the model parameter values needed to fit the data generated in an experiment with different conditions is assessed and maximised by the experiment design. Since no a priori knowledge on the "best" model is available at the moment the experiment design is performed, the rival models have to be treated on an equal basis. The design calculations are identical to the abovementioned procedure but as an objective function to be maximised the parameter sets for the first (real) experiment and the second (simulated) experiment are compared for both models. For this, the Mahalanobis distance between the parameter vectors is calculated and the smallest distance is maximised. The Mahalanobis distance is a measure of the difference between two vectors which takes the estimation accuracy into account.

3.4.2   Application: Real-time OED/SC in a Respirometer

In the respirometer case study developed throughout this chapter, the possibility exists to adjust the experimental conditions applied to the bioreactor. The aim is to maintain the quality of the identification of models describing the biodegradation processes. In this section attention will be given to a specific method that enables OED/SC even under the real-time constraints imposed by the fact that a new optimal experiment for structure characterisation must be designed within the 30 minute interval between consecutive experiments. It was shown that the methods described above could not handle this constraint due to the excessive computational burden and, therefore, a dedicated method had to be developed [261]. It is important to note that, as for every OED/SC method, the objective function is closely related to the method of structure characterisation applied during the identification stage. In the case study, the a priori structure characterisation method based on the number of inflection points is used.

It is good to recall that to determine the number of inflection points from the respirograms and their reliability, the second derivative (the curvature) must be calculated. To estimate its value and to determine whether it is significant, a moving window regression with window width n is applied and both a straight line and a parabola are fitted from points j up to $j + n$.

Whether the parabolic fit is significantly better, and therefore, whether the estimated value of the curvature is significantly different from zero, is tested by [13]:

$$\frac{(SSR_1 - SSR_2)/1}{SSR_2/(n - 2)} \simeq F(1; n - 2) \tag{3.32}$$

where SSR_1 and SSR_2 are the residuals sum of squares of the linear and parabolic regression respectively. $SSR_1 - SSR_2$ is the extra sum of squares due to the inclusion of the curvature in the regression. The null hypothesis that the parabolic fit is not significantly better than the linear one can be tested by referring this ratio of mean squares to the F-distribution with 1 and $n - 2$ degrees of freedom. The 1 in the numerator is the difference in degrees of freedom between a straight line and a parabola. One should note that, basically, the approach used to discriminate between the parabola and the linear regression is based on the F-test that was presented as one of the traditional structure characterisation methods.

If the parabolic fit is significantly better, the highest order coefficient (the curvature) is returned, otherwise its value is set to zero. An inflection point is defined

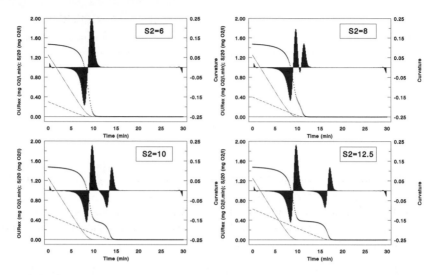

FIG. 3.8. Curvature (filled curve) of respirograms (top dashed lines starting horizontally) obtained with different ratios of two substrates (two dashed lines starting linearly decreasing): $S_1(0) = 25$, $S_2(0) = 6, 8, 10, 12.5$.

by a point where the curvature crosses zero. The results for simulated (noise free) example respirograms are displayed in Figure 3.8.

To define the reliability of an inflection point, it is to be noticed that, as illustrated in Figure 3.8, an inflection point is surrounded with two *pulses*, a positive one and a negative one. These two pulses can be used in several ways to define the reliability $r(f)$ of an inflection point f, e.g.:

- The surface of both pulses
- The total height of both pulses
- The height of the smallest pulse

The first approach will be used in the sequel. In order to increase the discriminative power of the experiments, the aim is to determine the inflection points with the highest reliability, and hence the aim of the optimal experiment design is to maximise $r(f)$.

The examples given will be restricted to the cases where the wastewater influent contains two substrates S_1 and S_2 (hence, a Double Monod model can be used) and the aim is to design an experiment such that the Double Monod model will be reliably selected. Pulse injection of such wastewater to the batch reactor of the respirometer results in initial substrate concentrations noted as $S_1(0)$ and $S_2(0)$. Two related problems will be treated:

OED/SC for Calibrations: In this type of experiment, the experimenter can add a chosen mixture of the two substrates. Since both the initial concentrations

Table 3.2 Two parameter sets used in the simulations for OED/SC

Parameter	Set 1	Set 2
X	4000 mg/l	4000 mg/l
μ_{max1}	5. e-4 /min	2.62 e-4 /min
K_{s1}	1.mg/l	0.226 mg/l
μ_{max2}	1. e-4 /min	2.85 e-4 /min
K_{s2}	0.2 mg/l	0.6 mg/l

$S_1(0)$ and $S_2(0)$ can be chosen by the experimenter, two degrees of freedom for the optimisation problem exist.

OED/SC for Wastewater: In this case, since the wastewater composition can not be altered, the ratio $\frac{S_1(0)}{S_2(0)}$ is fixed. Only the amount of wastewater injected is variable and, hence, only one degree of freedom is left.

OED/SC for Calibrations.   In the respirometer case study, calibrations are regularly performed, mainly to verify the correct operation of the measuring device. As shown by Vanrolleghem and Verstraete [266], this calibration can, however, also be used to independently characterise the two main groups of aerobic organisms in activated sludge, i.e. heterotrophs and nitrifiers. To this end, a calibration mixture of ammonia and a readily biodegradable carbon source such as acetate is injected. The optimal experiment design is then aimed at finding the amount of each substrate such that the resulting OUR_{ex}-curve allows the extraction of the three inflection points with the highest reliability and within a short experimentation time. This can be done by maximising the reliability of the three inflection points, or by maximising the least reliable inflection point. The first approach has been chosen and hence the following optimisation problem can be formulated:

$$\max_{S_1(0), S_2(0)} r(f_1) + r(f_2) + r(f_3) \tag{3.33}$$

where $S_1(0)$ and $S_2(0)$ are the initial concentrations of the calibration substrates and $r(f_i)$ is the area of the positive and negative pulses that determine the ith inflection point (see above).

This optimisation problem can be approximately solved by computing the sum of the reliabilities for each substrate combination on the grid $S_1(0) = 5(5)50$ (from 5 to 50 in steps of 5) and $S_2(0) = 5(5)50$ mg/l. One obtains a response surface that points to the optimal substrate combination. In Table 3.2, the two sets of biokinetic parameters that were used in the simulations are summarised. For Set 1, the results are schematised in Figure 3.9. On the left, the sum of surfaces is given as a contour-plot. Black indicates the experimental conditions to avoid and the lighter areas result in more reliable inflection points. The length of the respirogram is depicted on the right side. An experimental condition resulting in a respirogram that takes longer than 30 mins to return to an $OUR_{ex} = 0$ is colored black, reflecting

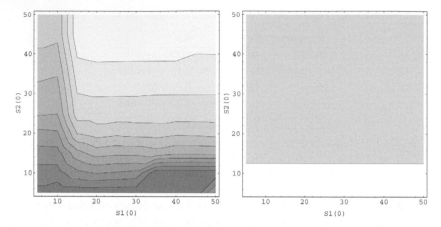

FIG. 3.9. Contourplot indicating the reliability (left) and the length (right) of the respirogram for parameter Set 1. Black indicates lower reliability or a violation of the real time constraint.

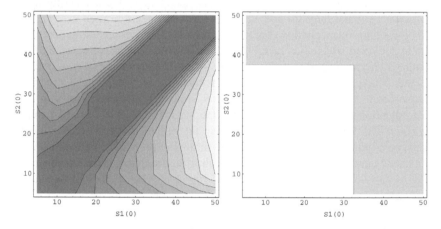

FIG. 3.10. Contourplot indicating the reliability (left) and the length (right) of the respirogram for parameter Set 2. Black indicates lower reliability or a violation of the real time constraint.

the importance given to the real-time constraint in this optimisation problem. From this it is clear that the line $S_2(0) = 5$ mg/l should be avoided and that $S_2(0)$ should be less than 12.5 mg/l.

In Figure 3.8 the OUR_{ex}-curve and corresponding substrate removal curves for four cases with increasing $S_2(0)$ are displayed ($S_1(0)$ is fixed at 25 mg COD/l). Note that the OUR_{ex}-profile in the top left case degenerates in a respirogram typical for a Single Monod model, implying that the parameters of the Double Monod

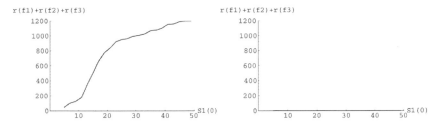

FIG. 3.11. The reliability of inflection point determination for a given $S_1(0)$ and a ratio $\frac{S_1(0)}{S_2(0)} = 1$ for parameter Set 1 (left) and Set 2 (right).

model have become practically unidentifiable from such experimental data [266].

It is very important to realise that this optimal experiment design for SC needs to be performed regularly, because changed biokinetic parameters (e.g. because of biomass adaptation) may result in a completely different advice. This is clearly illustrated by the OED/SC results (Figure 3.10) for the second set of parameter values of Table 3.2. For these sludge characteristics, the substrate concentrations to avoid, i.e. the line $S_1(0) = S_2(0)$, are clearly different compared to the ones obtained for the first parameter set (Figure 3.9).

This result emphasises that the need may exist for certain applications to perform OED/SC on-line. It also stresses the requirement for a priori SC methods so as to meet real-time constraints for timely OED.

OED/SC for Wastewater.   The OED/SC for wastewater can be derived from the OED/SC for Calibrations. Suppose the wastewater composition is characterised by a ratio $r = \frac{S_1(0)}{S_2(0)} = \alpha$. This implies that only initial concentrations lying on the line $S_1(0) = \alpha S_2(0)$ need to be considered. This line can be drawn on the given contour plots (Figure 3.9 and Figure 3.10), reducing the two-dimensional contourplot to a one-dimensional plot. One of these is shown in Figure 3.11(left) for the first set of parameters and for a fixed ratio $\frac{S_1(0)}{S_2(0)} = 1$. The reliability of the inflection point $r(f)$ is plotted versus $S_1(0)$, while $S_2(0)$ can be computed from the known ratio. Figure 3.11(left) illustrates that the reliability increases monotonically with increasing $S_1(0)$. The increase, however, is variable, implying that the gain in reliability is not constant. No maximum is found, but $S_1(0)$ is limited by the same real-time constraints as in Figure 3.9, limiting $S_1(0)$ to a maximum of about 12.5 mg/l.

The same wastewater composition (i.e. $S_1(0) = S_2(0)$) with an activated sludge characterised by parameter Set 2, results in respirograms which do not allow selection of the correct (Double Monod) model as illustrated in Figure 3.11(right): no $S_1(0)$ can be found for which the resulting respirogram will yield significant inflection points and, hence, no Double Monod model will be selected from data collected from such an experiment, although the process may intrinsically be a Double Monod type process. Only increasing the degrees of freedom of the experiment design may help to solve the identification problem. This may, for instance, be achieved by an artificial change of the wastewater's composition by deliberate addition of one of the substrate's concentrations (e.g. ammonia).

3.5 Conclusions

In this chapter we have introduced the basic concepts for Structure Characterisation (SC) and illustrated them via a set of different model structure candidates used in respirometry.

We have first made a distinction between a priori and a posteriori structure characterisation methods. Most of the proposed methods use different statistical tests. We have then proposed an approach for the optimal design of experiments in order to obtain the most reliable structure characterisation possible in the context of WWTP models.

The optimal experiment design for structure characterisation will be further developed in Section 5.5 where it will be combined with parameter estimation (OED for SC/PE).

Finally, as an extension of the methods introduced in Chapter 7, we shall introduce another approach for model selection based on the Key Transformation introduced in Chapter 2 (Section 2.8) that allows the selection of a reaction scheme independently of knowledge of the kinetics.

4

Structural Identifiability

4.1 Introduction

The identification of the dynamical models describing wastewater processes is characterised by two important features:

1. The models are most often highly complex, they are usually high-order non linear systems incorporating a large number of state variables and parameters. For instance, the IWA Activated Sludge Model No. 1 ([120], see also Chapter 2) contains 13 state variables and 19 parameters.
2. There is, generally speaking, a lack of cheap and reliable sensors and techniques for measurement of the key state variables. Despite considerable efforts, measurement methodology is still considered to be the weakest part in process modelling and control [111], [265].

Both problems are common to all biotechnological processes, although particularly crucial in wastewater treatment processes, because of the inherent particularly complex nature of these processes, involving for instance many different microbial populations which are often difficult to reliably identify with the available instrumentation.

The values of the parameters in a model are to be inferred from a priori knowledge and experimental data. The quality of the estimation of parameters will depend on the amount and quality of (real-time) data that is available to the identification algorithm. Besides these limitations on the available information, another problem in the identification of the process model appears: the model parameters may exhibit considerable correlation.

Because of the model complexity and the scarcity of (on-line) sensors, the identifiability study of the dynamical models, prior to any identification, is certainly a key question. The central question of the identifiability analysis is the following:

Assume that a certain number of the state variables are available for measurement; on the basis of the model structure (structural identifiability) or on the type and quality of available data (practical identifiability), can we expect to give via parameter estimation a unique value to the model parameters?

Simply speaking, one would wonder what is the use of trying to calibrate the parameters of a model which is, structurally or practically, unidentifiable. The above formulation is quite crude, but the answer to the identifiability analysis is often more subtle: it is not just a "yes or no" answer, but when it results in some conclusions (what is not a priori obvious with nonlinear models), these may indicate that some subset or combinations of the model parameters are a priori identifiable.

The goal of Chapters 4 and 5 is to study both the structural and practical identifiability of models used in biological wastewater treatment processes. Our intention is to give an introduction to identifiability in the context of wastewater treatment processes. Because the book is basically dedicated to WWTP engineers and not to mathematicians, we have decided to concentrate on the basic concepts without giving a fully rigourous and involved mathematical description. The interested reader can anyway be referred to a number of very good books and papers, e.g. [100], [206], [279], [254], [99]. For illustrative purposes, we shall use different examples including models employing Monod type limitation kinetics. Yet our objective is to deal with the identifiability in a sufficiently general way so as to allow the extension of the proposed study to other practical situations.

The chapter is organised as follows. The theoretical framework of the identification study will be briefly addressed in Section 4.2. It will be further developed in Section 4.3, where some important definitions are reviewed and basic concepts for the structural identifiability tests are introduced. The structural identifiability of the models is studied in Sections 4.4 and 4.5. Six different techniques (Laplace transform, Taylor series expansion, generating series, local state isomorphism, transformation of the nonlinear models, and Lyapunov based analysis) have been considered. The first five are introduced in Sections 4.3 and 4.4 and illustrated on the basis of a simple nonlinear model (yet linear in the parameters). The sixth method (Lyapunov based analysis) is introduced in Section 4.5 in a historical perspective by considering the structural identifiability of the Monod model. Section 4.6 illustrates the structural identifiability concept and tests with respirometer-based models. Finally, Section 4.7 introduces the notion of overparametrisation illustrated with an anaerobic digestion model.

4.2 Theoretical Framework

The notion of structural identifiability is related to the possibility to give a unique value to each parameter of a mathematical model. In simple words, the question of structural identifiability of a model can be formulated as follows (a rigourous definition can be found e.g. in [100]): given a model structure and perfect (i.e. that corresponds perfectly to the model) data of model variables, are all the parameters of the model identifiable? From the structural identifiability analysis one may conclude that only combinations of the model parameters are identifiable. If the number of resulting combinations is lower than the original model parameters, or if there is not a one-to-one relationship between both parameter sets, then a priori knowledge about some parameters may be required to achieve identifiability of each individual parameter. A simple example may illustrate this: in the model $y = ax_1 + bx_2 + c(x_1 + x_2)$, only the parameter combinations $a + c$ and $b + c$ are structurally identifiable (and not the three parameters a, b, c); two parameters (e.g. a and b) will be identifiable if the value of a third one (here, c) is known a priori.

For linear systems, the structural identifiability is rather well understood, and besides classical identifiable models (like dynamical models in canonical form [234], [159], [90]), there exists a number of tests for the identifiability (e.g. Laplace transform method, Taylor series expansion of the observations, Markov parameter matrix approach, modal matrix approach,..., see e.g. [100]). However, for models that are nonlinear in the parameters (like the models studied in this book), the problem is much more complex. Several structural identifiability tests also exist, but they are usually very complex (they typically require the (very helpful) use of symbolic software packages [206], as will be illustrated below).

Practical identifiability on the other hand is related to the quality of the data and their "information" content: are the available data informative enough for identifying the model parameters and for giving accurate values? In the model $y = ax_1 + bx_2$ the parameters are structurally identifiable but they will not be practically identifiable if the experimental conditions are such that the independent variables x_1 and x_2 are always proportional ($x_1 = \alpha x_2$) (then only the combination $a\alpha + b$ is identifiable). This topic (practical identifiability) will be the object of Chapter 5.

4.3 Notion of Structural Identifiability of Linear Systems

4.3.1 A Simple Example

Let us start with a simple example. Let us consider a CSTR with one reaction:

$$A \longrightarrow B \tag{4.1}$$

described by first order kinetics. The dynamical mass balance of the substrate A is given by the following equation:

$$\frac{dC}{dt} = -DC + DC_{in} - k_0 C \tag{4.2}$$

with C the concentration of reactant A, C_{in} its influent concentration, D the dilution rate, and k_0 the kinetic constant.

Assume that C and C_{in} are accessible for measurement and that k_0 and D are unknown and constant, i.e. these will be parameters whose value has to be determined.

These two parameters are structurally identifiable. Indeed, if for instance you apply a step of C_{in} to the reactor (Figure 4.1(bottom)), then the step response (Figure 4.1(top)) is equal to:

$$C(t) = C_0 + \frac{D}{D + k_0} \Delta C_{in} (1 - e^{-(D+k_0)t}) \tag{4.3}$$

with C_0 the initial value (before the step) of the reactant concentration C and ΔC_{in} the amplitude of the influent concentration step. From Figure 4.1(top), we note that the amplitude A of the step response of $C(t)$ is equal to the difference between the initial value and the final response value (i.e. after a sufficiently long time). Looking at equation (4.3), we see that A is equal to $\frac{D}{D+k_0} \Delta C_{in}$ (= $C(t = \infty)$ - C_0). Figure 4.1(top) can also be used to compute the time constants corresponding to the dynamics of $C(t)$. Indeed, from equation (4.3), we can deduce that 95 % of the final value of $C(t)$ is reached at a time $t_r = \frac{3}{D+k_0}$ after the time step has been applied. This corresponds to three times the time constant $\tau = \frac{1}{D+k_0}$ (as given in equation (4.3)). In conclusion, from the graphical representation of the step response for equations (4.2), we can deduce the amplitude $A = \frac{D}{D+k_0} \Delta C_{in}$, and the time response $t_r = \frac{3}{D+k_0}$. The parameters D and k_0 can therefore be readily calculated from the values of the amplitude $A = \frac{D}{D+k_0} \Delta C_{in}$ and the time response $t_r = \frac{3}{D+k_0}$.

Let us now consider that the measuring device for the reactant concentration gives a signal y which is proportional to the concentration C:

$$y = y_C C \tag{4.4}$$

And assume that the proportionality coefficient y_C is unknown, i.e. in this context y_C is an additional parameter. The step response of y will be very similar to that of C in Figure 4.1, except that the amplitude A is now equal to $y_C \frac{D}{D+k_0} \Delta C_{in}$. And it is now impossible to uniquely determine the values of the parameters y_C, D and k_0 from the values of the amplitude and time response of the step response. This means that the above three parameters are not identifiable. More precisely, only the following combinations of parameters are identifiable: $\theta_1 = D + k_0$ (from the time response $\frac{3}{D+k_0}$), and $\theta_2 = y_C D$ (= $A(D + k_0)$ = $A\theta_1$).

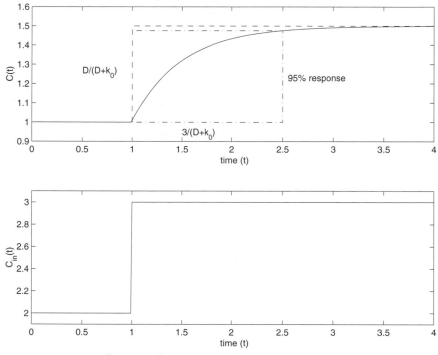

FIG. 4.1. Step response of a first order system.

4.3.2   *The Laplace Method*

The above simple example illustrates the structural identifiability of the two parameters k_0 and D, i.e. the structural (i.e. based on the model structure) property
to determine uniquely the value of these parameters in "ideal" conditions. Conversely, the above example shows the lack of structural identifiability of the three
parameters y_C, D and k_0, i.e. the structural property to determine non uniquely the
value of these parameters in "ideal" conditions. Let us try to generalise the results
of the above example.

Consider a dynamical model $\mathcal{M}(\theta)$ with the parametrisation $\theta = [\theta_1, \theta_2,...,$
$\theta_p]^T$, with q outputs y_i (i = 1 to q) (i.e. here measured variables) and m inputs
u_j (j = 1 to m). If the model is linear, its dynamics will be described either by state
space equations:

$$\frac{dx}{dt} = A(\theta)x(t, \theta) + B(\theta)u(t), \qquad x(0, \theta) = x_0(\theta) \qquad (4.5)$$

$$y(t, \theta) = C(\theta)x(t, \theta) \qquad (4.6)$$

or by a transfer function between the input vector $u(t)$ and the output vector
$y(t, \theta)$, $H(s, \theta)$, which for the above state space model will be equal to:

$$H(s, \theta) = C(\theta)(sI - A(\theta))^{-1} B(\theta) \tag{4.7}$$

This transfer function for a SISO (Single Input Single Output) system is characterised by the ratio of two polynomials of order n (n is the number of states x), i.e. it has (maximum) $2n$ parameters.

Note that the transfer functions between C_{in} and C and, C_{in} and y for equations (4.2), (4.4) can be formally written as follows:

$$H(s) = \frac{b}{s + a} \tag{4.8}$$

with:

$$a = D + k_0 \tag{4.9}$$

$$b = D \text{ or } y_C D \tag{4.10}$$

In the above example, we note that D and k_0 can be readily obtained from the values of a and b in the first case ($D = b$, $k_0 = a - b$, while in the second case, it is not possible to distinguish between y_C and D from the value of b. This suggests the following possible generalisation. Indeed, the $2n$ parameters of the transfer function $H(s)$ are structurally identifiable from measurements of $y(t, \theta)$. The analysis can be based on the impulse response of this system, which is written as follows:

$$y(t, \theta) = \sum_{i=1}^{n} c_i(\theta) e^{\lambda_i(\theta)t} \tag{4.11}$$

The $2n$ coefficients c_i and λ_i can be determined from data of the output $y(t, \theta)$. Therefore the relationship between these coefficients and the transfer function parameters are very important for the structural identifiability. If we consider the Laplace transform for instance for n = 2 and distinct eigenvalues λ_i, we obtain:

$$H(s) = \frac{c_1}{s + \lambda_1} + \frac{c_2}{s + \lambda_2} \tag{4.12}$$

$$= \frac{(c_1 + c_2)s - (c_2\lambda_1 + c_1\lambda_2)}{s^2 + (\lambda_1 + \lambda_2)s + \lambda_1\lambda_2} \tag{4.13}$$

$$= \frac{\beta_1 s + \beta_2}{s^2 + \alpha_1 s + \alpha_2} \tag{4.14}$$

with $\beta_1 = c_1 + c_2$, $\beta_2 = -(c_2\lambda_1 + c_1\lambda_2)$, $\alpha_1 = \lambda_1 + \lambda_2$ and $\alpha_2 = \lambda_1\lambda_2$. This suggests that if there are no common factors in the numerator and denominator polynomials, the $2n$ parameters α_i and β_i can be determined from the output data just as the $2n$ coefficients c_i and λ_i.

Note that the identifiability property is typically valid for *almost* all parameter values: indeed for example there may be instances for which particular parameter combinations or particular input functions give rise to pole-zero cancellation in the transfer function. However, these do not invalidate the general analysis (see also Section 4.3.4 here below).

4.3.3   *Some Generalisations and Definitions*

The above line of reasoning can lead to the following rather formal and abstract definitions of identifiability (see also [100] and [279] for a more rigourous treatment).

- *Definition # 1:* the parameter θ_i is structurally globally identifiable for the input class $\mathcal{U}$ if and only if for almost any value of the parameter vector θ (i.e. of the admissible parametric space $\mathcal{P}$) one has

$$\left.\begin{array}{l} \hat{\theta} \in \mathcal{P} \\ \hat{y}(\hat{\theta}, t) = y(\theta, t), \forall t > 0, \forall u \in \mathcal{U} \end{array}\right\} \Rightarrow \hat{\theta}_i = \theta_i$$

  i.e. in other words, the structurally global identifiability of the parameter θ_i means that if there exists another parameter vector $\hat{\theta}$ belonging to the admissible space for the parameters $\mathcal{P}$, and if the outputs y with both parameter vectors θ and $\hat{\theta}$ are equal for all time t and all input u, then the parameter θ_i (of the vector θ) and the parameter $\hat{\theta}_i$ (of the vector $\hat{\theta}$) are equal.

- *Definition # 2:* the parameter θ_i is structurally locally identifiable for the input class $\mathcal{U}$ if and only if for almost any value of the parameter vector θ (i.e. of the admissible parametric space $\mathcal{P}$) there exists a neighbourhood $V(\theta)$ such that

$$\left.\begin{array}{l} \hat{\theta} \in V(\theta) \subset \mathcal{P} \\ \hat{y}(\hat{\theta}, t) = y(\theta, t), \forall t > 0, \forall u \in \mathcal{U} \end{array}\right\} \Rightarrow \hat{\theta}_i = \theta_i$$

  The above definition is very similar to the preceding one, except that now we limit the domain of validity of the identifiability property to a subdomain (a neighbourhood $V(\theta)$) of the whole admissible space $\mathcal{P}$.

- *Definition # 3:* the model $\mathcal{M}(\theta)$ is structurally globally (locally) identifiable if and only if all the parameters θ_i are structurally globally (locally) identifiable.

The above definitions are obviously very important for defining the concept of structural identifiability, but they are not very useful for testing the identifiability of models. Before introducing different identifiability tests for nonlinear models, let us first further illustrate the concept via a second-order example in order to introduce some more generalisation (for state-space models).

4.3.4   *A Second-Order Example: The Two Interconnected CSTRs Model*

Let us go back to the model of the two interconnected CSTRs introduced in Chapter 2 (Section 2.5.1, Figure 2.12) with the following modifications:

1. S is the reactant of a first order chemical reaction;
2. S is fed in both tanks.

Let us note $F_{1,in}$, $S_{1,in}$, $F_{2,in}$, $S_{2,in}$ the influent flow rates and the influent concentration of S in tank 1 and 2, respectively, and k_0 the kinetic constant. The dynamics are then given by the following equations:

$$\frac{dS_1}{dt} = \frac{F_{1,in}}{V_1} S_{1,in} + \frac{F_1 - F_{1,in}}{V_1} S_2 - \frac{F_1}{V_1} S_1 - k_0 S_1 \tag{4.15}$$

$$\frac{dS_2}{dt} = \frac{F_{2,in}}{V_2} S_{2,in} + \frac{F_1}{V_2} S_1 - \frac{F_1 + F_{2,in}}{V_2} S_2 - k_0 S_2 \tag{4.16}$$

If we consider that the two influent concentrations are the process inputs, the above equations can be rewritten in the following matrix form:

$$\frac{d}{dt}\begin{pmatrix} S_1 \\ S_2 \end{pmatrix} = \begin{pmatrix} a_{11} & a_{12} \\ a_{21} & a_{22} \end{pmatrix} \begin{pmatrix} S_1 \\ S_2 \end{pmatrix} + \begin{pmatrix} b_1 & 0 \\ 0 & b_2 \end{pmatrix} \begin{pmatrix} u_1 \\ u_2 \end{pmatrix} \tag{4.17}$$

with:

$$a_{11} = -\frac{F_1}{V_1} - k_0, \quad a_{12} = \frac{F_1 - F_{1,in}}{V_1} \tag{4.18}$$

$$a_{21} = \frac{F_1}{V_2}, \quad a_{22} = -\frac{F_1 + F_{2,in}}{V_2} - k_0 \tag{4.19}$$

$$u_1 = S_{1,in}, \quad u_2 = S_{2,in} \tag{4.20}$$

$$b_1 = \frac{F_{1,in}}{V_1}, \quad b_2 = \frac{F_{2,in}}{V_2} \tag{4.21}$$

Assume now that the concentration of S can possibly be measured in both tanks, i.e.:

$$\begin{pmatrix} y_1 \\ y_2 \end{pmatrix} = \begin{pmatrix} c_1 & 0 \\ 0 & c_2 \end{pmatrix} \begin{pmatrix} S_1 \\ S_2 \end{pmatrix} \tag{4.22}$$

The transfer function *matrix* is readily derived by using the Laplace transform:

$$H(s) = \begin{pmatrix} \frac{Y_1(s)}{U_1(s)} & \frac{Y_1(s)}{U_2(s)} \\ \\ \frac{Y_2(s)}{U_1(s)} & \frac{Y_2(s)}{U_2(s)} \end{pmatrix} \tag{4.23}$$

$$= \frac{1}{\Delta(s)} \begin{pmatrix} c_1 b_1 (s - a_{22}) & c_1 b_2 a_{12} \\ c_2 b_1 a_{21} & c_2 b_2 (s - a_{11}) \end{pmatrix} \tag{4.24}$$

with:

$$\Delta(s) = s^2 - (a_{11} + a_{22})s + a_{11}a_{22} - a_{12}a_{21} \tag{4.25}$$

Let us now examine the identifiability for different situations.

$u_1 \neq 0$, $u_2 = 0$, *and only* y_1 *measured.*   The transfer function $\frac{Y_1(s)}{U_1(s)}$ can be written in the format (4.14) with:

$$\beta_1 = c_1 b_1, \quad \beta_2 = -c_1 b_1 a_{22} \tag{4.26}$$

$$\alpha_1 = -(a_{11} + a_{22}), \quad \alpha_2 = a_{11} a_{22} - a_{12} a_{21} \tag{4.27}$$

From the structural identifiability of α_1, α_2, β_1 and β_2, and the above relationships, we can readily conclude that only four parameter combinations out of the six parameters of $\frac{Y_1(s)}{U_1(s)}$, i.e. $c_1 b_1$, a_{22}, a_{11} and $a_{12} a_{21}$ are identifiable. In other words, the six parameters $a_{11}, a_{22}, a_{12}, a_{21}, c_1$ and b_1 are *not* identifiable while the four "parameters" $\theta_1 = c_1 b_1, \theta_2 = a_{22}, \theta_3 = a_{11}$ and $\theta_4 = a_{12} a_{21}$ are:

$$\alpha_1, \ \alpha_2, \ \beta_1, \ \beta_2 \ \Rightarrow \ \begin{cases} \theta_1 = \beta_1 \\ \theta_2 = -\frac{\beta_2}{\beta_1} \\ \theta_3 = -\alpha_1 + \frac{\beta_2}{\beta_1} \\ \theta_4 = -\alpha_2 - \frac{\beta_2}{\beta_1}\left(-\alpha_1 + \frac{\beta_2}{\beta_1}\right) \end{cases}$$

Now, can we expect to have more identifiable parameters if we consider the second input and/or the second output?

$u_1 \neq 0$, $u_2 = 0$, y_1 *and* y_2 *measured.*   We have here an additional measurement. However the denominators of $\frac{Y_1(s)}{U_1(s)}$ and $\frac{Y_2(s)}{U_1(s)}$ are the same. So any improvement in the number of identifiable parameters may only come from the numerator of $\frac{Y_2(s)}{U_1(s)}$, i.e. $b_1 c_2 a_{12}$. We note that a priori nothing is gained (we have one more parameter c_2 and one more identifiable parameter combination $b_1 c_2 a_{12}$) except if there is some a priori information about c_2. A priori information may be, for instance, the equality between c_1 and c_2 ($c_1 = c_2$) (this case is even the most probable in our example), then the numerator of $\frac{Y_2(s)}{U_1(s)}$ is equal to $b_1 c_1 a_{12}$. Since $b_1 c_1$ is already identifiable (see the preceding case), then a_{12} is identifiable. Now, since a_{11} and a_{22} are also identifiable (see above), then we can deduce from the definition of α (4.27) that a_{12} is also identifiable.

$u_1 \neq 0$, $u_2 \neq 0$, *and only* y_1 *measured.*   Let us evaluate what is gained by considering both inputs u_1 and u_2 (in practice via test-inputs applied at different time instants so that both responses can be completely distinguished from each other, so that $u_1(t) \neq \gamma u_2(t)$ at each time time instant). A conclusion similar to the one of the previous paragraph can be drawn, i.e. there is one extra parameter (b_2) and one extra parameter combination ($c_1 b_2 a_{12}$), and nothing is gained in terms of identifiability, except if there is some a priori information about b_2.

Now, if we consider the particular case when the same inputs are applied at the same time (i.e. $u_1 = u_2$), then the transfer function $\frac{Y_1(s)}{U_1(s)}$ is equal to:

$$\frac{Y_1(s)}{U_1(s)} = \frac{c_1 b_1 s + c_1(b_2 a_{12} - b_1 a_{22})}{\Delta(s)} \tag{4.28}$$

We are then in the particular situation when some input combinations may lead to a loss of identifiability (only 4 (instead of 5) parameters are then identifiable). This illustrates the introduction of the term *for almost any value of the parameters* in the above definitions, and also points to the fact that the choice of appropriate experimental conditions can have a great influence on the *practical* identifiability of the system (but this is the topic of Chapter 5).

$u_1 \neq 0$, $u_2 = 0$, *only* y_2 *measured, and* $a_{12} = 0$. Let us finally consider the case when $a_{12} = 0$, i.e. when we consider a sequence of two CSTRs without any possible flow back from tank 2 to tank 1 (i.e. $F_2 = 0$, or more precisely $F_1 = F_{1,in}$, see also Figure 2.12). Then the transfer function $\frac{Y_2(s)}{U_1(s)}$ is equal to:

$$\frac{Y_2(s)}{U_1(s)} = \frac{c_2 b_1 a_{21}}{(s - a_{11})(s - a_{22})} \tag{4.29}$$

The parameters a_{11} and a_{22} are identifiable, but only locally. Indeed it is not possible to distinguish between both parameters, since they have both two possible (interchangeable) values for the above configuration. In other words, this means that in the present example, we cannot distinguish between the two volumes V_1 and V_2, i.e. we cannot say if the first volume is small and the second is large or the reverse.

4.4 Methods for Testing Structural Identifiability of Nonlinear Systems

In this section, we introduce different approaches to test the structural identifiability of models. These represent the most largely used methods for structural identifiability. But the list is not exhaustive. Our objective is indeed to provide tools that can be helpful for testing structural identifiability, not to write a full monograph on the topic (see e.g. [100], [279], [254] for a more rigourous approach and technical mathematical details). Note that, in addition to the methods discussed in this section, we shall consider another method illustrated with the first example (Monod model) in the next section dedicated to illustrative examples.

The Laplace transform method has already been introduced in Section 4.2. One of the most important difficulties with the Laplace transform is that strictly speaking it only applies to linear models, and that results obtained from linearised models of nonlinear models may be difficult to interpret. The results obtained for the linearised model are only sufficient conditions: this means that the (combinations of) parameters that are identifiable for the model linearised around some steady state are also identifiable for the nonlinear model around that steady state only.

The basic concepts of each method are first presented. Their application is then illustrated on a simple example (nonlinear model linear in the parameters) in Section 4.4.5.

4.4.1 *Taylor Series Expansion*

Consider that the model equations are written under the following form:

$$\frac{dx}{dt} = f(x, u, \theta), \quad x(0) = x_0(\theta) \tag{4.30}$$

$$y(t, \theta) = h(x, \theta) \tag{4.31}$$

where x, u, y and θ represent the state vector, the input vector, the output (measured variable) vector, and the (unknown) parameter vector, respectively.

The method is based on a Taylor series expansion of the observations y(t) around time t=0:

$$y(t) = y(0) + t\frac{dy}{dt}(0) + \frac{t^2}{2!}\frac{d^2 y}{dt^2}(0) + ... \tag{4.32}$$

and consists of looking at the successive derivatives to check if they contain information about the parameters to be identified. More precisely, $y(0)$ and the successive derivatives of $y(t)$ at time t = 0 can be expressed from the model equations (4.30)(4.31) as functions of the unknown parameters $\theta^T = [\theta_1, \theta_2,..., \theta_p]$:

$$y(0) = \gamma_0(\theta) \tag{4.33}$$

$$\frac{dy}{dt}(0) = \gamma_1(\theta) \tag{4.34}$$

$$\vdots \tag{4.35}$$

$$\frac{d^q y}{dt^q}(0) = \gamma_q(\theta) \tag{4.36}$$

The second step consists of trying to invert the above expressions (4.33)-(4.36) so as to express the parameters θ_i (i = 1 to p) as functions of only $y(0)$, its successive derivatives and the input u, i.e.:

$$\theta_1 = \beta_1(y(0), \frac{dy}{dt}(0), ..., \frac{d^q y}{dt^q}(0), u) \tag{4.37}$$

$$\theta_2 = \beta_2(y(0), \frac{dy}{dt}(0), ..., \frac{d^q y}{dt^q}(0), u) \tag{4.38}$$

$$\vdots \tag{4.39}$$

$$\theta_p = \beta_p(y(0), \frac{dy}{dt}(0), ..., \frac{d^q y}{dt^q}(0), u) \tag{4.40}$$

If such a set of equations exist, this means that the parameters θ_i (i = 1 to p) are structurally identifiable. But it may also happen that the above set of equations can be written only for combinations of the parameters θ_i, and then only these combinations are identifiable.

Note that generally speaking the number of successive derivatives (= q) used for the identifiability analysis is not equal to the number of unknown parameters (= p). q is typically at least equal to p (because at least p relations are needed to invert and obtain expressions for the p parameters θ_i), but it may be larger in order to obtain new expressions that are independent of the preceding ones and therefore susceptible to introduce new information for the analysis.

Remark: for simplicity, we had chosen one time instant (t = 0) in the expansion (4.32). But other time instants may be used because these can be helpful in order to simplify the analysis by considering different (simpler) submodels (see e.g. the example in Section 4.6.3).

4.4.2   Generating Series

The Generating Series method is based on nonlinear control theory concepts, basically on the Lie derivatives and its link to observability of nonlinear systems. Without entering into the details, the method can be briefly summarised as follows:

Let us consider that the system equations can be written as follows[4]:

$$\frac{dx}{dt} = f_0(x, \theta) + \sum_{i=1}^{m} u_i(t) f_i(x, \theta), \quad x(0) = x_0(\theta) \qquad (4.41)$$

$$y(t, \theta) = h(x, \theta) \qquad (4.42)$$

The analysis is based on the output functions $h(x, \theta)$ and its successive Lie derivatives $L_{f_{j0}} ... L_{f_{jk}} h(x, \theta)$ evaluated at t=0. The Lie derivative along the vector field f_i is defined as follows:

$$L_{f_i} = \sum_{j=1}^{n} f_{j,i}(x, \theta) \frac{\partial}{\partial x_j} \qquad (4.43)$$

with $f_{j,i}$ the jth component of f_i. As a matter of illustration, the Lie derivative of h and of L_{f_k} along the vector field f_i are equal to:

$$L_{f_i} h(x, \theta) = \sum_{j=1}^{n} f_{j,i}(x, \theta) \frac{\partial}{\partial x_j} h(x, \theta) \qquad (4.44)$$

$$L_{f_i} L_{f_k} = \sum_{j=1}^{n} f_{j,i}(x, \theta) \frac{\partial}{\partial x_j} L_{f_k} \qquad (4.45)$$

Similarly to the Taylor series expansion method, we look at the successive generated Lie derivatives evaluated at time t = 0 (and assumed "known") to check if they contain information about the parameters to be identified.

---

[4]Note that the model (4.41) is linear in $u(t)$.

4.4.3 Local State Isomorphism

Let us consider that the system equations can be written as follows (it is only for convenience of the presentation of the method that we consider here a state representation different from the above equation (4.41)):

$$\frac{dx}{dt} = f(x, \theta) + u^T(t)g(x, \theta), \quad x(0) = x_0(\theta) \tag{4.46}$$

$$y(t, \theta) = h(x, \theta) \tag{4.47}$$

Let us denote $\bar{x}$ and $\tilde{x}$ two states corresponding to two different sets of parameter values, $\bar{\theta}$ and $\tilde{\theta}$ respectively. The models corresponding to each parameter set will have the same input-output behaviour (and $\bar{\theta}$ and $\tilde{\theta}$ are therefore not distinguishable) for any input u up to a time $t_1 > 0$ if and only if there exists a local state isomorphism:

$$\lambda : V \to R^n, \bar{x} \to \tilde{x} = \lambda(\bar{x}), \quad V \text{ is an open neighbourhood of } \bar{x}(0) \tag{4.48}$$

such that for any $\bar{x}$ in the neighbourhood $V(\bar{x}(0))$ the following conditions are satisfied:

$$\lambda \text{ is a diffeomorphism}: rank \frac{\partial \lambda(x = \bar{x})}{\partial x^T} = n \tag{4.49}$$

$$\text{the initial states correspond}: \lambda(\bar{x}(0)) = \tilde{x}(0) \tag{4.50}$$

$$\text{the drift terms correspond}: f(\lambda(\bar{x}), \tilde{\theta}) = \frac{\partial \lambda(x = \bar{x})}{\partial x^T} f(\lambda(\bar{x}), \bar{\theta}) \tag{4.51}$$

$$\text{the input terms correspond}: g(\lambda(\bar{x}), \tilde{\theta}) = \frac{\partial \lambda(x = \bar{x})}{\partial x^T} g(\lambda(\bar{x}), \bar{\theta}) \tag{4.52}$$

$$\text{the observations correspond}: h(\lambda(\bar{x}), \tilde{\theta}) = h(\bar{x}, \bar{\theta}) \tag{4.53}$$

One can test the structural identifiability by looking at all the solutions for $\bar{\theta}$ and λ of the above equations (4.49)-(4.53). If for almost any $\tilde{\theta}$ the only possible solution is $\bar{\theta} = \tilde{\theta}$, $\lambda(\bar{x}) = \tilde{x}$, then the model is uniquely identifiable.

Remark: make sure that you don't confuse between the two different notations, e.g. $\tilde{\theta}$ and $\bar{\theta}$, or $\tilde{x}$ and $\bar{x}$!

4.4.4 Transformation of Nonlinear Models

Another way to analyse the structural identifiability is to transform the nonlinear model into a model linear in the parameters, and then look at the identifiability of the linear model. A better understanding of the approach can be drawn from the examples here below in Sections 4.4.5 and 4.6.3.

4.4.5 *A Simple Example*

Let us illustrate the above identifiability tests with a simple nonlinear model linear in three parameters ($\theta_1, \theta_2, \theta_3$):

$$\frac{dx_1}{dt} = -\theta_1 x_1 - \theta_2 x_1 + \theta_3 x_1 x_2 + u, \quad x_1(0) = 1 \tag{4.54}$$

$$\frac{dx_2}{dt} = \theta_2 x_1 - \theta_3 x_1 x_2, \quad x_2(0) = 0 \tag{4.55}$$

$$y(t, \theta) = x_1 \tag{4.56}$$

This model holds for instance for a bioprocess in a batch reactor where x_2 is the concentration of a reactant and x_1 the concentration of an autocatalyst (e.g. microorganisms), and where there are three reactions: autocatalysis of x_1 with first order kinetics with respect to the reactant ($r_1 = \theta_3 x_1 x_2$), a decomposition (e.g. lysis) of x_1 into x_2 ($r_2 = \theta_2 x_1$) and a third reaction which may be a mortality reaction of x_1 ($r_3 = \theta_1 x_1$). Beside y, the input u is assumed to be known (in the context of the above example, u could be interpreted as an external addition of the autocatalyst x_1 in quantities small enough to keep the variations of the fermenter volume negligible).

The choice of the above example is motivated by our will to give an illustration of the above identifiability analysis tools that remains as simple and clear as possible. This has led to the present choice, i.e. a model linear in the parameters and nonlinear in the state variables, yet rather simple and with a potential connection with models considered in WWTP. This choice was a difficult one for us, because we are conscious that the linearity in the parameters may be interpreted as a limitation of the applicability of the results presented here. But starting with a nonlinear example would have led us to considerations that may have hidden the basic remarks that we feel are necessary to be understood to apply the proposed analysis tools. Other examples of models nonlinear in the parameters are therefore dealt with in the rest of the chapter, more precisely in Section 4.6.

Taylor series expansion. Let us start with the Taylor series expansion of $y(t, \theta)$:

$$\frac{dy}{dt}(0) = -(\theta_1 + \theta_2) + u \tag{4.57}$$

$$\frac{d^2 y}{dt^2}(0) = (\theta_1 + \theta_2)^2 - (\theta_1 + \theta_2)u + \theta_2 \theta_3 \tag{4.58}$$

$$\frac{d^3 y}{dt^3}(0) = -(\theta_1 + \theta_2)^3 + (\theta_1 + \theta_2)^2 u + 2\theta_2 \theta_3 u - 2\theta_2 \theta_3 (\theta_1 + \theta_2) - \theta_2 \theta_3^2 \tag{4.59}$$

Let us denote:

$$z_i = \frac{d^i y}{dt^i}(0) \tag{4.60}$$

Basically the z_i can be considered to be variables that have a known values since they can be readily obtained from (ideal) measurements y. The above equations can be rewritten as follows:

$$z_1 = -(\theta_1 + \theta_2) + u \tag{4.61}$$

$$z_2 = (\theta_1 + \theta_2)^2 - (\theta_1 + \theta_2)u + \theta_2\theta_3 \tag{4.62}$$

$$z_3 = -(\theta_1 + \theta_2)^3 + (\theta_1 + \theta_2)^2 u + 2\theta_2\theta_3 u - 2\theta_2\theta_3(\theta_1 + \theta_2) - \theta_2\theta_3^2 \tag{4.63}$$

We have three unknown parameters θ_1, θ_2, θ_3 and three equations in z_i ($i = 1, 2, 3$). The difficult task can now start, i.e. to see if it is possible to invert the above equations (4.61) (4.62) (4.63) to obtain expressions for the θ_i that are only functions of the z_i and of the input u. One possible way to proceed is the following. A look at the above equations shows that there are three groups of parameters: $\theta_1 + \theta_2$, $\theta_2\theta_3$ (last term of the second equation), and $\theta_2\theta_3^2$ (last term of the third equation). $\theta_1 + \theta_2$ can be expressed as a function of z_1 and u from the first equation (4.61). The second equation (4.62) can be used to express $\theta_2\theta_3$ as a function of z_2, u and z_1 (via $\theta_1 + \theta_2$) only. Finally the same procedure can be followed in the third equation (4.63) to express $\theta_2\theta_3^2$ as a function of z_1, z_2, z_3 and u:

$$\theta_1 + \theta_2 = u - z_1 \tag{4.64}$$

$$\theta_2\theta_3 = z_2 - (u - z_1)^2 + (u - z_1)u \tag{4.65}$$

$$\theta_2\theta_3^2 = z_3 + (u - z_1)^3 + (u - z_1)^2 u$$
$$-2(u - (u - z_1))(z_2 - (u - z_1)^2 + (u - z_1)u) \tag{4.66}$$

It is now possible to compute θ_3 (as the ratio of $\theta_2\theta_3^2$ and $\theta_2\theta_3$), θ_2 and θ_1, successively:

$$\theta_1 = u - z_1$$
$$-(z_2 - 2u^2 + z_1^2 + 3uz_1)^2(2u^3 - 10u^2 z_1 + 6uz_1^2 + z_1^3 + 2z_1 z_2 - z_3) \tag{4.67}$$

$$\theta_2 = (z_2 - 2u^2 + z_1^2 + 3uz_1)^2(2u^3 - 10u^2 z_1 + 6uz_1^2 + z_1^3 + 2z_1 z_2 - z_3) \tag{4.68}$$

$$\theta_3 = \frac{2u^3 - 10u^2 z_1 + 6uz_1^2 + z_1^3 + 2z_1 z_2 - z_3}{z_2 - 2u^2 + z_1^2 + 3uz_1} \tag{4.69}$$

This shows that the three parameters θ_1, θ_2 and θ_3 can be formally computed from the values of z_1, z_2, z_3, and u, i.e. they are structurally identifiable.

Generating series.   Let us now look at the generating series test. We know that:

$$f_0 = \begin{pmatrix} -\theta_1 x_1 - \theta_2 x_1 + \theta_3 x_1 x_2 \\ \theta_2 x_1 - \theta_3 x_1 x_2 \end{pmatrix}, \quad f_1 = \begin{pmatrix} 1 \\ 0 \end{pmatrix}, \quad h = x_1 \tag{4.70}$$

The Lie derivatives are equal to:

$$L_{f_0} = [-\theta_1 x_1 - \theta_2 x_1 + \theta_3 x_1 x_2]\frac{\partial}{\partial x_1} + [\theta_2 x_1 - \theta_3 x_1 x_2]\frac{\partial}{\partial x_2} \tag{4.71}$$

$$L_{f_1} = \frac{\partial}{\partial x_1} \tag{4.72}$$

Let us consider the following three successive Lie derivatives of h evaluated at time $t = 0$ (some others could have been written down too, e.g. $L_{f_1} L_{f_0} h(0)$ ($= \theta_3 x_1(0) = 0$), but are unnecessary here):

$$z_1 = L_{f_0} h(0) = -\theta_1 - \theta_2 \tag{4.73}$$

$$z_2 = L_{f_0} L_{f_0} h(0) = (\theta_1 + \theta_2)^2 + \theta_2 \theta_3 \tag{4.74}$$

$$z_3 = L_{f_0} L_{f_0} L_{f_0} h(0)$$
$$= -(\theta_1 + \theta_2)[(\theta_1 + \theta_2)^2 + 2\theta_2\theta_3 - \theta_2\theta_3(\theta_1 + \theta_2) - \theta_2\theta_3^2] \tag{4.75}$$

By using the same argument as for the Taylor series approach (inversion of the expressions), we can rewrite the three parameters in terms of the successive Lie derivatives, and therefore the three parameters θ_1, θ_2 and θ_3 are structurally identifiable.

Note the resemblance of the above expressions (4.73)-(4.75) with those obtained with the Taylor series expansion approach (4.61)-(4.63), but without the terms in the input u. The present approach allows the separation of both types of terms, i.e. the terms depending on the state variables and those depending on the inputs u. This offers the advantage of handling possibly simpler expressions (compare equations (4.73)-(4.75) and (4.61)-(4.63)).

Local state isomorphism. Let us now apply the local state isomorphism. Now we have:

$$f = \begin{pmatrix} -\theta_1 x_1 - \theta_2 x_1 + \theta_3 x_1 x_2 \\ \theta_2 x_1 - \theta_3 x_1 x_2 \end{pmatrix}, \quad g = \begin{pmatrix} 1 \\ 0 \end{pmatrix}, \quad h = x_1 \tag{4.76}$$

First note that the dimension of λ is 2:

$$\lambda = \begin{pmatrix} \lambda_1 \\ \lambda_2 \end{pmatrix} \tag{4.77}$$

i.e.:

$$\begin{pmatrix} \tilde{x}_1 \\ \tilde{x}_2 \end{pmatrix} = \begin{pmatrix} \lambda_1(\bar{x}_1, \bar{x}_2) \\ \lambda_2(\bar{x}_1, \bar{x}_2) \end{pmatrix} \tag{4.78}$$

From the relations (4.48), we have here:

$$\tilde{x}_1 = \lambda_1(\bar{x}_1) \tag{4.79}$$

And the condition (4.53) gives:

$$\lambda_1(\bar{x}_1) = \bar{x}_1 \tag{4.80}$$

This gives:

$$\tilde{x}_1 = \lambda_1(\bar{x}_1) = \bar{x}_1 \tag{4.81}$$

This means that:

$$\frac{\partial \lambda_1}{\partial x_1} = 1, \quad \frac{\partial \lambda_1}{\partial x_2} = 0 \tag{4.82}$$

If we consider now the relation (4.52), we have:

$$\begin{pmatrix} 1 \\ 0 \end{pmatrix} = \frac{\partial \lambda}{\partial x^T} \begin{pmatrix} 1 \\ 0 \end{pmatrix} = \begin{pmatrix} \frac{\partial \lambda_1}{\partial x_1} \\ \frac{\partial \lambda_2}{\partial x_1} \end{pmatrix} \tag{4.83}$$

This implies that:

$$\frac{\partial \lambda_2}{\partial x_1} = 0 \tag{4.84}$$

Let us now look at the relation (4.51), which specialises here as follows:

$$\begin{pmatrix} -\tilde{\theta}_1 \tilde{x}_1 - \tilde{\theta}_2 \tilde{x}_1 + \tilde{\theta}_3 \tilde{x}_1 \tilde{x}_2 \\ \tilde{\theta}_2 \tilde{x}_1 - \tilde{\theta}_3 \tilde{x}_1 \tilde{x}_2 \end{pmatrix} = \begin{pmatrix} 1 & 0 \\ 0 & \frac{\partial \lambda_2}{\partial x_2} \end{pmatrix} \begin{pmatrix} -\bar{\theta}_1 \bar{x}_1 - \bar{\theta}_2 \bar{x}_1 + \bar{\theta}_3 \bar{x}_1 \bar{x}_2 \\ \bar{\theta}_2 \bar{x}_1 - \bar{\theta}_3 \bar{x}_1 \bar{x}_2 \end{pmatrix} \tag{4.85}$$

Since $\tilde{x}_1 = \bar{x}_1$ and $\tilde{x}_2 = \lambda_2(\bar{x})$, the first row implies that:

$$-\tilde{\theta}_1 \bar{x}_1 - \tilde{\theta}_2 \bar{x}_1 + \tilde{\theta}_3 \bar{x}_1 \lambda_2(\bar{x}) = -\bar{\theta}_1 \bar{x}_1 - \bar{\theta}_2 \bar{x}_1 + \bar{\theta}_3 \bar{x}_1 \bar{x}_2 \tag{4.86}$$

since $\bar{x}_1$ and $\bar{x}_2$ are independent (they are solutions of the two ordinary differential equations (4.54)(4.55)), we have:

$$\tilde{\theta}_1 + \tilde{\theta}_2 = \bar{\theta}_1 + \bar{\theta}_2 \tag{4.87}$$

$$\lambda_2(\bar{x}) = \frac{\bar{\theta}_3}{\tilde{\theta}_3} \bar{x}_2 \tag{4.88}$$

Therefore, the derivative of λ_2 with respect to x_2 is equal to:

$$\frac{\partial \lambda_2}{\partial x_2} = \frac{\bar{\theta}_3}{\tilde{\theta}_3} \tag{4.89}$$

The second row of (4.85) then becomes:

$$\tilde{\theta}_2 \bar{x}_1 - \tilde{\theta}_3 \bar{x}_1 \tilde{x}_2 = \bar{\theta}_2 \frac{\bar{\theta}_3}{\tilde{\theta}_3} \bar{x}_1 - \bar{\theta}_3 \frac{\bar{\theta}_3}{\tilde{\theta}_3} \bar{x}_1 \bar{x}_2 \tag{4.90}$$

which implies that:

$$\tilde{\theta}_2 \tilde{\theta}_3 = \bar{\theta}_2 \bar{\theta}_3, \quad \tilde{\theta}_3 = \bar{\theta}_3 \tag{4.91}$$

The condition (4.50) is immediate, and the relation (4.49) simply implies that $\tilde{\theta}_3$ must be different from zero.

We can then conclude, specifically from (4.87) and (4.91) that the three parameters θ_1, θ_2 and θ_3 are here also structurally identifiable.

Transformation of nonlinear models. The objective of the method is to rewrite the model in input-output format linear in the parameters. This is performed here by the elimination of x_2 via differentiating the output x_1 twice.

x_2 can be put in evidence from equation (4.54):

$$x_2 = \frac{1}{\theta_3 x_1}(\frac{dx_1}{dt} + \theta_1 x_1 + \theta_2 x_1 - u) \tag{4.92}$$

By differentiating the output y ($= x_1$) twice with respect to time t, we obtain:

$$\frac{d^2 y}{dt^2} - \frac{1}{y}(\frac{dy}{dt})^2 + \frac{u}{y} = (\theta_1 + \theta_2)\frac{dy}{dt}(\frac{dy}{dt} - 1) + \theta_2 \theta_3 y^2 - \theta_3 y(\frac{dy}{dt} - u) - \theta_3(\theta_1 + \theta_2)y\frac{dy}{dt} \tag{4.93}$$

It can be rewritten in the usual linear regression format $Y = \theta^T \Phi$ with:

$$Y = \frac{d^2 y}{dt^2} - \frac{1}{y}(\frac{dy}{dt})^2 + \frac{u}{y}, \quad \theta = \begin{pmatrix} \theta_1 + \theta_2 \\ \theta_2 \theta_3 \\ \theta_3 \\ \theta_3(\theta_1 + \theta_2) \end{pmatrix}, \quad \Phi = \begin{pmatrix} \frac{dy}{dt}(\frac{dy}{dt} - 1) \\ y^2 \\ y(\frac{dy}{dt} - u) \\ y\frac{dy}{dt} \end{pmatrix} \tag{4.94}$$

θ is identifiable if the components of the regressor vector Φ are independent (see e.g. [14] [165] for more details). It is then obvious that it is structurally possible to reconstruct the parameters θ_1, θ_2 and θ_3 from the parameter vector θ.

4.5   The Lyapunov-Based Method: An Historical Perspective with the Monod Model

The Monod model is largely used in biotechnological process applications, and in particular in biological wastewater treatment, to characterise growth kinetics. It has been the object of many (structural and practical) identifiability studies since the seventies (see e.g. [2], [126], [276], [100], [161]). In this section we present the first structural identifiability analysis performed on the Monod model. It has been published by Aborhey and Williamson in 1978 ([2]). The original aspect of the proposed analysis is that it is based on a Lyapunov function, a concept largely used to analyse the stability of dynamical systems. In that sense, this approach can be viewed as another method to analyse the structural identifiability, although it is not very popular so far.

Let us start by rewriting the mass balance equations of a simple microbial growth process with Monod kinetics in a CSTR:

$$\frac{dX}{dt} = \frac{\mu_{max} S X}{K_S + S} - DX \tag{4.95}$$

$$\frac{dS}{dt} = -\frac{1}{Y}\frac{\mu_{max} S X}{K_S + S} + DS_{in} - DS \tag{4.96}$$

In the above model, there are 3 parameters: μ_{max}, K_S, and Y.

Assume now that X and S are accessible for on-line measurement, as well as D and S_{in}. The structural identifiability of the 3 parameters can be deduced from the existence of an estimation algorithm that is shown to be theoretically convergent.

In their paper, Aborhey and Williamson ([2]) propose the following estimation scheme[5]:

$$\frac{dz_1}{dt} = X[\hat{\mu} - D - g_1(z_1 - X)] \tag{4.97}$$

$$\frac{dz_2}{dt} = X[\hat{\alpha} - g_2(z_2 - S)] + DS_{in} - DS \tag{4.98}$$

$$\frac{d\hat{\mu}_{max}}{dt} = -\lambda_1 XS(z_1 - X) \tag{4.99}$$

$$\frac{d\hat{K}_S}{dt} = X[\lambda_2\hat{\mu}(z_1 - X) + \lambda_3\hat{\alpha}(z_2 - S)] \tag{4.100}$$

$$\frac{d\hat{\alpha}_m}{dt} = -\lambda_4 XS(z_2 - S) \tag{4.101}$$

with:

$$\alpha_m = -\frac{\mu_{max}}{Y}, \ \alpha = -\frac{\mu}{Y}, \ \hat{\mu} = \frac{\hat{\mu}_{max} S}{\hat{K}_S + S} \tag{4.102}$$

Let us choose the estimator design parameters g_i (i = 1, 2) and λ_i (i = 1 to 4) such that:

$$g_i > \frac{DS_{in}}{2(S + K_S)X} \tag{4.103}$$

$$\lambda_i > 0 \tag{4.104}$$

It is obvious that the estimate of Y is readily derived from the estimates of μ_{max} and α_m:

$$\hat{Y} = \frac{\hat{\mu}_{max}}{\hat{\alpha}_m} \tag{4.105}$$

Let us now consider the following Lyapunov (positive definite: $V > 0$) candidate function:

$$V = (S + K_S)(l_1 e_1^2 + l_2 e_2^2) + \sum_{i=3}^{5} l_i e_i^2, \ l_i > 0, \quad \text{for i = 1 to 5} \tag{4.106}$$

where e_i are the following error terms:

$$e_1 = z_1 - X \tag{4.107}$$

[5] As the reader may detect either by personal experience or by looking at the chapter on state observation (Chapter 7), the following structure resembles somewhat to that of a classical observer like the extended Luenberger observer, but the choice of the observer gains is different (it is based on a Lyapunov function).

$$e_2 = z_2 - S \tag{4.108}$$

$$e_3 = \hat{\mu}_{max} - \mu_{max} \tag{4.109}$$

$$e_4 = \hat{K}_S - K_S \tag{4.110}$$

$$e_5 = \hat{\alpha}_m - \alpha \tag{4.111}$$

The estimation algorithm (4.97)-(4.101) will be convergent, i.e. the estimates of μ_{max}, K_S, and Y will converge to their true values, if the time derivative of V along the solutions of (4.95)(4.96)(4.97)-(4.101) is negative definite. It is what we are going to check in the following paragraphs.

Let us first write the dynamics of the error terms e_i (i = 1 to 5) from (4.95), (4.96), (4.97)-(4.101), (4.107)-(4.111). This gives:

$$\frac{de_1}{dt} = -g_1 X e_1 - X(\hat{\mu} - \mu) \tag{4.112}$$

$$\frac{de_2}{dt} = -g_2 X e_2 - X(\hat{\mu} - \mu) \tag{4.113}$$

$$\frac{de_3}{dt} = -\lambda_1 X S e_1 \tag{4.114}$$

$$\frac{de_4}{dt} = X[\lambda_2 \hat{\mu} e_1 + \lambda_3 \hat{\alpha} e_2] \tag{4.115}$$

$$\frac{de_5}{dt} = -\lambda_4 X S e_2 \tag{4.116}$$

Let us compute dV/dt:

$$\frac{dV}{dt} = l_1 e_1^2 [\frac{dS}{dt} - 2g_1(S + K_S)X] + l_2 e_2^2 [\frac{dS}{dt} - 2g_2(S + K_S)X]$$
$$+ 2[e_1 e_3(l_1 - l_3\lambda_1)XS + e_1 e_4(l_4\lambda_2 - l_1)X$$
$$+ e_2 e_4(l_4\lambda_3 - l_2)X + e_2 e_5(l_2 - l_5\lambda_4)XS] \tag{4.117}$$

Now we choose l_i (i = 1 to 5) such that:

$$l_1 = l_3\lambda_1, \quad l_4\lambda_2 = l_1, \quad l_4\lambda_3 = l_2, \quad l_2 = l_5\lambda_4 \tag{4.118}$$

Then the last four terms of dV/dt are equal to zero, i.e.:

$$\frac{dV}{dt} = l_1 e_1^2 [\frac{dS}{dt} - 2g_1(S + K_S)X] + l_2 e_2^2 [\frac{dS}{dt} - 2g_2(S + K_S)X] \tag{4.119}$$

By using (4.103), it is straighforward that dV/dt is negative, since $dS/dt \leq DS_{in}$. And if $dV/dt = 0$ for some time t between t_1 and t_2, this means that:

$$e_1 = e_2 = 0 \quad \text{and} \quad \frac{de_1}{dt} = \frac{de_2}{dt} = 0 \tag{4.120}$$

Let us introduce the above equalities (4.120) in equations (4.112) and (4.113). Let us only consider the calculations with equation (4.112) here (the argument is completely similar with (4.113)). $e_1 = 0$ and $de_1/dt = 0$ imply that equation (4.112) becomes:

$$\frac{\mu_{max} S}{K_S + S} = \frac{\hat{\mu}_{max} S}{\hat{K}_S + S} \qquad (4.121)$$

Let us multiply both sides of the above equation by $(K_S + S)$ and then substract (also on both sides) $\hat{\mu}_{max} S$.

We obtain (after changing signs on both sides):

$$\hat{\mu}_{max} S - \mu_{max} S = \hat{\mu}_{max} S - \frac{\hat{\mu}_{max} S(K_S + S)}{\hat{K}_S + S} \qquad (4.122)$$

If we multiply and divide the first term of the right hand side by $(\hat{K}_S + S)$, the right hand side becomes:

$$\frac{\hat{\mu}_{max} S}{\hat{K}_S + S}(\hat{K}_S - K_S) \qquad (4.123)$$

Thus equation (4.122) becomes:

$$e_3 S = \hat{\mu} e_4 \qquad (4.124)$$

Similarly we obtain from equation (4.113):

$$e_5 S = \hat{\alpha} e_4 \qquad (4.125)$$

For the last three equations (4.114)(4.115)(4.116), we readily obtain by using equation (4.120):

$$\frac{de_3}{dt} = \frac{de_4}{dt} = \frac{de_5}{dt} = 0 \qquad (4.126)$$

Since S, $\hat{\mu}$ and $\hat{\alpha}$ will generally speaking be varying independently from each other, this implies that:

$$e_3 = e_4 = e_5 = 0 \qquad (4.127)$$

on the interval $[t_1, t_2]$ provided that $S(t)$ is not constant on this interval.

Therefore the time derivative of the candidate Lyapunov function V is negative definite. This implies the convergence of the estimation algorithm, and in consequence, the identifiability of the parameters of the Monod model. Or in other words, if we are able to build an estimation algorithm that is mathematically guaranteed to give estimates that converge to their true values, this means that a fortiori, the parameters that are considered in the estimator are structurally identifiable.

4.6 Example #2: Respirometer-based Models

Let us perform the analysis of the structural identifiability of the four models introduced in Chapter 3 (Section 3.2) based on the on-line measurement of the oxygen uptake rate via a respirometer (see also [74]). We shall consider the following approaches for the different models:

1. first order kinetics (type 1): Laplace transform;
2. single Monod model (type 2): Taylor series expansion;
3. double Monod model (type 3): nonlinear transformation;
4. modified ASM1 model (type 4): nonlinear transformation and generating series.

4.6.1 Identifiability of the First Order Kinetics Model (Laplace Transform)

Recall that the equations of the model are the following:

$$\frac{dS_1}{dt} = -\frac{k_{max1}X}{Y_1}S_1 \tag{4.128}$$

$$y(t) = OUR_{ex}(t) = -(1 - Y_1)\frac{dS_1}{dt} \tag{4.129}$$

The parameters for which we would like to check the structural identifiability are here Y_1, μ_{max1}, X, and $S_1(0)$ (since we have a priori no idea of the initial value of S_1 at the beginning of the respirometric experiment).

First note that considering the initial value of S_1 as a parameter is an extension of the cases considered in the preceding sections (yet already suggested by considering $x(0) = x_0(\theta)$ in equations (4.30), (4.41) and (4.46)). This case appears to be of great interest in several applications (like the one presented here).

The structural identifiability of the first model is rather straightforward.

Since the model is linear in the state variable $S_1(t)$ and of the output $OUR_{ex}(t)$, we can use the Laplace transform to perform the identifiability analysis. The Laplace transform $\mathcal{L}(s)$ applied to equations (4.128) and (4.129) gives:

$$\mathcal{L}(S_1) = \frac{S_1(0)}{s + \frac{k_{max1}X}{Y_1}} \tag{4.130}$$

$$\mathcal{L}(y) = \frac{(1 - Y_1)k_{max1}X}{Y_1}\frac{S_1(0)}{s + \frac{k_{max1}X}{Y_1}} \tag{4.131}$$

We have a first order equation similar to the Laplace transform of equation (4.2). From the arguments developed in Section 4.3.2, we know that only two parameter combinations (corresponding to the numerator coefficient and to the denominator coefficient) will be identifiable. Indeed the inverse Laplace transform of (4.131) gives the following time evolution for $y(t)$:

$$y(t) = \frac{(1 - Y_1)k_{max1}X}{Y_1}S_1(0)e^{-\frac{k_{max}Xt}{Y_1}} \tag{4.132}$$

The initial value $y(0)$ gives the amplitude $\frac{(1-Y_1)k_{max1}X}{Y_1}S_1(0)$, and the time response (decrease of 95 % after three time constants, i.e. at $t = \frac{3Y_1}{k_{max}X}$) gives the time constant $\tau = \frac{Y_1}{k_{max}X}$.

Therefore we see that only the two parameter combinations $\theta_1 = (1 - Y_1)S_1(0)$ and $\theta_2 = \frac{k_{max}X}{Y_1}$ are identifiable. Note that Y_1 is identifiable if $S_1(0)$ is known.

Note also that we could have equivalently written the model equations in a linear regression form by considering for instance the integral of $OUR_{ex}(t)$ as the output (see also [74]):

$$y'(t) = \int_O^t OUR_{ex}(\tau)d\tau \qquad (4.133)$$

This means that:

$$\frac{dy'}{dt} = OUR_{ex}(t) \qquad (4.134)$$

From equation (4.129), we know that the integral of OUR_{ex}, $y'(t)$, is equal to:

$$y'(t) = -(1 - Y_1)(S_1(t) - S_1(0)) \qquad (4.135)$$

By combining equations (4.128) and (4.129), we can write $S_1(t)$ as a function of $OUR_{ex}(t)$:

$$S_1(t) = \frac{Y_1}{(1 - Y_1)k_{max1}X} OUR_{ex}(t) \qquad (4.136)$$

By introducing (4.133), (4.134) and (4.136) into equation (4.135), we obtain:

$$\frac{dy'}{dt} = \beta_1 y' + \beta_2 \qquad (4.137)$$

with:

$$\beta_1 = -\frac{k_{max1}X}{Y_1}, \quad \beta_2 = \frac{(1 - Y_1)k_{max1}X}{Y_1}S_1(0) \qquad (4.138)$$

Observe that the two parameter combinations β_1 and β_2 are structurally identifiable from the data of OUR_{ex} and their time integral (4.133). Let us illustrate the structural identifiability concept via (real-life) data (Figure 4.2) with an initial substrate concentration $S_1(0)$. Figure 4.3 shows the data pairs $(OUR_{ex}, \int_o^t OUR_{ex}(\tau)d\tau)$ corresponding to the $(OUR_{ex}(t), t)$ data presented in Figure 4.2: β_2 is given by the initial value of OUR_{ex}, and β_1 is the slope.

4.6.2   Identifiability of the Single Monod Model (Taylor Series Expansion)

Let us consider the identifiability properties of the second model (Single Monod with one substrate):

$$\frac{dS_1}{dt} = \frac{\mu_{max1}X}{Y_1}\frac{S_1}{K_{S1} + S_1} \qquad (4.139)$$

Let us now use the Taylor series expansion method. This means that we look at the series expansion of $OUR_{ex}(t)$ around time t=0:

$$OUR_{ex}(t) = OUR_{ex}(0) + t\frac{dOUR_{ex}}{dt}(0) + \frac{t^2}{2!}\frac{d^2OUR_{ex}}{dt^2}(0) + ... \qquad (4.140)$$

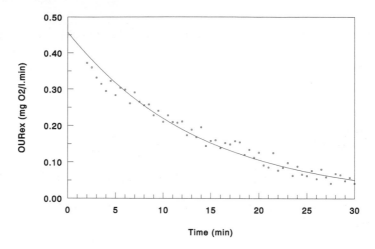

FIG. 4.2. OUR data corresponding to the exponential model.

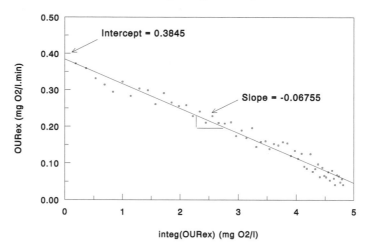

FIG. 4.3. Transformation of the OUR data into a linear regression form.

Let us compute the successive derivatives of $OUR_{ex}(t)$ at t = 0. The first terms will be written as follows:

$$OUR_{ex}(0) = \frac{\mu_{max1}X(1-Y_1)}{Y_1} \frac{S_1(0)}{K_{S1}+S_1(0)} \qquad (4.141)$$

$$\frac{dOUR_{ex}}{dt}(0) = -\frac{\mu_{max1}^2 X^2(1-Y_1)}{Y_1^2} \frac{K_{S1}S_1(0)}{(K_{S1}+S_1(0))^3} \qquad (4.142)$$

$$\frac{d^2 OU R_{ex}}{dt^2}(0) = \frac{\mu_{max1}^3 X^3 (1 - Y_1)}{Y_1^3} \frac{K_{S1} S_1(0)(K_{S1} - 2S_1(0))}{(K_{S1} + S_1(0))^5} \qquad (4.143)$$

$$\frac{d^3 OU R_{ex}}{dt^3}(0) = -\frac{\mu_{max1}^4 X^4 (1 - Y_1)}{Y_1^4} \frac{K_{S1} S_1(0)}{(K_{S1} + S_1(0))^7}$$
$$(K_{S1}^2 - 8K_{S1} S_1(0) + 6S_1(0)^2) \qquad (4.144)$$

$$\frac{d^4 OU R_{ex}}{dt^4}(0) = \frac{\mu_{max1}^5 X^5 (1 - Y_1)}{Y_1^5} \frac{K_{S1} S_1(0)}{(K_{S1} + S_1(0))^9}$$
$$(K_{S1}^3 - 22K_{S1}^2 S_1(0) + 58K_{S1} S_1(0)^2 - 24S_1(0)^3)(4.145)$$

The number of parameters has now increased: there are five parameters to be identified: Y_1, μ_{max1}, X, K_{S1} and $S_1(0)$.

The key question is then the following: are they all structurally identifiable, or only combinations of them?

Let us first note that the following parameter combinations:

$$\theta_1 = \frac{\mu_{max1} X (1 - Y_1)}{Y_1}, \quad \theta_2 = (1 - Y_1) S_1(0), \quad \theta_3 = (1 - Y_1) K_{S1} \qquad (4.146)$$

are combined in the first three derivatives. Indeed by noting:

$$z_i = \frac{d^i OU R_{ex}}{dt^i}(0), \quad i = 0, \ 1, \ 2, \ ... \qquad (4.147)$$

equations (4.141), (4.142) and (4.143) can be rewritten under the following (equivalent) form:

$$z_0 = \frac{\theta_1 \theta_2}{\theta_2 + \theta_3} \qquad (4.148)$$

$$z_1 = -\frac{\theta_1^2 \theta_2 \theta_3}{(\theta_2 + \theta_3)^3} \qquad (4.149)$$

$$z_2 = \frac{\theta_1^3 \theta_2 \theta_3 (\theta_3 - 2\theta_2)}{(\theta_2 + \theta_3)^5} \qquad (4.150)$$

Therefore the "parameters" θ_1, θ_2 and θ_3 can be formally calculated from the values of z_i (which can be theoretically calculated from a $(OUR(t), t)$ dataset) by inverting the above expressions, i.e.:

$$\theta_1 = \frac{z_0(z_0 z_2 - 3z_1^2)}{z_0 z_2 - z_1^2} \qquad (4.151)$$

$$\theta_2 = -\frac{2z_0^2 z_1}{z_0 z_2 - 3z_1^2} \qquad (4.152)$$

$$\theta_3 = -\frac{4z_0^2 z_1^3}{(z_0 z_2 - 3z_1^2)(z_0 z_2 - z_1^2)} \tag{4.153}$$

The question is then the following: can we expect to increase the number of identifiable parameters by considering higher order derivatives?

If we look at the additional derivatives for $i \geq 3$ (e.g. 4.145), the above parameter combinations are still combined basically in the same way as for the lower derivative terms, without any possibility to put in evidence other parameter combinations which could lead to a larger set of identifiable parameters. The conclusion appears to remain the same (we can never be sure!) if we consider even higher order derivatives: only the above parameter combinations θ_1, θ_2 and θ_3 are structurally identifiable.

The above set of parameter combinations is not the only one that fits in the above identifiability analysis. Other combinations can also be considered (e.g. θ_1 and θ_2 as in (4.146), and $\theta_3 = (1 - Y_1)(K_{S1} + S_1(0))$, but they are combinations of the above parameter combinations (4.146), and therefore result basically in the same conclusions to the one given above. Note also that the method using the transformation of the nonlinear model has also been applied to the Single Monod model and leads to the same conclusions.

Finally it is worth noting that symbolic manipulation software has been used to compute the successive derivatives and, once a set of parameter combinations was chosen, to perform the subsequent computations (e.g. (4.148), (4.149), (4.150) and (4.151), (4.152), (4.153) above).

4.6.3 Identifiability of the Double Monod Model (Nonlinear Transformation)

Let us now consider the Double Monod model (two pollutants simultaneously degraded without mutual interaction, (k=2)):

$$\frac{dS_1}{dt} = -\frac{\mu_{max1} X}{Y_1} \frac{S_1}{K_{S1} + S_1} \tag{4.154}$$

$$\frac{dS_2}{dt} = -\frac{\mu_{max2} X}{Y_2} \frac{S_2}{K_{S2} + S_2} \tag{4.155}$$

Let us here find a transformation of the nonlinear model into a model linear in the parameters. The line of reasoning is basically similar to the one considered for the exponential model above (development (4.134)-(4.138)). Recall that here the oxygen uptake rate $OU R_{ex}$ is the sum of the contribution of two substrates S_1 and S_2:

$$OU R_{ex} = -(1 - Y_1)r_{S_1} - (1 - Y_2)r_{S_2} \tag{4.156}$$

A typical $OU R_{ex}$ profile is shown in Figure 4.4. In the following, we assume (as it is suggested in Figure 4.4) that one substrate (S_1) is completely eliminated from the mixed liquor after the first part of the experiment (note that there is only one pathological case when this assumption does not hold: when S_1 and S_2 are

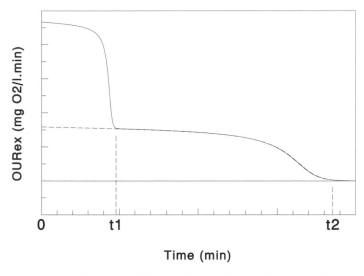

FIG. 4.4. Conceptual OUR profile with double Monod kinetics.

eliminated at exactly the same time). With this assumption the oxygen uptake rate can be subdivided in two parts corresponding with the degradation of each substrate. Hence, the identifiability analysis reduces to the analysis of the Single Monod model performed in two steps: for $0 \leq t < t_1$ for the first Monod model, and for $t_1 \leq t \leq t_2$ for the second one.

Let us insist that the present procedure is only valid in the presence of fully decoupled degradation of the substrates S_1 and S_2.

Let us first proceed for the first step and denote the first term of the right hand side of (4.156) by OUR_1:

$$OUR_1 = -(1 - Y_1)\frac{dS_1}{dt} \tag{4.157}$$

The integration of the above equation gives:

$$S_1(t) = S_1(0) - \frac{1}{1 - Y_1} \int_0^t OUR_1(\tau)d\tau \tag{4.158}$$

By introducing equation (4.154), the oxygen uptake rate equation (4.157) can be rewritten as follows:

$$OUR_1 = \frac{1 - Y_1}{Y_1} \frac{\mu_{max1} S_1}{K_{S1} + S_1} X \tag{4.159}$$

By multiplying both sides of the above equation by $(K_{S1} + S_1)$, and by considering (in order to have more compact notations) the variable $y_1(t)$ defined as follows:

$$y_1(t) = \int_0^t OUR_1(\tau)d\tau \tag{4.160}$$

equation (4.159) after some manipulations becomes:

$$y_1\frac{dy_1}{dt} = -\alpha_1 - \alpha_2 y_1 + \alpha_3\frac{dy_1}{dt} \tag{4.161}$$

where the parameters α_i (i = 1, 2, 3) are indeed combinations of the parameters θ_i (i = 1, 2, 3) defined in (4.146):

$$\alpha_1 = \theta_1\theta_2, \quad \alpha_2 = \theta_1, \quad \alpha_3 = \theta_2 + \theta_3 \tag{4.162}$$

There is clearly a one-to-one relation between these two sets of parameters:

$$\theta_1 = \alpha_2, \quad \theta_2 = \frac{\alpha_1}{\alpha_2}, \quad \theta_3 = \alpha_3 - \frac{\alpha_1}{\alpha_2} \tag{4.163}$$

Then we can conclude that with independent data of $y_1\frac{dy_1}{dt}$, y_1 and $\frac{dy_1}{dt}$ (generated via an appropriate experiment design), the parameters $\alpha_1, \alpha_2, \alpha_3$, and therefore the parameters $\theta_1, \theta_2, \theta_3$, are identifiable.

This result corresponds to the one obtained in the preceding section for the Single Monod model (for which we used the Taylor series expansion approach).

We can proceed similarly for the second step $t_1 \leq t \leq t_2$, and using similar definitions for OUR_2 and $y_2(t)$, it is straightforward that the parameters:

$$\theta_4 = \frac{\mu_{max2}X(1 - Y_2)}{Y_2}, \quad \theta_5 = (1 - Y_2)S_2(0), \quad \theta_6 = (1 - Y_2)K_{S2} \tag{4.164}$$

are identifiable. This means that only six combinations of the nine original parameters (Y_1, $S_1(0)$, μ_{max1}, K_{S1}, Y_2, $S_2(0)$, μ_{max2}, K_{S2}, and X) are structurally identifiable.

4.6.4 Identifiability of a Modified ASM1 Model (Nonlinear Transformation and Generating Series)

The analysis carried out for the modified ASM1 model is based on the following model equations (3 pollutants, 2 hydrolysed into the first substrate which is used for growth according to the Monod kinetics):

$$\frac{dS_1}{dt} = -\frac{\mu_{max1}X}{Y_1}\frac{S_1}{K_{S1} + S_1} + k_r X_r + k_s X_s \tag{4.165}$$

$$\frac{dX_r}{dt} = -k_r X_r \tag{4.166}$$

$$\frac{dX_s}{dt} = -k_s X_s \tag{4.167}$$

and the oxygen uptake rate is written as follows:

$$OU R_{ex} = \frac{\mu_{max1} X (1 - Y_1)}{Y_1} \frac{S_1}{K_{S1} + S_1} \tag{4.168}$$

As in Section 4.6.3, the first step in the analysis consists of considering that during a part of the experiment, the concentration of the rapidly hydrolysable substrate X_r should be approximately zero.

The effects of the two substrates cannot be decoupled, unlike in the Double Monod model where this was possible due to the saturation in the kinetics of S_2. Yet we can assume that the rapidly biodegradable substrate X_r is completely consumed after a time $t = t_1$. Therefore we can perform the analysis in two steps.

1. Step 1: $t_1 \leq t \leq t_2$ when X_r is assumed to be equal to zero. This means that in the first step only equations (4.165), (4.167) and (4.168) with $X_r = 0$ are considered.

2. Step 2: $0 \leq t < t_1$, with all equations (4.165), (4.166), (4.167) and (4.168) and the knowledge given.

Let us consider here two approaches: the nonlinear transformation, and the generating series.

Nonlinear transformation. The analysis is performed as follows:

1. Integration of equations (4.167) and (4.168). This gives the following relation for $S_1(t)$:

$$
\begin{aligned}
S_1(t) &= S_1(0) - \tfrac{1}{1-Y_1} \int_0^t OU R_{ex}(\tau) d\tau + \int_0^t k_s X_s(\tau) d\tau \quad (4.169) \\
&= S_1(0) - \tfrac{1}{1-Y_1} \int_0^t OU R_{ex}(\tau) d\tau + k_s X_s(0)(1 - e^{-k_s t}) (4.170)
\end{aligned}
$$

2. Linearisation of the exponential term $e^{-k_s t}$ around t = 0 (in order to carry out the analysis with a model linear in the parameters):

$$e^{-k_s t} \cong 1 - k_s t + \frac{k_s^2 t^2}{2} \tag{4.171}$$

(We stop the series expansion at the second order term since the additional terms do not add extra useful information for the analysis.)

3. Introduction of these results in equation (4.168) and rewriting by multiplying both sides by $(1 - Y_1) K_{m1} + (1 - Y_1) S_1(0) - \int_0^t OU R_{ex}(\tau) d\tau +$

$(1 - Y_1)k_s X_s(0)(k_s t - \frac{k_s^2 t^2}{2})$ and considering $y(t)$ defined as the integral of $OU R_{ex}(t)$, as in (4.133):

$$y\frac{dy}{dt} = \beta_1 + \beta_2 y + \beta_3 \frac{dy}{dt} + \beta_4 t \frac{dy}{dt} - \beta_5 \frac{t^2}{2}\frac{dy}{dt} - \beta_6 t + \beta_7 \frac{t^2}{2} \quad (4.172)$$

$$\beta_1 = -\theta_1\theta_2, \quad \beta_2 = \theta_1, \quad \beta_3 = \theta_2 + \theta_3 \quad (4.173)$$
$$\beta_4 = \theta_7\theta_8^2, \quad \beta_5 = \theta_7\theta_8^3, \quad \beta_6 = \theta_1\beta_4, \quad \beta_7 = \theta_1\beta_5 \quad (4.174)$$

with $\theta_1, \theta_2, \theta_3$ as defined in (4.146), and θ_7 and θ_8 defined as follows:

$$\theta_7 = (1 - Y_1)X_s(0), \quad \theta_8 = k_s \quad (4.175)$$

Among the seven parameters β_i, only five are independent (β_6 and β_7 are related to β_4, β_5 and β_2). Therefore five parameter combinations are identifiable, i.e. $\beta_1, \beta_2, \beta_3, \beta_4$ and β_5, or equivalently $\theta_1, \theta_2, \theta_3, \theta_7$ and θ_8.

The second step for $0 \le t \le t_1$ considers that the dynamics are given by equations (4.165), (4.166), (4.167) and (4.168) with $X_r \neq 0$, and that the values of the parameters β_i (i = 1 to 7) are already available from the first step, i.e. from data for times between t_1 and t_2. Then by following the same line of reasoning as in step 1, one obtains two more identifiable parameter combinations: $(1 - Y_1)X_r(0)$ and k_r.

Generating series. The equations (4.41) and (4.42) specialise here as follows:

$$x = \begin{pmatrix} S_1 \\ X_s \end{pmatrix}, \quad f_0 = \begin{pmatrix} -\frac{\mu_{max1}X}{Y_1}\frac{S_1}{K_{S1}+S_1} + k_s X_s \\ -k_s X_s \end{pmatrix} \quad (4.176)$$

$$h(x, \theta) = (1 - Y_1)\frac{\mu_{max1}X}{Y_1}\frac{S_1}{K_{S1} + S_1} - (1 - Y_1)k_s X_s \quad (4.177)$$

$$x(0) = x_0(\theta) = \begin{pmatrix} S_1(0) \\ X_s(0) \end{pmatrix} \quad (4.178)$$

From the knowledge that we have gained from the preceding examples, we can suspect that the individual parameters μ_{max1}, X and Y_1 will not be identifiable, but only the combination $\frac{\mu_{max1}X}{Y_1}$. For simplifying the rest of the developments, let us consider it as one parameter. Therefore we can see that we have six parameters:

$$\theta_1 = \frac{\mu_{max1}X}{Y_1} \quad (4.179)$$
$$\theta_2 = (1 - Y_1) \quad (4.180)$$
$$\theta_3 = K_{S1} \quad (4.181)$$
$$\theta_4 = k_s \quad (4.182)$$

$$\theta_5 = S_1(0) \tag{4.183}$$

$$\theta_6 = X_s(0) \tag{4.184}$$

Let us now calculate $h(0, \theta)$ and the successive Lie derivatives.

Since the calculations are quite complex and the resulting equations are very much involved, we shall concentrate on the first four elements calculated in the first step. (Let us recall that at least six elements are necessary to complete the analysis.) Let us first define these first four terms as follows:

$$h(0, \theta) = z_1 \tag{4.185}$$

$$L_{f_0} h(0, \theta) = z_2 \tag{4.186}$$

$$L_{f_0} L_{f_0} h(0, \theta) = z_3 \tag{4.187}$$

$$L_{f_0} L_{f_0} L_{f_0} h(0, \theta) = z_4 \tag{4.188}$$

Then the calculations give:

$$z_1 = \theta_2 \theta_1 \frac{\theta_5}{\theta_3 + \theta_5} - \theta_2 \theta_4 \theta_6 \tag{4.189}$$

$$z_2 = -\frac{\theta_1 \theta_3}{(\theta_3 + \theta_5)^2} z_1 + \theta_2 \theta_4^2 \theta_6 \tag{4.190}$$

$$z_3 = -\frac{\theta_1 \theta_3}{\theta_2 (\theta_3 + \theta_5)^3} z_1^2 - \frac{\theta_1 \theta_3}{(\theta_3 + \theta_5)^2} z_2 - \theta_2 \theta_4^3 \theta_6 \tag{4.191}$$

$$
\begin{aligned}
z_4 = {}& -\frac{6\theta_1 \theta_3}{\theta_2^2 (\theta_3 + \theta_5)^4} z_1^3 + \frac{6\theta_1^2 \theta_3^2}{\theta_2 (\theta_3 + \theta_5)^5} z_1^2 \\
& -\frac{\theta_1^3 \theta_3^3}{(\theta_3 + \theta_5)^6} z_1 - \frac{3\theta_1 \theta_3}{\theta_2 (\theta_3 + \theta_5)^3} z_1 z_2 \\
& -4\theta_4^2 \frac{\theta_1 \theta_3 \theta_6}{(\theta_3 + \theta_5)^3} z_1 - \theta_4^2 \frac{\theta_1 \theta_2 \theta_3 \theta_6}{(\theta_3 + \theta_5)^2} \left(\frac{\theta_1 \theta_3}{(\theta_3 + \theta_5)^2} - \theta_4 \right) + \theta_2 \theta_4^4 \theta_6
\end{aligned} \tag{4.192}
$$

Although we do not have enough information from the above equations to conclude, these already contain useful hints about the identifiability.

- First note that the last term of the right hand side of each equation (4.189), (4.190), (4.191) and (4.192) is a combination of three parameters (θ_2, θ_4 and θ_6). Among the three, only θ_4 appears with an increasing power. This suggests that θ_4 ($= k_s$) might be identifiable (*hint #1*). For the other two parameters, we have to investigate further.
- Let us now look at z_1 (4.189). We note that θ_2 appears linearly in both terms. In other words, the parameters θ_1, θ_2 and θ_6 appear to be linked in the following combinations $\theta_2 \theta_1$ and $\theta_2 \theta_6$ (*hint #2*).
- For z_2 (4.190), we note that the term multiplying z_1 is a fraction where the degree of the numerator with respect to the parameters is equal to the degree

of the denominator (numerator: degree 1 for θ_1, degree 1 for θ_3; denominator: degree 2 for $\theta_3 + \theta_5$).

- Let us look at all the terms but the last one (already considered here above) in each equation (4.189), (4.190), (4.191) and (4.192). We note that θ_2 appears whenever the degree of the numerator is different from the degree of the denominator. More precisely, it appears with a power equal to the difference between both degrees. (Equivalently each term can be rewritten with θ_2 appearing both at the numerator and the denominator with a degree equal to the total degree of the parameters of the numerator and the denominator, respectively.) This suggests that the four parameters θ_1, θ_3, θ_5 and θ_6 cannot be separated from θ_2, or in other words, that only the combinations $\theta_1\theta_2$, $\theta_2\theta_3$, $\theta_2\theta_5$, $\theta_2\theta_6$ (*hint #3*) can.

From the above comments, we have hints of what we may expect in terms of structural identifiability. These are in line with the results that we had obtained with the preceding examples (first order kinetics, single Monod, double Monod). Let us insist that this preliminary analysis does not allow us to make a conclusion. Only a complete analysis (with at least two extra generating series terms) will allow us to give results. This full analysis has indeed been performed. It shows that five parameter combinations

$$\theta_4 = k_s \tag{4.193}$$

$$\theta_1\theta_2 = \frac{\mu_{max1} X (1 - Y_1)}{Y_1} \tag{4.194}$$

$$\theta_2\theta_3 = (1 - Y_1) K_{S1} \tag{4.195}$$

$$\theta_2\theta_5 = (1 - Y_1) S_1(0) \tag{4.196}$$

$$\theta_2\theta_6 = (1 - Y_1) X_s(0) \tag{4.197}$$

are identifiable in step 1 (since we can write them as functions of the z_i). This confirms the results obtained with the nonlinear transformation.

4.6.5   *Summary of the Results and Discussion*

The identifiability results for the four models are summarised in Table 4.1.

The examples presented in Sections 4.6.1, 4.6.2 and 4.6.3 are illustrative of the advantages and drawbacks of the considered methods. The implementation of the series expansion method has the advantage of being systematic in the sense that it follows a clearly identified route. The above example is illustrative of the potential difficulties with the series expansion:

- How many derivatives of OUR_{ex} are needed to obtain conclusive results? For certain models the question may rise whether one can achieve better identifiability properties by considering more terms in the expansion. Here

Table 4.1 Identifiable parameter combinations of the four models

Exponential	Single Monod	Double Monod	Modified ASM1
$(1 - Y_1)S_1(0)$ $\frac{k_{max1}X}{Y_1}$	$(1 - Y_1)S_1(0)$ $\frac{\mu_{max1}X(1-Y_1)}{Y_1}$ $(1 - Y_1)K_{S1}$	$(1 - Y_1)S_1(0)$ $\frac{\mu_{max1}X(1-Y_1)}{Y_1}$ $(1 - Y_1)K_{S1}$ $(1 - Y_2)S_2(0)$ $\frac{\mu_{max2}X(1-Y_2)}{Y_2}$ $(1 - Y_2)K_{S2}$	$(1 - Y_1)S_1(0)$ $\frac{\mu_{max1}X(1-Y_1)}{Y_1}$ $(1 - Y_1)K_{S1}$ $(1 - Y_1)X_r(0)$ k_r $(1 - Y_1)X_s(0)$ k_s

we found that the additionally evaluated terms did not yield additional information. Generally speaking the approach may imply more and more symbolic computations, and yet not lead to conclusive results (as experienced for the Double Monod model).

- How can we derive the right combinations of identifiable parameters? There is indeed no general systematic rule for selecting these combinations, and therefore the procedure may look a little tricky. However the structure of the different terms of the expansion is often a source for good initial guesses. For instance, the choice of θ_1 $(= -\frac{\mu_{max1}X(1-Y_1)}{Y_1})$ looks quite obvious from equations (4.141), (4.142) and (4.143).

On the other hand, the nonlinear transformation may suffer from the difficulty to easily find out the transformation that will a priori suit the problem (although in the proposed example, the choice of the transformation (multiplication by the denominator of the Monod model) is rather straightforward).

Let us also make some comments about the obtained identifiability results.

First note that the yield coefficient(s) Y_1 (Y_2) is present in all the parameter combinations (except in k_r and k_s in the modified ASM1 model). This is not surprising since, on the basis of OUR_{ex} data only, one can have no idea which quantity of substrate has been transformed into biomass. This explains why the term $(1 - Y_1)S_1(0)$ $((1 - Y_2)S_2(0))$, i.e. the fraction of substrate which is oxidised, appears as a parameter combination. The same remark applies to $(1 - Y_1)K_{S1}$ $((1 - Y_2)K_{S2})$ which can be viewed as no more than a rescaling of the substrate affinity constant.

Secondly the parameter combination $\frac{\mu_{max1}X(1-Y_1)}{Y_1}$ $(\frac{\mu_{max2}X(1-Y_2)}{Y_2})$ is an expression of the total activity of the sludge, and with that respect can be considered as giving information different from that of the individual parameters.

Finally let us point out that a priori information about some individual parameters (e.g. the yield coefficient(s) Y_1 (Y_2) values obtained via separate experiments) can be incorporated in the parameter evaluation procedure. Then individual parameters (e.g. $S_1(0)$ or K_{S1}) can be estimated. As a matter of example, in the single Monod model, if $S_1(0)$ is known, then Y_1 is identifiable, and consequently, K_{S1} is also identifiable.

4.7   General Structural Identifiability Results for the ASM-type Models

Petersen *et al.* [194] propose a generalisation of the structural identifiability results for the ASM1. The proposed generalisation applies to models like the ASM1 where the kinetics are either first order kinetics or Monod-type models. It is based on measured variables that are either concentrations of components appearing explicitly in the mass balance models or directly related to these (like the OUR, which has been linked to the substrate concentration in the respirometer-based examples of the preceding section). Finally the results apply to one reaction at a time: in that sense, they apply to the ASM1 under the assumption that each reaction can be decoupled from the others (an assumption that obviously has to be validated, see also the preceding section for examples).

The generalisation results are summarised in Table 4.2, for two cases: when one measurement is available and when two measurements are available. They are based on the tabular form of the ASM1 (Section 2.3.3). The terms v correspond to the yield coefficients, the terms K correspond to the saturation constant in the kinetic Monod models, the indices i and j correspond to the number of the measured component (column) in the ASM1 table, and the number of the process (reaction) (row) in the same table. More precisely, j corresponds to one or two columns in the ASM1 table (depending on the number of measured components), and i corresponds to the rows in the ASM1 tables, corresponding to the processes (reactions) considered in the dynamical model.

It is important to note that so far there is no technical (or mathematical) proof that the proposed results represent a generalisation. On the other hand, all the examples that have been treated so far give identifiablity results that follow the proposed generalisation.

4.8   Overparametrisation: An Illustrative Example

So far, we have introduced different methods for testing the structural identifiability and illustrated them on several examples. Typically, when the number of identifiable parameters is lower than the original number of parameters, the model is said to be overparametrised. Overparametrisation can indeed be detected in many instances via a preliminary analysis based for instance on a transformation of the original model formulation into an equivalent one in which the number of parameters may be lower. Formally this means that, if we consider the following system dynamics:

Table 4.2 Identifiable parameter combinations of the ASM-type model: generalisation for one and two measurements

One measurement (j)		Two measurements (j)	
No growth	Growth	No growth	Growth
$\mid v_{i,j} \mid \mu_{max,j} X$	$\mu_{max,j}$	$\mid v_{i,j} \mid \mu_{max,j} X$	$\mu_{max,j}$
$\mid \dfrac{v_{i,j}}{v_{k,j}} \mid K_j$	$\mid v_{i,j} \mid X(0)$	$\mid \dfrac{v_{i,j}}{v_{k,j}} \mid K_j$	$\mid v_{i,j} \mid X(0)$
$\mid \dfrac{v_{i,j}}{v_{k,j}} \mid S_k(0)$	$\mid \dfrac{v_{i,j}}{v_{k,j}} \mid K_j$	$\mid \dfrac{v_{i,j}}{v_{k,j}} \mid S_k(0)$	$\mid \dfrac{v_{i,j}}{v_{k,j}} \mid K_j$
	$\mid \dfrac{v_{i,j}}{v_{k,j}} \mid S_k(0)$		$\mid \dfrac{v_{i,j}}{v_{k,j}} \mid S_k(0)$
		$\mid \dfrac{v_{i(1),j}}{v_{i(2),j}} \mid$	$\mid \dfrac{v_{i(1),j}}{v_{i(2),j}} \mid$

$$\frac{dx}{dt} = f(\theta, x, u) \tag{4.198}$$

$$y = h(\theta, x) \tag{4.199}$$

and if we find a state transformation:

$$x' = g(x) \tag{4.200}$$

such that the new model formulation:

$$\frac{dx'}{dt} = f'(\theta', x', u) \tag{4.201}$$

$$y = h'(\theta', x') \tag{4.202}$$

contains less parameters:

$$dim(\theta') < dim(\theta) \tag{4.203}$$

then the original model is overparametrised, and only the parameters θ' can possibly be structurally identifiable.

Let us illustrate this with an example. Let us consider the following dynamical model of an anaerobic digestion process ([15]):

$$\frac{dS_0}{dt} = -DS_0 + \alpha D S_{in} - k_0 S_0 X_1 \tag{4.204}$$

$$\frac{dS_1}{dt} = -DS_1 + k_0 S_0 X_1 - \frac{1}{Y_1} \frac{\mu_{max1} S_1}{K_{S1} + S_1} X_1 \tag{4.205}$$

$$\frac{dX_1}{dt} = -DX_1 + \frac{\mu_{max1} S_1}{K_{S1} + S_1} X_1 - k_{d1} X_1 \tag{4.206}$$

$$\frac{dS_2}{dt} = -DS_2 + Y_3 \frac{\mu_{max1} S_1}{K_{S1} + S_1} X_1 - \frac{1}{Y_2} \frac{\mu_{max2} S_2}{K_{S2} + S_2} X_2 \tag{4.207}$$

$$\frac{dX_2}{dt} = -DX_2 + \frac{\mu_{max2} S_2}{K_{S2} + S_2} X_2 - k_{d2} X_2 \tag{4.208}$$

$$Q_{CH_4} = Y_4 \frac{\mu_{max2} S_2}{K_{S2} + S_2} X_2 \tag{4.209}$$

i.e. a model with three steps: solubilisation of organic compounds S_0 (equation (4.204)), acidification of solubilised substrate S_1 (equations (4.205)(4.206)), and methanisation of volatile fatty acids S_2 (4.207)(4.208)(4.209). Note that in this model compared to the one presented in Chapter 2, a solubilisation step has been added. X_1, X_2, S_{in} and Q_{CH_4} are the concentrations of acidogenic bacteria and of methanogenic bacteria, the concentration of nonsolubilised organic matter in the influent, and the methane gas outflow rate, respectively. Y_i (i=1 to 4) are yield coefficients, α is an availability coefficient, and k_0 is the kinetic constant of the solubilisation reaction. μ_{maxi}, K_{Si} and k_{di} (i=1, 2) are the maximum specific growth rates, the affinity constants, and the death coefficients related to acidogenesis and methanisation, respectively.

The above model contains 12 parameters (Y_1, Y_2, Y_3, Y_4, α, k_0, μ_{max1}, K_{S1}, k_{d1}, μ_{max2}, K_{S2}, k_{d2}). In the identifiability study performed in [15], the measured variables were the dilution rate D, the influent substrate concentration S_{in}, and the methane gas flow rate Q_{CH_4}. The dilution rate D and the influent substrate concentration S_{in} are typically actions (*inputs*) on the process, while the methane gas flow rate Q_{CH_4} is typically a result (*output*) of the process operation. This means that the output and input are defined as follows:

$$u = \begin{bmatrix} D \\ S_{in} \end{bmatrix}, \quad y = Q_{CH_4} \tag{4.210}$$

Let us define the following state transformation:

$$\tilde{X}_1 = k_0 X_1 \tag{4.211}$$

$$\tilde{X}_2 = Y_4 X_2 \tag{4.212}$$

$$\tilde{S}_0 = \frac{S_0}{K_{S1}} \tag{4.213}$$

$$\tilde{S}_1 = \frac{S_1}{K_{S1}} \tag{4.214}$$

$$\tilde{S}_2 = \frac{S_2}{K_{S2}} \tag{4.215}$$

If we consider the following parameter combinations:

$$\tilde{Y}_1 = k_0 K_{S1} Y_1 \tag{4.216}$$

$$\tilde{Y}_2 = K_{S2} Y_4 Y_2 \tag{4.217}$$

$$\tilde{\alpha} = \frac{\alpha}{K_{S1}} \tag{4.218}$$

$$\tilde{Y}_3 = \frac{Y_3}{k_0 K_{S2}} \tag{4.219}$$

and by using the above state transformation (4.211)(4.215), the dynamical equations (4.204)-(4.209) can be represented by the following set of equations:

$$\frac{d\tilde{S}_0}{dt} = -D\tilde{S}_0 + \tilde{\alpha} D S_{in} - \tilde{S}_0 \tilde{X}_1 \tag{4.220}$$

$$\frac{d\tilde{S}_1}{dt} = -D\tilde{S}_1 + \tilde{S}_0 \tilde{X}_1 - \frac{1}{\tilde{Y}_1} \frac{\mu_{max1} \tilde{S}_1}{1 + \tilde{S}_1} \tilde{X}_1 \tag{4.221}$$

$$\frac{d\tilde{X}_1}{dt} = -D\tilde{X}_1 + \frac{\mu_{max1} \tilde{S}_1}{1 + \tilde{S}_1} \tilde{X}_1 - k_{d1} \tilde{X}_1 \tag{4.222}$$

$$\frac{d\tilde{S}_2}{dt} = -D\tilde{S}_2 + \tilde{Y}_3 \frac{\mu_{max1} \tilde{S}_1}{1 + \tilde{S}_1} \tilde{X}_1 - \frac{1}{\tilde{Y}_2} \frac{\mu_{max2} \tilde{S}_2}{1 + \tilde{S}_2} \tilde{X}_2 \tag{4.223}$$

$$\frac{d\tilde{X}_2}{dt} = -D\tilde{X}_2 + \frac{\mu_{max2} \tilde{S}_2}{1 + \tilde{S}_2} \tilde{X}_2 - k_{d2} \tilde{X}_2 \tag{4.224}$$

$$Q_{CH_4} = \frac{\mu_{max2} \tilde{S}_2}{1 + \tilde{S}_2} \tilde{X}_2 \tag{4.225}$$

The above formulation is equivalent to the original one (4.204)-(4.209) with respect to the inputs (D, S_{in}) and the output Q_{CH_4}. The model contains now only eight parameters $(\tilde{Y}_1, \tilde{Y}_2, \tilde{Y}_3, \tilde{\alpha}, \mu_{max1}, k_{d1}, \mu_{max2}, k_{d2})$: this means that only these eight parameters are possibly structurally identifiable.

4.9   Conclusions

In this chapter, we have introduced the concept of structural identifiability. As stated in the introduction, the notion of structural identifiability may be essential in the study of wastewater treatment processes and the use of dynamical models for numerical simulation, process design and/or control design, because the structural identifiability analysis will tell a priori if there is any chance that a candidate model is identifiable, i.e. if its parameters can be given unique values. This property is essential for the reliability of the model. If you have an unidentifiable model, this means that any numerical values of its parameters (as long as they correspond to a unique value of the identifiable parameter combinations[6]) can be given. How can you then have any confidence in such a model whether it is used for simulation, process design or control? And what kind of interpretation can you then give to the (physical) parameters of the model, if there exists an infinite possibility for choosing their values?

So far, we have introduced several tools to test the structural identifiability. These are:

1. Laplace transform,
2. Taylor series expansion,
3. generating series,
4. local state isomorphism,
5. transformation of nonlinear models,
6. Lyapunov-based observer analysis.

The first one was introduced in Section 6.3 and is only valid for linear models. However since most models in WWTP are nonlinear (not only in the state, but also in the parameter), it is important to propose tests that can be used to handle the structural identifiability of nonlinear models. These were introduced in Section 6.4 and 6.5. The method introduced in Section 6.5 (Lyapunov-based method) is quite a special one and difficult to generalise, especially for non-experts in system analysis and automatic control. Moreover, the motivation to introduce it was also a historical perspective with the first identifiability analysis of the Monod model. This motivated our choice to put it in a separate section. The methods introduced in Section 6.4 are indeed of a more general use, and are basically applicable to any model available in the literature on WWTP.

The use of these tools has been illustrated with several examples:

1. two interconnected tanks,
2. two reaction models,
3. Monod model,
4. respirometer-based models.

The basic features of these tools can be summarised as follows.

1. First of all, it should be noted that it is very difficult to a priori select the best method to test the structural identifiability of a dynamical model. It may happen that one method is much easier to apply to one model, and

---

[6]Assume that you have a model where $\theta_1 + \theta_2$ is identifiable (but not each parameter individually). It is obvious that any combination of values for θ_1 and θ_2 such that the sum is equal to a specific constant will give the same result in the model behaviour.

becomes completely cumbersome with another. This probably explains why there are (at least) six different structural identifiability methods available in the literature (see e.g. [55]).

2. Most of the structural identifiability tests only give local structural identifiability results for nonlinear models (except the local state isomorphism approach, and the Lyapunov approach if the domain of validity covers the whole physical space). The results given by these tests have been obtained for some specified time t (typically, t = 0) for the Taylor series expansion, or equivalently at the initial values for the generating series: therefore the results obtained are strictly speaking only valid for these values. In order to become "global", the approach should cover the whole physical state space via the computation of the different terms required for the test in this space for *all* admissible state variable values. For the transformation of nonlinear models, one should be particularly careful at some singularities (typically possible division by zero, for instance) that may arise during the transformation and back-transformation processes.

3. The methods may give sufficient or necessary identifiability results. The Taylor series expansion gives a *sufficient* [7] structural identifiability condition (see [197]) (because there exists no upper bound on the number of coefficients to be considered in the test). Generally speaking, the generating series method also gives sufficient conditions, but it results in necessary and sufficient conditions for bilinear and polynamial models.

4. The use of symbolic software can be very helpful to apply identifiability tests to the studied models. Very quickly the computation burden may become enormous, and without symbolic manipulation software, the computation may become impossible to handle in practice. The computation burden may be less important with the generating series approach than with the Taylor series approach. Yet this is not the panacea so far and symbolic software also exhibits its limitations: in several applications, it appears that the complexity of the required computations is such that the symbolic softwares that we have been using were not able to solve the problems (but of course, we can hope that this will improve in the future...).

Finally we have briefly introduced the notion of overparametrisation of dynamical models, and illustrated it with an anaerobic digestion model.

Different structural identifiability studies dedicated to water and wastewater treatment processes can be found in the literature.

---

[7]The notion of necessary and sufficient conditions is essential in mathematics. A condition C is sufficient means that if the condition C is fulfilled, the result R follows (in mathematical terms, C $\Rightarrow$ R). Yet this does not mean that if the condition C is not fulfilled, the result R is wrong or false. A necessary condition is the reciprocal proposal (R $\Rightarrow$ C): the result R is true or correct *only* if the condition C is fulfilled, and R may be false or wrong even if C is fulfilled.

We have presented here part of the results of the paper by Dochain *et al.* [74], illustrating the structural identifiability of different respirometer-based models was stabilised by considering various approaches including the Taylor series expansion and nonlinear transformations.

We have also mentioned the generalization effort done in Petersen [193] and Petersen *et al.* [194] for ASM1 type models (with first order kinetics and Monod kinetics) by considering respirometric measurements as well as combined respiro-metric-titrimetric measurements.

In Bourrel *et al.* [35], the authors analyse the identifiability of a denitrifying biofilter model. The process dynamics are described by partial differential equations (PDEs): this is one of the main original aspects of this work, especially with respect to what has been presented in the present chapter. There were seven measurement points along the column. The available data corresponds to different steady states: the authors have therefore studied the steady state equations (ordinary differential equations) for the identifiability analysis.

In Chen and Bastin [57], the authors consider the structural identifiability of the yield coefficients independently of the reaction rate model parameters. This analysis is made possible by considering the state transformation that we have considered in Chapter 2, and is based on the structural properties of the General Dynamical model.

The structural identifiability of the ASM model No. 1 and of a reduced-order version of this model is presented in Julien *et al.* (1992) [139], and Julien *et al.* (1998) [138], respectively. The model consists indeed in two submodels: one for aerobic conditions, the other one for anoxic conditions. The model was applied to an alternating operation of a WWTP. In the aerobic phase, the dynamical model is composed of three differential equations, while two differential equations describe the dynamics in anoxic conditions. The methods considered by the authors are local state isomorphism, and the transformation of the nonlinear model into a linear one (for the anoxic model in [138]).

In Keesman *et al.* [142], the authors study the identifiability of a model for endogenous respiration in an activated sludge in a batch reactor in the absence of dissolved oxygen limitation. This model is derived from the ASM1 model and contains six parameters $(\mu_m, Y, K_S, k_h, f_p, b)$. The structural identifiability is studied numerically by identifying the model parameters θ from a "thought-experiment" with a selected parameter vector θ^*, and by computing the gradient and the Hessian of the output prediction error in order to check that the solution $\theta = \theta^*$ is a local minimum. The authors obtain the following results. If only the endogenous respiration rate is measured, then k_h, f_p, and combinations of μ_m and K_S, and of Y, K_S and b (i.e. four parameters) are identifiable. If, in addition, measurements of the volatile suspended solids in the mixed liquor are available, then five parameters are identifiable: Y, k_h, f_p, b, and a combination of μ_m and K_S.

5

Practical Identifiability and Optimal Experiment Design for Parameter Estimation (OED/PE)

5.1 Introduction

In the preceding chapter, we discussed the notion of structural identifiability, which is related to the possibility of giving a unique value to each parameter of a mathematical model. The question that we addressed was the following: given a model structure and perfect (i.e. that fits perfectly to the model) data of model variables, are all the parameters of the model identifiable? A structural identifiablity study may result in the following conclusions. First it is possible that only combinations of the model parameters are identifiable. Moreover, if the number of resulting combinations is lower than the number of original model parameters, or if there is not a one-to-one relationship between both parameter sets, then a priori knowledge about some parameters may be required to achieve identifiability of each individual parameter.

In this chapter, we would like to discuss the notion of practical identifiability, which is the important complement to the structural identifiability in order to guarantee reliability of the calibration of the model parameters from available experimental data. Practical identifiability is indeed related to the quality of the data and their "information" content: are the available data informative enough for identifying the model parameters and, more specifically, for giving accurate values? For instance in the model $y = ax_1 + bx_2$ the parameters a and b are structurally identifiable but they will not be practically identifiable if the experimental conditions are such that the independent variables x_1 and x_2 are always proportional ($x_1 = \alpha x_2$) (then only the *combination* $a\alpha + b$ is practically identifiable).

While the structural identifiability is studied under the assumption of perfect, i.e. noiseless, data, the problem with highly correlated parameters arises when a limited set of experimental, noise-corrupted data is used for parameter estimation. Under such conditions the uniqueness of parameter estimates predicted by the structural analysis may no longer be guaranteed because a change in one parameter can be compensated almost completely by a proportional shift in another, still producing a satisfying fit between experimental data and model predictions. In addition, the numerical algorithms that perform the nonlinear parameter estimation (presented in Chapter 6) show poor convergence when faced with this type of ill-conditioned optimisation problem, the estimates being very sensitive to the initial parameter values given to the algorithm [126], [171]. Consequently, the estimated parameters may vary over a broad range and little physical interpretation can be given to the parameter values obtained.

The Monod-model (μ_{max} is the maximum specific growth rate (min^{-1}), K_S is the saturation constant (mg/L)),

$$\mu(S) = \frac{\mu_{max} S}{K_S + S} \qquad (5.1)$$

is probably the best-known example in biological systems of a model in which parameter estimates may be highly correlated [41],[126],[179]. In many cases the experiments provide only sufficient information to estimate the ratio between both parameters in this model, μ_{max}/K_S. A simple example may illustrate this (Figure 5.1): if only growth rates are available for low substrate concentrations (in the example of Figure 5.1, these range between 0 and 0.1 mg/L), no distinction can be made between different parameter sets, i.e. the Monod model is practically unidentifiable. In order to overcome this problem, it has been proposed to use additional a priori information (e.g. a known maximum growth rate), to impose parameter bounds [126], or to sample more frequently in defined periods of the experiment in order to increase the informative content of the collected data [276]. Evidently, measuring at higher substrate concentrations (see Figure 5.1, right) also allows unique, reliable parameter estimates to be obtained.

The chapter is organised as follows: we shall first discuss the concept of practical identifiability and the related notions of confidence intervals and sensitivity functions in Section 5.2. Section 5.3 will be devoted to optimal design of experiments in order to obtain the most reliable parameter values possible. The optimal experiment design will be illustrated in Section 5.4 with a respirometry-based

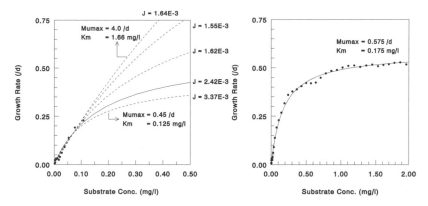

FIG. 5.1. Practical identifiability of the Monod model parameters (J represents the sum of the squared errors for different parameter sets).

model that we have already considered in Chapters 3 (Section 3.2) and 4 (Section 4.6). Finally the question of the optimal experiment design for the dual problem of structure characterisation and parameter estimation will be briefly discussed in Section 5.5.

5.2 Practical Identifiability

5.2.1 Theoretical Framework

The question addressed in this section is the following: with the available experimental data, what is the accuracy we can obtain for the parameter estimates, or, in other words, if a small deviation in the parameter set occurs, does this have a considerable decrease of the fit as a consequence. Mathematically, this can be formalized as follows [180].

Let us recall first (see also Chapter 6) that parameter estimation can often be formulated as the minimisation of the following quadratic objective functional by optimal choice of the parameters θ [180]:

$$J(\theta) = \sum_{i=1}^{N}(y_i(\hat{\theta}) - y_i)^T Q_i(y_i(\hat{\theta}) - y_i) \tag{5.2}$$

in which y_i and $y_i(\hat{\theta})$ are vectors of N measured values and model predictions at times t_i (i = 1 to N) respectively, and Q_i is a square matrix with user-supplied weighting coefficients. The expected value of the objective functional for a parameter set slightly different from the optimal one is given by [179]:

$$E[J(\theta + \delta\theta)] \cong \delta\theta^T [\sum_{i=1}^{N}(\frac{\partial y}{\partial \theta}(t_i))^T Q_i(\frac{\partial y}{\partial \theta}(t_i)]\delta\theta + \sum_{i=1}^{N} tr(C_i Q_i) \tag{5.3}$$

in which C_i represents the measurement error covariance matrix (Q_i is typically chosen as C_i^{-1} and the second term reduces to a scalar). An important consequence of (5.3) is that in order to *optimise the practical identifiability* one has to *maximise the term between brackets [.]* in equation (5.3). By doing so, one maximizes the difference between $J(\theta + \delta\theta)$ and $J(\theta)$ or in other words, one ensures that the fit of a parameter set that is slightly different from the best parameter set is significantly worse. The term between brackets in equation (5.3) is the so-called Fisher Information Matrix and expresses the information content of the experimental data [159]:

$$F = \sum_{i=1}^{N} (\frac{\partial y}{\partial \theta}(t_i))^T Q_i (\frac{\partial y}{\partial \theta}(t_i)) \tag{5.4}$$

This matrix is indeed the inverse of the parameter estimation error covariance matrix of the best linear unbiased estimator [99]:

$$V = F^{-1} = (\sum_{i=1}^{N} (\frac{\partial y}{\partial \theta}(t_i))^T Q_i (\frac{\partial y}{\partial \theta}(t_i)))^{-1} \tag{5.5}$$

The terms $\frac{\partial y}{\partial \theta}$ are the output sensitivity functions. These quantify the dependence of the model predictions on the parameter values. The evaluation of the sensitivity functions is a central task in the practical identifiability study and is dealt with in the next section.

As it will be discussed in detail in Chapter 6, the approximation (5.3) of the objective function allows one to draw lines of constant objective functional J values in the parameter space, and the delimited regions give confidence regions around the best parameter estimates for different confidence levels. In case a two-parameter problem is addressed, these lines form ellipses. As it is pointed out in Munack [179], the axes of the ellipses are given by the eigenvectors of the Fisher Matrix, and their lengths are proportional to the square root of the inverse of the corresponding eigenvalues. Hence, the ratio of the largest to the smallest (in absolute value) eigenvalue is a measure of the shape of the objective functional J close to the optimal parameter estimates.

It is important to note that many numerical optimization algorithms (needed to solve these non-linear parameter estimation problems) have difficulties in finding a global optimum in such valley-like functionals (for more details, see Chapter 6). The need to invert the Fisher Matrix in many of these algorithms is important in this respect [218]. Indeed, the above mentioned ratio of eigenvalues equals the Fisher Matrix's condition number which is a measure for the reliability by which the inversion can be made. Hence, if an appropriate experiment design could be found that alleviates this problem, increased estimation accuracy would result.

Petersen [193] studied this problem in more detail and concluded that these numerical problems can sometimes be solved by changing the units of the parameters to be estimated. Indeed, the eigenvalues of the Fisher Matrix are unit-dependent. In case numerical problems are to be expected during parameter estimation, rescaling of the parameter units can be sufficient to alleviate these problems.

5.2.2 Confidence Region of the Parameter Estimates

A rather important result of a practical identifiability study is the possibility to determine the parameter estimation error. It can be stated that reporting parameter estimates without the corresponding parameter variance is meaningless as no confidence can be given to the parameter estimates. If the covariance matrix V (5.5) is available, and the matrix Q_i was defined as the inverse of the measurement error covariance matrix for calculation of the Fisher Matrix, approximate standard errors for the parameters can be calculated as:

$$\sigma(\theta_i) = \sqrt{V_{ii}} \tag{5.6}$$

Confidence intervals for the parameters are then obtained as:

$$\theta \pm t_{\alpha;N-p}\sigma(\theta_i) \tag{5.7}$$

for a confidence level specified as $100(1 - \alpha)\%$ and t-values obtained from the Student-t distribution.

It should be mentioned though that these confidence intervals are too optimistic (too small) as they do not consider modelling errors. Indeed, only the measurement errors are included in the matrix Q_i.

In case only a single variable is measured, and fitted to, a more realistic estimate of the parameter confidence can be obtained by evaluating the residual mean square

$$s^2 = \frac{J_{opt}(\theta)}{N - p} \tag{5.8}$$

with p the number of parameters in the model and $J_{opt}(\theta)$ as defined in (5.2) and with Q_i a $p \times p$ identity matrix. Approximate standard errors for the parameters can then be calculated as:

$$\sigma(\theta_i) = s\sqrt{V_{ii}} \tag{5.9}$$

In this special case the standard errors are closer to the real ones since modelling errors are also included in $\sigma(\theta_i)$ since the J_{opt} contains both.

5.2.3 Sensitivity Functions

The output sensitivity $\partial y/\partial \theta$ equations are central to the evaluation of practical identifiability as they are a major component of the Fisher Information Matrix, and hence, also of the parameter estimation covariance matrix. If the sensitivity equations are proportional, the covariance matrix becomes singular and the model is not practically identifiable [218]. However, exceptions to this seem to exist. Petersen *et al.* [195] reported that certain parameters were practically identifiable despite the fact that the sensitivity functions are proportional. It was argued that the non-linearity of the estimation problem was the reason for this. Evidently, it was not possible to calculate the parameter estimation error covariance matrix since inversion of the (singular) Fisher Matrix was impossible. However, other (exploratory) methods introduced in Chapter 6 allow the confidence region to be obtained.

Overall, however, for many models used to describe biological phenomena, the sensitivity equations are nearly proportional, resulting in parameter estimates that are highly correlated. This is also visualised in the error functional J that looks like a valley, i.e. several combinations of parameters may describe the same data (almost) equally well.

Therefore, an easy way to study the practical identifiability of a model is to plot the sensitivity equations. In the literature numerous studies can be found in which this study is performed, especially for the Single Monod model considering measurements of both biomass and substrate concentrations [126], [127], [170], [195], [201], [218], [276].

To obtain a particular sensitivity function $\frac{\partial y_j}{\partial \theta_i}$ different approaches are possible. The most accurate is the analytical derivation of the sensitivity function. For somewhat more complex models it quickly becomes necessary to use symbolic manipulation software to minimise the errors that would certainly slip into a manual derivation.

Alternatively a numerical approximation is possible. It basically requires additional evaluations of the model for parameter values that are slightly different from the nominal ones. Typically one parameter θ_i will be perturbed at a time with a properly chosen perturbation value $\Delta \theta_i$. The sensitivity of output y_j to θ_i is then easily calculated as

$$\frac{\partial y_j}{\partial \theta_i} = \frac{y_j(\theta_i) - y_j(\theta_i + \Delta \theta_i)}{\Delta \theta_i} \tag{5.10}$$

For illustrative purposes, the sensitivity equations are deduced for the Single Monod and modified IWA ASM model (see Chapter 3, Section 3.2) with OUR_{ex} measurements as the only source of information for the identification of the biokinetic parameters. The sensitivity of OUR_{ex} with respect to μ_{max1} is:

$$\frac{\partial OUR_{ex}}{\partial \mu_{max1}} = \frac{\partial}{\partial \mu_{max1}} \left(-(1 - Y_1) \frac{dS_1}{dt} \right) = -(1 - Y_1) \frac{d}{dt} \left(\frac{\partial S_1}{\partial \mu_{max1}} \right) \tag{5.11}$$

in which the state sensitivity $\frac{\partial S_1}{\partial \mu_{max1}}$ is obtained by integration of the differential equation (with zero initial value):

$$\frac{d}{dt} \left(\frac{\partial S_1}{\partial \mu_{max1}} \right) = \frac{\partial}{\partial \mu_{max1}} \left(-\frac{\mu_{max1} X}{Y_1} \frac{S_1}{K_{s1} + S_1} \right)$$

$$= -\frac{X}{Y_1}\left(\frac{S_1}{K_{s1}+S_1} + \frac{\mu_{max1}K_{s1}\frac{\partial S_1}{\partial \mu_{max1}}}{(K_{s1}+S_1)^2}\right) \qquad (5.12)$$

where the substrate concentration S_1 is calculated by integration of the substrate dynamic model (note that X is assumed to be constant in the present model):

$$\frac{dS_1}{dt} = -\frac{\mu_{max1}X}{Y_1}\frac{S_1}{K_{s1}+S_1} \qquad (5.13)$$

Simultaneous solution of the differential equations (5.12) and (5.13) allows the output sensitivities (5.11) to be calculated. One can proceed similarly for the sensitivity of OUR_{ex} with respect to K_{s1}. The following relations are obtained:

$$\frac{\partial OUR_{ex}}{\partial K_{s1}} = -(1-Y_1)\frac{d}{dt}\left(\frac{\partial S_1}{\partial K_{s1}}\right) \qquad (5.14)$$

$$\frac{d}{dt}\left(\frac{\partial S_1}{\partial K_{s1}}\right) = -\frac{\mu_{max1}X}{Y_1}\left(\frac{K_{s1}\frac{\partial S_1}{\partial K_{s1}} - S_1}{(K_{s1}+S_1)^2}\right) \qquad (5.15)$$

The equations show that the sensitivities of the Single Monod model are dependent on the parameter values. This is a general characteristic of nonlinear models that has even been used to define nonlinearity [80]. Consequently the Fisher Information Matrix (5.4) depends on the parameter values and this feature has important implications for the optimal experiment design (see below).

An example of an OUR_{ex} profile with the corresponding sensitivity function evolutions is given in Figure 5.2 (left). One observes that the sensitivity functions for K_{s1} and μ_{max1} are nearly proportional, a well-known characteristic of the Monod model. Intuitively, the sensitivity functions express the dependence of the output or state variable on a change in the parameters. Hence, the sensitivity functions indicate conditions where the dependence is the largest and therefore, under which conditions the most information can be gathered on the parameters. In the example of Figure 5.2 (left) these conditions prevail when the substrate concentration has dropped to a level close to the affinity constant K_{s1}. From this one can deduce a first approach to increase the information content of an experiment: choose the sampling times when the parameters are influent, i.e. in the high sensitivity zone [276].

The output sensitivities for the IWA ASM model as modified by Sollfrank and Gujer [233] are deduced in a similar manner (in case OUR_{ex} is the only measured variable and the biokinetic parameters μ_{max1}, K_{s1}, k_r and k_s are to be inferred):

$$\frac{\partial OUR_{ex}}{\partial k_r} = (1-Y_1)\frac{\partial}{\partial k_r}\left(-\frac{dS_1}{dt} + k_r X_r + k_s X_s\right)$$

$$= (1-Y_1)\left(X_r + k_r\frac{\partial X_r}{\partial k_r} - \frac{d}{dt}\left(\frac{\partial S_1}{\partial k_r}\right)\right) \qquad (5.16)$$

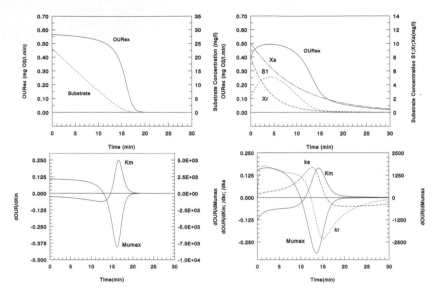

FIG. 5.2. Left: Output sensitivities (bottom) for a Single Monod-type OUR_{ex}-profile (top). Right: Output sensitivities (bottom) for an IWA ASM-type OUR_{ex}-profile (top).

$$\frac{\partial OUR_{ex}}{\partial k_s} = (1 - Y_1)\left(X_s + k_s \frac{\partial X_s}{\partial k_s} - \frac{d}{dt}\left(\frac{\partial S_1}{\partial k_s}\right)\right) \tag{5.17}$$

State sensitivities needed for the calculation of the output sensitivities (5.16)(5.17) are:

$$\frac{d}{dt}\left(\frac{\partial X_r}{\partial k_r}\right) = \frac{\partial}{\partial k_r}(-k_r X_r) = -\left(X_r + k_r \frac{\partial X_r}{\partial k_r}\right) \tag{5.18}$$

$$\frac{d}{dt}\left(\frac{\partial X_s}{\partial k_s}\right) = -\left(X_s + k_s \frac{\partial X_s}{\partial k_s}\right) \tag{5.19}$$

The output and state sensitivities for μ_{max1} and K_{s1} are identical to (5.11)(5.12) and (5.14)(5.15) respectively.

In Figure 5.2(right) the practical identifiability of the modified ASM model is studied by checking the output sensitivities for a short term batch experiment. No clear proportionality between sensitivity functions is observed. Stronger evidence can be obtained, however, by calculation of the rank of the Fisher Information Matrix. If no linear dependency exists, it should be full rank. This is indeed the case for this example. The condition number of the Fisher Matrix, or equivalently, the ratio of the largest to the smallest eigenvalue, indicates whether the sensitivities are nearly linearly dependent: the higher the condition number, the lower the practical identifiability.

So far, the initial conditions of the model variables were not included in the practical identifiability study, though they may be a highly desired outcome of parameter estimation. For the Single Monod model for instance, one can write:

$$\frac{\partial OUR_{ex}}{\partial S_1(0)} = -(1 - Y_1)\frac{d}{dt}\left(\frac{\partial S_1}{\partial S_1(0)}\right)$$

$$= (1 - Y_1)\frac{\mu_{max1}X}{Y_1}\frac{\partial}{\partial S_1(0)}\left(\frac{S_1}{K_{s1} + S_1}\right) \qquad (5.20)$$

To solve this, one must introduce the initial condition, using the relationship:

$$S_1(t) = S_1(0) - \frac{\int_0^t OUR_{ex}(\tau)d\tau}{1 - Y_1} \qquad (5.21)$$

yielding:

$$\frac{\partial OUR_{ex}}{\partial S_1(0)}$$

$$= \frac{(1 - Y_1)\mu_{max1}X}{Y_1}\frac{\partial}{\partial S_1(0)}\left(\frac{(1 - Y_1)S_1(0) - \int_0^t OUR_{ex}(\tau)d\tau}{(1 - Y_1)(K_{s1} + S_1(0)) - \int_0^t OUR_{ex}(\tau)d\tau}\right)$$

and the final equation:

$$\frac{\partial OUR_{ex}}{\partial S_1(0)}$$

$$= \frac{(1 - Y_1)^2\mu_{max1}XK_{s1}}{Y_1}\left(\frac{(1 - Y_1) - \int_0^t \frac{\partial OUR_{ex}(\tau)}{\partial S_1(0)}d\tau}{\left((1 - Y_1)(K_{s1} + S_1(0)) - \int_0^t OUR_{ex}(\tau)d\tau\right)^2}\right)$$

5.3 Optimal Experiment Design for Parameter Estimation (OED/PE)

We have already introduced the basic concepts of optimal experiment design (OED) in Chapter 1. Let us now concentrate on OED applied to parameter estimation. The quality of a set of parameter estimates can be assessed in different ways, e.g. in the way they allow a model to make good predictions of process behaviour. However, in most cases, one specifies parameter estimation quality by providing information on the parameter estimation errors (confidence intervals as given in (5.7)) or more generally, by providing the covariance matrix (5.5) altogether. Clearly the quality of parameter estimation is directly related to the practical identifiability of parameters.

If the objective of an experiment design exercise is to improve parameter estimation, it is evident that this covariance matrix (5.5) or elements thereof, or its inverse, the Fisher Information Matrix (5.4), are a central component.

Although we will extensively discuss the methods built around these matrices, other methods to deal with practical identifiability problems may be more appropriate. Indeed, sometimes the route of improving the experimental data is not followed to improve identifiability. For instance, it is sometimes possible to transform the model into an equivalent form that is numerically more tractable leading to more reliable estimation [208]. Alternatively, one may take a more drastic step and leave the initial model structure for a reduced order model that is less "data-hungry" and with improved practical identifiability [134]. It has also been proposed to use additional a priori information – such as a known maximum growth rate – to impose parameter bounds and in this way improve the identifiability of the other parameters [179].

One may also try to circumvent a practical identifiability problem by changing the goal of the modelling exercise. Indeed, sometimes it may suffice only to give a reasonable description ("curve fit") of the experimental data. Dedicated estimation algorithms, such as the set membership [143] or GLUE [28] methods have been developed that yield sets of parameters that allow description of process "behaviour". The goal of the modelling exercise is then no longer to find unique estimates but one is content with a set of good parameter values.

5.3.1 *Theoretical Background of OED/PE*

If one aims at designing experiments for optimal parameter estimation (OED/PE), it is illustrative to recall that the variable describing the reliability of a parameter estimate, the approximate standard error of the parameter, is given by:

$$\sigma(\theta_i) = s\sqrt{V_{ii}} \qquad (5.22)$$

in case a single output variable is measured for parameter estimation purposes.

One observes two terms that can be manipulated to increase the parameter estimation accuracy. The first one, the residual mean square s^2 is readily calculated from the sum of squared errors between N model predictions and experimental data J_{opt}:

$$s^2 = \frac{J_{opt}(\theta)}{N - p} \qquad (5.23)$$

To a certain extent this term can be decreased by increasing the number of experimental data N, e.g. by repeating the experiment. This is especially useful when p is not negligible compared to N. When N is already large, the increase of the denominator will be proportional to the increase of the objective functional J_{opt} as this is typically a sum of N squared errors.

Alternatively one may aim at reducing the parameter estimation error covariance matrix V. The methods to decrease V by optimal experiment design are the focus of attention in this section.

Different strategies have been developed to design experiments in such a way that the measurement data allow unique determination of the (combinations of) parameters that were shown to be structurally identifiable, i.e. produce "informative" experiments. The Fisher Information Matrix F or, equivalently, the covariance matrix V are the cornerstones of the optimal experiment design procedures because these matrices summarise the information content of an experiment or the precision of the parameter estimates. Depending on the requirements imposed by the application different scalar measures of these matrices are optimised [180]:

$$A - optimal\ design\ criterion: \qquad min[tr(F^{-1})] \qquad (5.24)$$

$$Modified\ A - optimal\ design\ criterion: max[tr(F)] \qquad (5.25)$$

$$D - optimal\ design\ criterion: \qquad max[det(F)] \qquad (5.26)$$

$$E - optimal\ design\ criterion: \qquad max[\lambda_{min}(F)] \qquad (5.27)$$

$$Modified\ E - optimal\ design\ criterion: min[\frac{\lambda_{max}(F)}{\lambda_{min}(F)}] \qquad (5.28)$$

in which $\lambda_{min}(F)$ and $\lambda_{max}(F)$ are the smallest and largest eigenvalue of the Fisher Information Matrix.

The following interpretation can be given to these optimal experiment design criteria [180]. The A- and D-optimal designs minimise the arithmetic and geometric mean of the identification errors respectively, while the E-criterion based experimental designs aim at minimising the largest error. Because in these criteria a maximisation of eigenvalues of the Fisher Information Matrix is pursued, they guarantee the maximisation of the distance from the singular (non-informative) case. The modified E criterion should be interpreted in the frame of the objective functional shape. The ratio of the largest to the smallest eigenvalue is an indication of this shape. The objective is to have eigenvalues as close as possible to each other: the shape is then circular. When $\lambda_{min}(F)$ is zero, this ratio is infinite, i.e. an infinite number of parameter combinations can be used to describe the experimental data and, hence, the experiment is non-informative. The Fisher Matrix is then singular, and, hence, the D- and E-criteria are zero while the A-criterion cannot be determined since inversion of F is impossible. This case also points to problems that can be encountered with the modified A-criterion: even if a non-informative and unidentifiable experiment is conducted, the modified A-criterion may still be maximised because one of the other eigenvalues has become large [103].

Petersen [193] pointed to a property of the Fisher Information Matrix that is very relevant to experiment design. Inherently the elements of the Fisher Information Matrix are dependent on the unit of the parameters. For instance, the unit of the diagonal elements of the Fisher Information Matrix is the square of the unit of the parameter this matrix element corresponds to. Consequently, the eigenvalues of the Fisher Matrix are unit dependent and can therefore be manipulated by rescaling the units. Henceforth, an experiment that is optimal for one particular set of units, may not be optimal for a rescaled parameter estimation problem. This is true for all experimental design criteria, except for the D-criterion: although the absolute value of this criterion is different for different parameter units, the optimal experiment remains the same. Hence, only for the D-criterion is the optimal experiment scale-invariant. To make her point particularly clear, Petersen [193] proved that the best attainable value of the Modified E criterion (Modified E=1) could be obtained simply by adequate rescaling of the parameters. This result has quite some implications for the experiment design methodology developed around such scale dependent criteria.

On the other hand, as mentioned before, the unit dependency of the Fisher Information Matrix can be used to the advantage of parameter estimation in case numerical problems occur with its inversion. Indeed, simple rescaling of the parameter units can change the condition number (which is equivalent to the Modified E criterion) and therefore the reliability with which inversion of the matrix can be done. This is particularly relevant for quite a number of numerical optimization algorithms used for parameter estimation (see Chapter 6).

Finally, it should be mentioned that other design criteria can be proposed, e.g. reducing the estimation error of a particular parameter can be obtained by designing experiments with this particular variance component as design criterion.

5.4   Examples of OED/PE

Below are a few studies which review the design of experiments allowed to collect more informative data. Vialas *et al.* [276] proposed to sample more frequently in defined periods of the experiment whereas Holmberg [126] showed that the practical identifiability of Monod parameters from batch experiments depends significantly on the initial substrate concentration. This author further stated that the optimal initial substrate concentration depends on the noise level and the sampling instants. It is also obvious from her results that the experiment design is dependent on the parameter values, which, in view of the changing nature of the process studied in the wastewater treatment case study, implies that the experiment design is time-varying. Munack [179] proposed different modifications to batch experiments and it was shown that important improvements in parameter confidences can be achieved by optimal experiment design techniques.

Vanrolleghem *et al.* [258] reported enhancements in estimation accuracy for the two parameters of a Monod type biodegradation model. By adding an addition pulse of substrate to a batch experiment the confidence interval of the parameters could be reduced by 25%. This variance reduction was well-balanced over both parameters μ_{max} and K_S which is a typical result of a D-criterion based OED. This example is dealt with in great detail in the subsequent section to illustrate the different aspects of optimal experiment design for parameter estimation.

Petersen [193] performed a similar study for a full-scale application in which a wastewater biodegradation model was to be identified. The model contained two submodels, one for nitrogen oxidation (nitrification) and one for carbon oxidation.

Here too an improvement in parameter estimation accuracy was obtained by complementing the wastewater sample with a designed amount of ammonia. For the carbon oxidation submodel parameters the confidence intervals were 20% smaller whereas they were 50% smaller for the nitrification kinetic parameters.

Baetens et al. [10] estimated six parameters in a biological phosphorus removal process. Optimal experiment designs were evaluated with a number of degrees of freedom. The D-criterion could be improved with a factor of at least 4, but this was completely attributed to a change in the reactor's acetate concentration, as no effect of changing the phosphate concentration could be observed in the system under study. This result corresponds with an average improvement of the parameter estimation confidence intervals with a factor 2. The evolution of the E-criterion with changing experimental conditions indicated that the longest axis of the confidence ellipsoid could be reduced with a factor 2 since the E-criterion could be increased with a factor 4 by increasing the dosage of acetate.

In relation to the degrees of freedom for experiment design, Petersen [193] used another approach. She evaluated the differences in parameter estimation accuracy when different sets of measured variables were used to identify a nitrification model. The following possibilities were evaluated:

- Single dissolved oxygen measurement S_O in an aerated batch reactor

- Two dissolved oxygen measurements S_O at in- and outlets of a closed respiration chamber

- Respiration rate OUR_{ex} calculated from two oxygen measurements

- Proton production rate H_p obtained from the pH controller

- Two dissolved oxygen measurements S_O and the proton production rate H_p.

Despite the fact that for the design options with oxygen concentration measurements the mass transfer parameters K_La and $S_{O,sat}$ need to be estimated in addition to the three nitrification parameters μ_{max}, K_{NH} and $S_{NH}(0)$, the (nitrification parameter) estimates one is interested in can be estimated much more accurately when dissolved oxygen measurements are used, i.e. the confidence regions are no less than 10 times smaller (or the variance is 100 times smaller). The larger noise on the respiration rate measurements is a partial explanation of this difference. This result points out that it is important to carefully reflect on the measured variable one is using for parameter estimation.

Munack [180] also evaluated the effect of different measurement set-ups that could lead to significantly different information contents of the data sets. He looked not only at the type of measurements, but also at the number of them and their location within an aerated column reactor.

Versyck et al. [273] reported on a quite remarkable result in terms of optimal experiment design. For the more accurate estimation of the parameters in a Haldane type microbial growth model, a Modified E criterion based design of the substrate feed profile was conducted. Starting from a feed profile that was optimal in the sense of process performance, the authors were able to reach the truly optimal value of the Modified E criterion, namely the value is 1.

Versyck and Van Impe [274] also reported on the design of a temperature profile to identify a temperature dependency model of microbial growth kinetics. Again the unicity of the Modified E criterion could be reached. In the work conducted by Versyck and co-workers attention was drawn to the fact that more than one experiment design leads to this truly optimal value. It is therefore possible to also take into account additional criteria to select among the different possible designs, for instance practical feasibility or model validity. The latter aspect was also pointed out by Baltes *et al.* [12].

5.5   Application: Real-time OED/PE in a Respirometer

5.5.1   *Degrees of Freedom and Constraints for OED/PE*

Designing identification experiments requires several choices, e.g. what outputs should be measured at what time instants and at what frequency, and what inputs to manipulate and how. In the case study the output (OUR_{ex}) and sampling frequency (6 min^{-1}) are no longer available to the experimenter since they are fixed by the respirometer hardware used in the study. The only degree of freedom left is the design of the input. Optimal experiment design therefore reduces, in this case study, to find the input functions $u(t)$ that lead to the most informative experiments.

First, if one only considers batch experiments, the only possibility to change the information content of the experiment is the initial condition as imposed by the pulse of wastewater sample injected at the start of the experiment [126]. However, considering the assumption that biomass is constant in the course of the experiment and the on-line character of the sensor (the maximum experimentation time is 40 minutes), a constraint is placed on the maximum initial substrate concentrations.

As Munack [179] pointed out, fedbatch experiments are superior to batch experiments with respect to the practical identifiability of model parameters. In the respirometer under study, this degree of freedom is available as well since the wastewater pumps can be activated at any time, providing additional wastewater pulses to the bioreactor. A constraint is imposed, however, on the amount of sample injected per pulse. Real-time constraints must again be taken into account for the experiment design.

In the sequel, four examples of OED/PE will be developed theoretically:

1. Optimal initial substrate
2. Optimal additional pulse with fixed initial substrate
3. Optimal additional pulse and initial substrate
4. Optimal design with multiple additional pulses

In addition, the first and second design option will be illustrated with real-life data.

5.5.2   OED/PE for the Single Monod Model

Introduction.   In Section 4.6, the structural identifiability of four kinetic models (Exponential, Single Monod, Double Monod and modified IWA ASM1) was studied, based on OUR_{ex} measurements. Here, we concentrate on the practical identifiability and the optimal experiment design for parameter estimation (OED/PE) of one of these models (the Single Monod model):

$$\frac{dS_1}{dt} = -\frac{\mu_{max1} X}{Y_1} \frac{S_1}{K_{s1} + S_1} \tag{5.29}$$

$$OUR_{ex} = -(1 - Y_1)\frac{dS_1}{dt} \tag{5.30}$$

The choice of the Monod model is, at least partially, motivated by its very large use in biotechnological applications. More specifically in the context of this study, this choice means that it is assumed that the experimental data are characterised by Single Monod kinetics, either because the real-life data are always characterised by this type of kinetics, or because a preliminary model structure characterisation has been performed, leading to the selection of Monod kinetics.

It was shown in Section 5.6 that three combinations ($\frac{\mu_{max1} X(1-Y_1)}{Y_1}$, $(1-Y_1)S_1(0)$, $(1 - Y_1)K_{s1}$) of the five original parameters (μ_{max1}, X, Y_1, $S_1(0)$, K_{s1}) are structurally identifiable. In order to have a presentation as pedagogical as possible (via e.g. the use of 3-D plots of the confidence regions), it is assumed here that the initial substrate concentration $S_1(0)$, the yield coefficient Y_1, and the biomass concentration X are known a priori (e.g. via some separate experiments). This leaves two parameters (μ_{max1}, K_{s1}) to be estimated.

The start-up of a batch experiment by pulse injection of wastewater is included in the model via the initial conditions $S_1(0)$ (which will be the degree of freedom in the optimal experiment design). An additional term in the mass balance is required to describe the fedbatch experiments that are also treated in this section. In order to prevent numerical problems, a pulse injection of wastewater in the course of an experiment is described by a Gauss-like function:

$$S_{puls}e^{-\frac{(t-t_{puls})^2}{\sigma}} \tag{5.31}$$

in which t_{puls} is the time instant at which the pulse is given, σ is the width of the pulse and $S_{puls}\sqrt{\pi\sigma}$ is the total amount of substrate injected.

Reference Data Set.   As a reference data set for the theoretical examples of optimal experiment design, a single Monod model simulated OUR_{ex} profile was calculated with the following parameter values:

$$X(0) = 4000 \; mg/l, \; S_1(0) = 23 \; mg/l \tag{5.32}$$

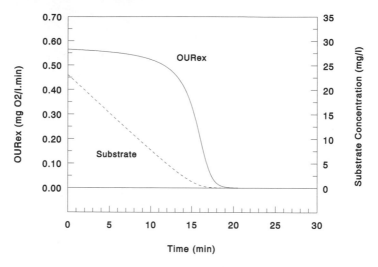

FIG. 5.3. Reference respirogram.

$$\mu_{max1} = 2.62 \ 10^{-4} \ min^{-1}, \ K_{s1} = 1 \ mg/l, \ Y_1 = 0.64 \qquad (5.33)$$

The resulting respirogram and corresponding substrate concentration trajectory are illustrated in Figure 5.3. The Fisher Information Matrix and the values of the different OED/PE criteria are equal to:

$$F = \begin{pmatrix} 3.456 \ 10^8 & -8182.2 \\ -8182.2 & 0.25702 \end{pmatrix}, \ V = \begin{pmatrix} 1.175 \ 10^{-8} & 3.742 \ 10^{-4} \\ 3.742 \ 10^{-4} & 15.802 \end{pmatrix} \ (5.34)$$

$$tr(V) = 1.857 \ 10^{-7}, \ tr(F) = 8.882 \ 10^{7}, \ Det(F) = 2.186 \ 10^{7}, \quad (5.35)$$

$$\lambda_{min} = 6.328 \ 10^{-2}, \ \frac{\lambda_{max}}{\lambda_{min}} = 5.46 \ 10^9 \qquad (5.36)$$

Theoretical Example 1: Initial Substrate. We have looked for an optimal initial substrate concentration by using the different experiment design criteria introduced above. Figure 5.4 shows the different criterion values as a function of the initial substrate concentration.

For comparative purposes the optimal concentrations proposed by the other criteria are indicated in each graph. All criteria except for the modified E criterion tend to a batch experiment with almost 60 mg S_1/l as the initial concentration.

When considering the first four criteria, experiments are proposed with the highest possible information content with the aim of decreasing the variances of the estimates. This practically implies that the exogenous oxygen uptake rate is different from zero for the longest possible time. Hence, substrate is added initially in such an amount that it is not depleted until the allowed experimentation time (in this example 40 minutes). Confidences in the estimates of μ_{max1} and K_{s1} improve

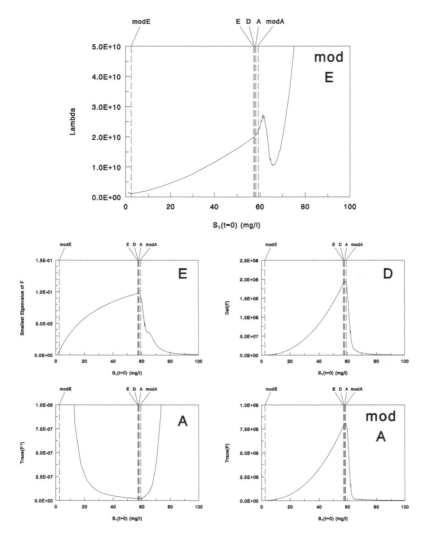

FIG. 5.4. Values of the OED/PE criteria with respect to the initial substrate concentration. Vertical lines indicate the optimal concentration for each criterion.

with a factor 2.4 and 1.25 respectively compared to the reference experiment. This indicates that experiment design with the initial substrate concentration as a degree of freedom is mostly beneficial to the estimation of the maximum growth rate.

The modified E criterion proposes an experiment with a very low substrate concentration of only 2.55 mg S_1/l. This can be interpreted as follows. In Figure 5.5 one can observe the flat valley for the parameter estimation problem of the reference OUR_{ex} profile. This flat valley is the main cause for the numerical

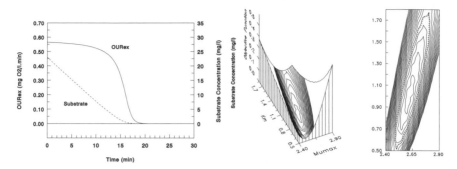

FIG. 5.5. 3D- (middle) and contour plot (right) of the objective function of the Monod
model parameters for the reference respirogram (left).

problems related to the parameter estimation of Monod-type models. To improve
the practical identifiability, the modified E criterion aims at OEDs where the objective functional's shape is as close as possible to a cone or funnel. The modified
E based experiment that is obtained starts with a substrate concentration which
is 10 times lower than the reference experiment. While the objective functional's
shape has improved (the eigenvalues ratio has decreased by 3.3), the variances of
the parameters indicate that this numerical advantage is at the expense of parameter estimation quality, i.e. the confidence regions have increased significantly,
for μ_{max1} by one order of magnitude and for K_{s1} with a factor 3. This "improvement" (with respect to the modified E criterion) has been achieved by lowering the
number of experimental data with a high sensitivity with respect to μ_{max1}.

Theoretical Example 2: One Additional Pulse. Let us now examine the effect of a
fedbatch experiment on the practical parameter identifiability. If one considers that
the pulse characteristics are fixed by the hardware used, i.e. pulse volume of the
sample pump and mixing intensity in the reactor, the only degree of freedom to be
evaluated here is the time of pulse addition, t_{puls}. In order to illustrate the increased
flexibility more clearly, the initial substrate concentration is chosen identical to the
reference case of the previous example.

Figure 5.6 illustrates the effect on the error functional's shape of an additional
pulse of 8 mg/l given at the optimal time in a fedbatch experiment, according to
the modified E criterion ($t_{puls} = 18.2$ min). The OUR_{ex} and substrate profiles of
this experiment are given as well. One observes that, although still very 'valley-
like', the properties of the error functional have been significantly improved (the
eigenvalue ratio has decreased by a factor 3.5). A closer look at the covariance
matrices V for the reference (no pulse) and optimal experiments for $S_1(t = 0) = 23$
mg/l:

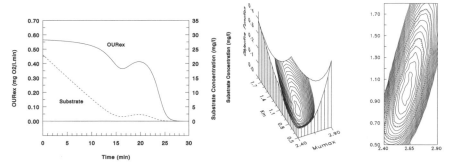

FIG. 5.6. 3D- (middle) and contour plot (right) of the objective function of the Monod
model parameters for the respirogram with optimal substrate pulse (of 8 mg/l) at t =
18.2 min (left).

$$V_{Reference} = \begin{pmatrix} 1.175 \ 10^{-8} & 3.742 \ 10^{-4} \\ 3.742 \ 10^{-4} & 15.802 \end{pmatrix} \tag{5.37}$$

$$V_{OED/PE} = \begin{pmatrix} 9.623 \ 10^{-9} & 1.752 \ 10^{-4} \\ 1.752 \ 10^{-4} & 6.735 \end{pmatrix} \tag{5.38}$$

shows that all the variances and covariances have improved, but in contrast with
the previous example, the OED with an additional pulse is especially attractive for
a more accurate estimation of the affinity constant. Indeed, while the confidence
interval for the μ_{max1} only decreased by 10%, the K_{s1} accuracy increased by more
than 50%. In addition, the results show that the covariance between both biokinetic
parameters is reduced almost to the same extent.

The study was also extended to evaluate the other OED/PE criteria. In this
overall study, however, a pulse amount of 2 mg/l was taken. In Figure 5.7 the
optimised OUR_{ex} and substrate profiles are summarised. As before the differences
among the design criteria are considerable.

1. To optimise the A- and modified A criteria, the experimental conditions
   where maximal substrate degradation takes place are prolonged, a feature
   which was also noticed in the previous example. This can probably be explained by the fact that these criteria try to minimise the mean variance of
   the parameters. It may well be that this can be achieved by improving only
   one of the variances, and potentially one could have designs in which one
   variance improves to such an extent that the variance deterioration of another
   variance is compensated. With the sludge properties (5.32)(5.33), the most
   important improvement seems to be possible for the μ_{max1} parameter and,
   consequently, experimental conditions are proposed that take advantage of
   this.

2. For the D- and E-criteria, experiments are proposed in which a fresh amount
   of substrate is injected only after the exogenous respiration has dropped
   completely.

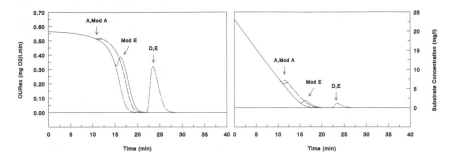

FIG. 5.7. OUR_{ex} (left) and substrate concentration (right) trajectories of fedbatch experiments with pulse additions (2 mg/l) at different injection times as proposed by different OED/PE criteria.

3. The modified E-criterion based OED results in a design which is in between both approaches.

With the D-, E- and modified E-criteria the substrate concentration is driven to remain for a longer period of time in the lower part of the Monod model. Consequently, additional information is obtained on the substrate range where the highest sensitivity with respect to the affinity constant is found.

From this observation it is clear why parameter accuracy has improved most for K_{s1} (see above). As a drawback to the D- and E-criteria it must be noted that the proposed experiments are significantly longer (approximately 30 %) than the other experiments, which should be considered in view of the real-time nature of the respirometer. Clearly, imposing a maximum experiment length will eliminate this problem but will result in suboptimal experiment designs that are a compromise between information content and experimentation time.

Theoretical Example 3: Additional Pulse + Initial Substrate. Let us now investigate whether the combination of the two degrees of freedom introduced above gives rise to an additional improvement in experimentation quality. This is clearly a two-dimensional optimisation problem: both the optimal initial substrate concentration and optimal time of pulse addition must be found within the time frame imposed by the real-time constraint.

To illustrate the results more clearly, the optimal $S_1(0)$ will first be sought for a pulse addition at 18.2 minutes, the optimal pulse addition time obtained for the case with $S_1(0) = 23$ mg/l (see above). At the end of this section some comments will then be given on the global 2-dimensional optimisation result.

Figure 5.8 gives the evolution of the different OED/PE criteria as function of the initial substrate concentration. The covariance matrices corresponding with the different optimal designs are equal to (note the large differences in the covariance values!):

$$V_{reference} = \begin{pmatrix} 1.175\ 10^{-8} & 3.742\ 10^{-4} \\ 3.742\ 10^{-4} & 15.802 \end{pmatrix}, \quad V_{Mod-E} = \begin{pmatrix} 8.257\ 10^{-8} & 1.050\ 10^{-3} \\ 1.050\ 10^{-3} & 16.72 \end{pmatrix}$$

$$V_{A,D,Mod-A} = \begin{pmatrix} 2.402\ 10^{-9} & 1.434\ 10^{-4} \\ 1.434\ 10^{-4} & 18.10 \end{pmatrix}, \quad V_E = \begin{pmatrix} 9.650\ 10^{-9} & 1.761\ 10^{-4} \\ 1.761\ 10^{-4} & 6.731 \end{pmatrix}$$

Again, the optimal experiment designs are significantly different. Figure 5.8 exhibits local extrema corresponding to conditions that are optimal for other criteria, especially for the E-optimal experiment designs. On one hand, this looks rather reassuring: if the wrong criterion is chosen, still suboptimal experiments are performed with respect to the other criteria. On the other hand, this does not seem to hold for the D- and modified A criteria where the E-based design gives rise to a local minimum in information quality. The modified E criterion has a rather different behaviour compared to the others: low initial substrate amounts (7 mg/l) are proposed to optimise this criterion. This is similar to the behaviour observed to an even higher extent in the case where only the substrate concentration was available for design. This deviation from the other criteria is probably due to the different underlying objective, i.e. to improve the numerical properties of the error functional shape.

Another interesting result concerns the substrate concentration of 23 mg/l for which the additional pulse was optimised (see above). The E-criterion keeps this value as the optimal one. This initial substrate concentration corresponds to a secondary (local) minimum for both the modified E and the A-criterion. However, 23 mg/l is considered as a poor experiment design value for both other criteria.

The presence of local extrema in the criterion profiles illustrates the problems that may arise in looking for the global optimal experiment design. While it has not been documented for the designs in which even more degrees of freedom are available, it can be expected that attaining the globally optimal design may be difficult. The 2-dimensional design problem that is treated next may give a first indication of the expected problems.

In order to get insight in the dependency of the OED-criteria on the design variables, a grid was evaluated of substrate concentrations ranging between 1 and 40 mg/l and a pulse addition at times between 1 and 40 minutes after the start of the experiment. A total of 40×40 combinations were simulated. The results are summarised in the 3D-plots of Figure 5.9. The substrate and time for pulse addition that are optimal according to a criterion are marked on these figures. In Table 5.1, the improvement of the criterion values is compared to the values for the reference respirogram (with $S_1(0) = 23$ mg/l) and the values for the experiments in which the time of pulse addition was the only design variable. The gains in criterion values are important, but depend on the type of considered criterion.

A more detailed analysis indicates that the initial substrate concentration is maximised within the limit of 40 mg/l as imposed by the grid choice, except for the modified E criterion which proposes lower substrate concentrations as before,

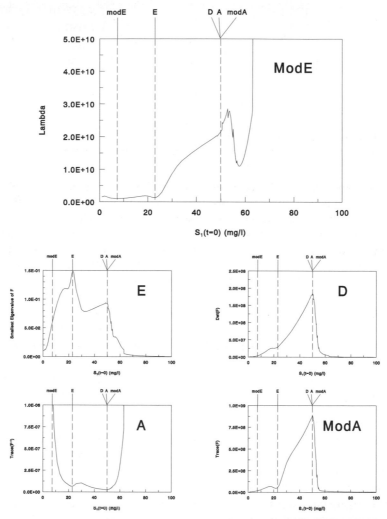

FIG. 5.8. Values of the OED/PE criteria with respect to the initial substrate concentration in a fedbatch experiment (t_{puls} = 18.2 min). Vertical lines indicate the optimal concentration for each criterion.

sacrificing μ_{max1} estimation accuracy to obtain a more cone-like error functional shape. All but the modified A criterion propose to inject the additional amount of substrate after 36 minutes. Clearly, this value is influenced by the 40 minute limit of the experiment, since to obtain the full information of the extra OUR_{ex} peak requires that the decreasing part of this peak finishes before data collection stops. This feature is visible in all 3D-plots where the criterion values decrease when t_{puls} exceeds 36 minutes.

Table 5.1 Optimal experiment design results with both the initial substrate concentration and time of pulse addition available for the design

	OED/PE		Gain in criterion values compared to	
Criterion	$S_1(0)$	t_{puls}	Reference	Optimised pulse
Modified E	4	36	7	2
E	40	36	2.05	1.65
D	40	36	5.8	1.65
A	40	36	7.75	2.2
Modified A	40	22	4.4	1.3

If one compares the 3D-plots (Figure 5.9) with the graphs in Figure 5.8 (that are in fact sections of the volume along the t_{puls} = 18.2 min line) the following observations can be made. A ripple on the surface (indicated with an arrow) can be found for all criteria. This corresponds to conditions in which the pulse addition is performed at the time the substrate initially present in the reactor is depleted. The experiments with "ripple conditions" result in OUR_{ex} profiles similar to the one presented in Figure 5.7, but with different lengths of the batch phase depending on the initial substrate concentration. For the modified E criterion surface, the valley is distinct but cannot be considered to be the minimum along any t_{puls} or $S_1(0)$ section. One can deduce it also from Figure 5.8: the minimum at 23 mg/l is only a secondary minimum. One finds for the ridge in the E criterion functional that the corresponding experiment designs are the optimum in the lower $S_1(0)$ range. At higher initial concentrations, however, the secondary (local) optimum becomes more pronounced and eventually takes over from the "ridge extremum".

Theoretical Example 4: Multiple Pulses.   A next evident optimisation step is to consider experiment designs with multiple pulses of substrate addition. The obvious question is then whether the quality of the data is consistently improving and to what extent the marginal increase decreases.

In Figure 5.10 the evolution of the modified E-criterion as a function of increasing experimental freedom is depicted. One observes the decreasing effect of adding another degree of freedom to the experiment design.

A remarkable result of this case study is that the design can be performed sequentially: first, the optimum time for the first addition is determined; then, the next pulse time is optimised with this 1-pulse experiment. The simulation results indicate that the alternative optimisation of both pulses in one step gives only a minor improvement of 1.1 % in criterion value. The same conclusion was deduced when the design of a 3-pulse experiment was performed in a single or three optimisation stages. The sequential design has the important advantage that the computational burden is considerably lower since only one-dimensional optimisation problems must be solved.

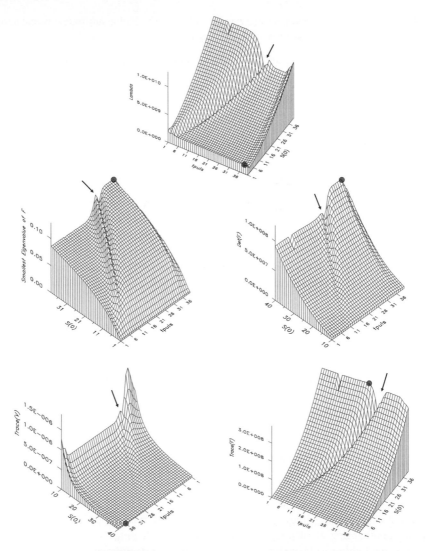

FIG. 5.9. Values of the OED/PE criteria with respect to the initial substrate concentration
and time of pulse addition. Optimal experimental conditions are indicated with a circle,
"ripple" conditions with an arrow.

This feature of the optimisation problem makes it even conceivable, to a certain extent, to adapt the experiments while they are still running, using the data
obtained so far to decide about the quality of the experiment and to possibly add
another pulse if necessary (see also [182]). Again one has to look for a compromise
between accuracy and real-time constraint.

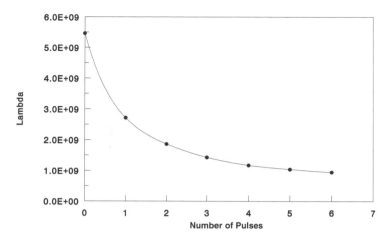

FIG. 5.10. Modified E criterion with respect to the increasing experimental flexibility.

A simulation of the optimal experiment obtained with six pulses is given in Figure 5.11. The numbers in the figure indicate the sequence in which the pulses are proposed by the OED method. One observes that the first two pulses are proposed to be injected during the decline phase of the OUR_{ex}. Adding two other degrees of freedom to the experiment design gives rise to pulses 3 and 4 that are initiated when the substrate is completely removed from the mixed liquor. If one allows two more pulses in the design (numbers 5 and 6), then these are scheduled such that the transients of pulse 3 and 4 are increased so as to enhance their information content.

Discussion of Theoretical Examples.   The results presented above can be interpreted and summarised as follows:

- The information quality of the experiments is highly dependent on the design and major improvements (see Table 5.1) can be achieved by changing initial substrate concentrations and extending the experiments to fedbatch operation (with injection of additional substrate at an optimal time in the course of the experiments).

- It was observed that the different OED/PE criteria mentioned above yield different OEDs. The constraint imposed by the desired real-time operation of the respirometer is shown to be necessary since all but the modified E criterion would lead to prohibitively long experiments.

- More than one pulse addition further improves practical identifiability but the benefits become marginal as the experiment complexity increases. One or two additional pulses are apparently enough in terms of practical identifiability improvement.

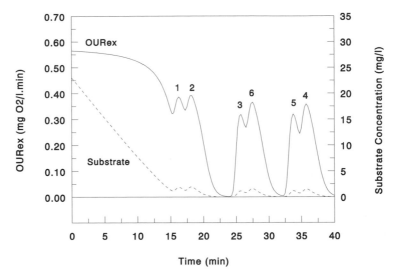

FIG. 5.11. OUR_{ex} and substrate concentration trajectories of fedbatch experiments with six pulse additions at injection times as proposed by the modified E criterion.

As a reasonable compromise of experimentation length and informative quality of the experimental data, it is proposed to perform respirometric experiments in which an additional pulse of substrate is injected at the time when the exogenous oxygen uptake rate is substantially decreasing, i.e. when the substrate has dropped to concentrations near to the affinity concentration. The amount of substrate at the beginning of the experiment is imposed by the allowable experimentation length.

5.5.3   *Experimental Validation of OED/PE for the Single Monod Model*

The above theoretical OED/PE results were checked in respirometric experiments performed with activated sludge that had slightly different properties than the sludge used in the theoretical OED/PE. First, a reference respirogram is commented and then the improvements obtained in parameter estimation accuracy are reported for improved initial substrate concentration and pulse additions.

Reference Experiment.   The reference experimental OUR_{ex} profile consists of a batch experiment with an initial acetate concentration of 20 mg COD/l. Figure 5.12 presents the collected experimental data and the fit of the Single Monod model. The OED/PE study was mainly directed to the improvement of the numerical properties of the optimisation problem via experiment designs based on the modified-E criterion. Therefore, the objective functional's shape was calculated for a grid of parameter combinations μ_{max1}, K_{s1} in the neighbourhood of the optimum. It should be emphasised that the surface and corresponding contour-plot depicted in Figure 5.12 is the result of systematic exploration of the error

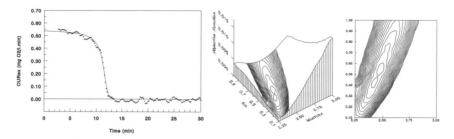

FIG. 5.12. 3D- (middle) and contour plot (right) of the objective function as a function of the Monod parameters for the validation reference respirogram (left).

functional in parameter space and is not a mere representation of the linearised objective functional around the optimum as it is often found in the literature ([161], see also Chapter 6).

This example of a flat valley in the parameter space may be the source of considerable problems to certain optimisation algorithms. The experience drawn from the cases studied so far tells that there exist adequate optimisation algorithms, such as the direction set method of Brent [42], which converge to the global minimum ($\mu_{max1} = 2.457 \ 10^{-4}$ /min; $K_{s1} = 0.456$ mg COD/l). Still, the valley is undesirable and the aim of the study was to see whether the proposed OED/PE methods would result in improved properties.

The Fisher Information Matrix corresponding with this experiment and the deduced values of the different OED/PE criteria are summarised here below:

$$F = \begin{pmatrix} 3.475 \ 10^8 & -12250.1 \\ -12250.1 & 0.57715 \end{pmatrix}, \ V = \begin{pmatrix} 1.148 \ 10^{-8} & 2.443 \ 10^{-4} \\ 2.443 \ 10^{-4} & 6.926 \end{pmatrix} \quad (5.39)$$

$$tr(F) = 2.005 \ 10^8, \ det(F) = 5.03 \ 10^7, \ tr(V) = 7.95 \ 10^{-8} \quad (5.40)$$

$$\lambda_{min}(F) = 0.145 \ 10^{-2}, \ \frac{\lambda_{max}}{\lambda_{min}}(F) = 2.39 \ 10^9 \quad (5.41)$$

Experimental Validation of Example 1 – Initial Substrate. As a first validation test, the effect of a change in initial concentration on the error functional shape and the estimation accuracy is assessed. For this purpose a batch experiment was conducted with half the initial concentration of the reference experiment (see Figure 5.13). The modified E-criterion value calculated from the experimental results was 2.42 times lower than the reference value, i.e. lower substrate concentrations give rise to batch experiments in which the error functional is more cone-like. However, it was already pointed out that this numerical improvement is at the expense of estimation accuracy. Indeed, if the parameter variances are calculated, it is found that the variances increase, especially for the μ_{max1} parameter (increased with a factor 3.82) and to a lesser extent also for the affinity constant K_{s1} (a fac-

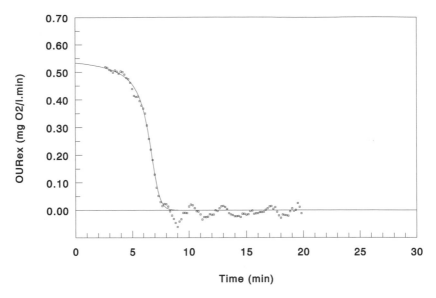

FIG. 5.13. Experimental respirogram in the initial substrate case.

tor 1.52). Let us note that OED based on the modified E criterion may sacrifice parameter estimation accuracy for improved numerical properties.

Experimental Validation of Example 2 – One Additional Pulse.   The effect of an additional pulse of substrate was validated with three experiments in which the substrate concentration in the bioreactor was increased at some time instant with 2 mg COD/l. Different injection times t_{puls} were tested in order to illustrate the effect of an optimal t_{puls}.

Suppose first that the data of the reference example are available and that an additional experiment has to be designed with the possibility of adding one more substrate pulse. The calculations result in curves of criterion values versus injection time, summarised in Figure 5.14. These graphs show the differences in optimum injection time for the different design criteria. The A, D and E criteria lead to an optimal substrate pulse after 14.6 minutes (after complete degradation of the initially present substrate). The modified A criterion based experiment consists of a prolonged batch-phase. And the modified E criterion OED results in a respirogram in which the oxygen uptake re-accelerates just before complete disappearance of the initial amount of substrate.

Three experiments were performed with injection times of 13, 14.1 and 14.6 minutes respectively. The resulting OUR_{ex} profiles are given in Figures 5.15, 5.16 and 5.17.

A first important observation is that the model extension for fedbatch operation is capable of simulating the behaviour remarkably well. The pulse is described very

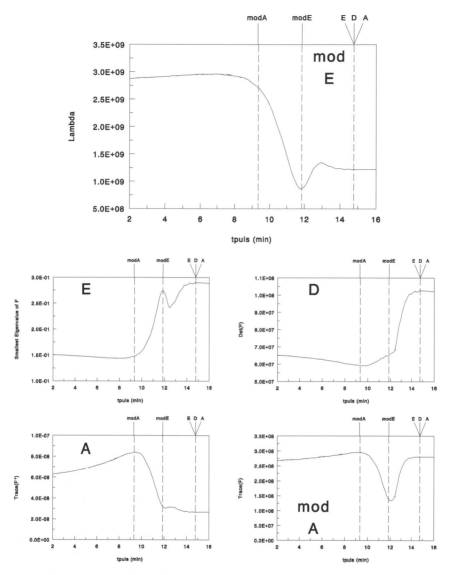

F<small>IG</small>. 5.14. Evolution of the different OED/PE criteria as a function of the pulse addition time (vertical lines = optimal time of addition for the different criteria).

well and microbial metabolism does not seem affected by the important transients imposed.

The following conclusions can be drawn when focusing on the effect of these fedbatch experiments on the error functional shape and parameter variances.

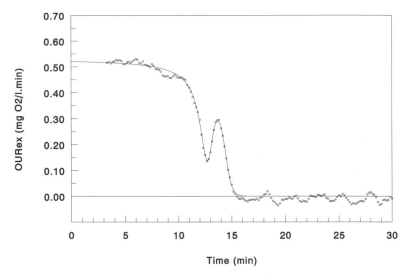

FIG. 5.15. Experimental respirogram obtained with a fedbatch experiment with additional pulse after 13 minutes.

Table 5.2 Dependence of the modified E criterion and of the parameter variances with respect to the pulse addition time

t_{puls}	Modified E	Var(μ_{max1})	Var(K_{s1})	Covariance
No pulse	1	1	1	1
13	0.676	0.4111	0.422	0.381
14.1	0.624	0.535	0.465	0.468
14.6	0.619	0.480	0.409	0.417

Although the expected values for the modified E criterion and the variances may change to a certain extent from the actually observed values due to changes in noise level, experimental error and biological changes, the trends set by the theoretical analysis are confirmed with these results. The predicted modified E criterion values for instance were approximately 20 % underestimated compared to the actual values. However, the data given in Table 5.2 clearly illustrate that a significant improvement in shape of the error functional is still obtained with fedbatch experiments. Moreover, as Figure 5.14 illustrates, the times of pulse addition that were evaluated were in the secondary minimum and more important effects could have been achieved if the substrate had been injected after 11.8 minutes.

A second conclusion concerns the variances. The experimental results confirm that significant improvements in parameter estimation accuracy can be obtained by this relatively small extension of the experiment. The variances have decreased with more than 50 % (Table 5.2). A similar effect on the parameter vari-

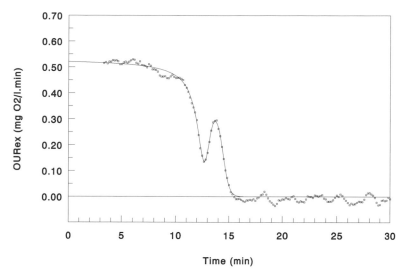

FIG. 5.16. Experimental respirogram obtained with a fedbatch experiment with additional
pulse after 14.6 minutes.

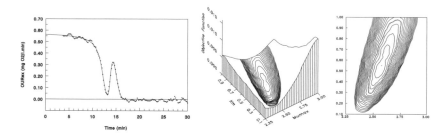

FIG. 5.17. 3D- (middle) and contour plot (right) of the objective function as a function of
the Monod parameters for the validation respirogram with additional pulse after 14.1
minutes (left).

ance could also be obtained by repeating the experiment twice, recall (5.22)(5.23).
However, this would double the experimentation time while the approach taken
here increases the experiment duration by only 3 minutes, i.e. 10 % of normal
operation.

5.6   Optimal Experiment Design for the Dual Problem of Structure Charac-
terisation and Parameter Estimation

As it was presented so far, the goal of experiment design was geared to *either* parameter estimation *or* structure characterisation. Frequently, however, investigators

want to perform experiments that will shed light on both questions simultaneously.

The simplest approach, which is often intuitively taken, is to first apply the discrimination criterion to select the best model and then use an OED/PE procedure for design of experiments for reliable parameter estimation [123]. However, one can imagine that during the structure characterisation experiments important information is already collected on parameter values, and that, conversely, during the experiments for parameter estimation additional data are gathered on the most appropriate model structure. Most importantly, it is likely that the same model accuracy can be reached with fewer experiments if a joint design criterion were used.

The more integrated approach presented by Hill $et\ al.$ [124] is based on a design criterion of the form

$$C = w_D D + w_E E \qquad (5.42)$$

where D is a measure for the discriminative power of an experiment and E reflects the parameter estimation accuracy. Such a joint design criterion should emphasise discrimination among rival models when there is a substantial uncertainty as to which model is the best. It should also emphasise parameter estimation accuracy when one particular model of the model set is definitely superior over the others. Hence, the values of the relative weights w_D and w_E should gradually change as experimentation proceeds to reflect the relative emphasis that is given to each of the objectives. A sequential design procedure where an $(n + 1)^{th}$ experiment is designed should therefore find the experimental conditions Ψ_{n+1} such that C is maximised. Hill $et\ al.$ [124] have proposed a particular form for this joint criterion which is written as follows for the general case of m rival models:

$$C = w_D \frac{D}{D_{max}} + (1 - w_D) \sum_{j=1}^{m} \Pi_{j,n} \frac{E_j}{E_{j,max}} \qquad (5.43)$$

where D_{max} and E_{max} are the maximum attainable values of D and E_j over the allowed Ψ-region. $\Pi_{j,n}$ is the probability of model j after n experiments, for which an iterative formula is given in [123]. The weighting w_D given to discrimination is suggested to be [124]:

$$w_D = \left(\frac{m}{m-1} \left(1 - \Pi_{max,n} \right) \right)^{\lambda} \qquad (5.44)$$

in which $\Pi_{max,n}$ is the largest probability of the $\Pi_{j,n}$ after n experiments. The tuning parameter λ gives a degree of freedom to the experimenter on the rate at which interest shifts from OED/SC to OED/PE.

Very few results have been reported on the performance of such criteria. The example given by Hill $et\ al.$ [124] indicates that no conflict arises as to the experimental conditions optimal for model selection or parameter estimation. In [123] on the other hand it is mentioned that design conditions that are best suited for discrimination may not be optimal for accurate parameter estimation. Clearly, some more experience with such joint criterion is necessary.

5.7 Conclusions

In this chapter, we have added an essential brick in the model building exercise: the practical identifiability analysis, which is mainly concerned with the informative quality of the experimental data used for parameter estimation. More precisely, the practical identifiability is the capability to produce reliable parameter estimates with the available data. The natural complement of this concept is the generation of sufficiently informative data for parameter estimation, which has been called "optimal experiment design" for parameter estimation.

To improve the accuracy (combinations) of parameters that can be estimated (from a theoretical point of view at least), it was indicated that two possibilities exist. One is to repeat experiments, while the other consists of improving the quality of the experiment(s) performed. While the former is a straightforward approach, the latter was paid a lot of attention and it was illustrated using the case study developed throughout the text that important improvements in parameter estimation accuracy can be expected.

A similar optimal experiment design methodology was introduced in Chapter 3 for the goal of model structure discrimination. It was indicated that the main problem with the mentioned techniques concerned the lengthy computations necessary in case a posteriori structure characterisation methods are used for the model selection. With the case study, for which specific a priori SC methods had been developed, it was shown that optimal experiment design does not necessarily have to violate the time constraints.

In the last section of the present chapter attention was paid to the application of a joint experiment design criterion for model discrimination and parameter estimation. Because its use may substantially reduce the necessary experiments in case of an overall model selection exercise, this approach warrants more attention than given so far.

6

Estimation of Model Parameters

6.1 Introduction

So far we have spent our time and the pages of the book to build models (Chapter 2) and to give tools for analysing these models, either for structure characterisation (Chapter 3), for structural (Chapter 4) and for practical identifiability (Chapter 5). Now we finally arrive at the following practical questions:

1. We have built and/or selected a model and we have a complete set of appropriate data (preferably after optimal experiment design (see Chapter 5): how do we estimate the parameters of this model ?

2. We have a model with estimated parameters, and we intend to use it for following the time evolution of (some of) the state variables (typically, process component concentrations) and/or parameters[8] (e.g. maximum specific growth rates) from available on-line measurements (typically, liquid and gaseous flow rates and some component concentrations): Which tools are we going to use?

The first topic (en bloc parameter estimation (identification or calibration)) will be the object of the present chapter. Chapter 7 will be concerned with algorithms

---

[8]Note that a parameter was defined in Section 1.2.1 as a model constituent whose value is constant for a particular system under study. However, here we will also consider that such a system may be characterised by time-varying behaviour which results in "time-varying parameters". Strict on the definition of a parameter, it means that the model is rebuilt every time the parameter changes.

to estimate (track) state variables and parameters on-line as new data become available.

Parameter estimation is defined as

"the determination of the "optimum" values of the parameters that arise in a mathematical description with the aid of experimental data, assuming that the structure of the process model, in other words the relationships between the variables and the parameters, are explicitly known."

This definition should be read with some caution as some overlap between parameter estimation and structure characterisation may occur. Indeed, for some particular parameter values (typically for zero values) a "branch" of a model structure may be deleted. Therefore, following [240], it is assumed that all parameters have non-zero (or other particular) parameter values.

The chapter is organised as follows. We shall first give two simple examples to introduce the two types of estimations mentioned above. In Section 6.3 we discuss some preliminary steps to be taken in a parameter estimation problem, such as the parameters that can be considered for estimation, linearisation and reparametrisation of the model, how to obtain initial guesses for the parameters to be estimated and the way to deal with constraints on parameter values. Section 6.4 focuses on the characteristics of measurement errors as they will determine which method is to be used for parameter estimation. The next section deals with the different objective functions that should be used for different parameter estimation problems. In Section 6.6, some linear and nonlinear parameter estimation algorithms are introduced in a rather qualitative manner. It is important to realise that most of the algorithms presented are basically optimisation algorithms that can also be useful in other tasks than parameter estimation, e.g. optimisation of an experiment design. Finally, in Section 6.7 methods are given that allow the assessment of the quality of the parameter estimation procedure and, hence, the resulting identified model. It also adresses the question of the quality of the parameter estimates themselves and the generation of confidence information.

6.2 Introductory Examples

As we shall be looking at a number of aspects of parameter estimation, two simple examples are given to help

- to differentiate between the two major types of estimation, i.e. en bloc versus recursive estimation and
- to get a feeling for the problems nonlinearities in the model cause in parameter estimation.

6.2.1 *Example 1: Estimating the Mean of a Data Set*

Probably the best known estimator is the determination of the mean of a random variable on the basis of N samples. With this example, Spriet and Vansteenkiste

[240] illustrate three basic computation schemes for estimation problems, the en bloc, the recursive and the iterative method.

En bloc method. This method is applied when all the measurements have been completed at the time the computation is initiated. Straightforward calculations allow us to obtain the en bloc estimate:

$$\bar{y}_N = \frac{1}{N} \sum_{i=1}^{N} y_i \qquad (6.1)$$

Recursive method. When one wants to be computationally efficient in case the current mean of a data set with increasing size is desired, the recursive approach is preferred. Indeed, this method allows the estimation of the mean of a growing data set with minor calculations each time a new data point is gathered. Computing recursively is basically characterised by two features:

1. The data set is expanding during computation;
2. Intermediate estimates of the quantity one desires are available and these values converge to the en bloc solution as the data set completes.

A recursive specification of the sample mean is the following:

$$\begin{aligned} \hat{a}_0 &= 0 \\ \hat{a}_k &= \hat{a}_{k-1} - \tfrac{1}{k}\left(\hat{a}_{k-1} - y_k\right) \end{aligned} \qquad (6.2)$$

which is calculated for k increasing from 0 to N.

Note that the algorithm shows the major features of recursivity. First, during computation the amount of data used, increases:
for $k = 1$: data base $\{y_1\}$

$$\vdots$$

for $k = 2$: data base $\{y_1, y_2\}$
for $k = N$: data base $\{y_1, y_2, \ldots, y_N\}$

Second, a sequence of numbers is obtained:

$$\hat{a}_1, \hat{a}_2, \hat{a}_3, \ldots, \hat{a}_N$$

However, the second feature asks for a proof that $\hat{a}_N$ converges to $\bar{y}_N$. We do not do this formally, but show how the recursive estimate can be deduced from the en bloc calculation procedure:

Write the en bloc solution for $k - 1$ and k samples

$$\hat{a}_{k-1} = \frac{1}{k-1} \sum_{i=1}^{k-1} y_i \qquad (6.3)$$

$$\hat{a}_k = \frac{1}{k} \sum_{i=1}^{k} y_i = \frac{1}{k} \left(\sum_{i=1}^{k-1} y_i + y_k \right) \tag{6.4}$$

and we readily obtain:

$$\hat{a}_k = \frac{k-1}{k} \hat{a}_{k-1} + \frac{1}{k} y_k = \hat{a}_{k-1} - \frac{1}{k} \left(\hat{a}_{k-1} - y_k \right) \tag{6.5}$$

The validity of the recursive procedure follows from induction. Although the en bloc (non-recursive and single iteration) method of estimating the mean is well-known, the recursive algorithm is relatively little known. And yet, the algorithm is significant in a number of ways: not only is it elegant and computationally attractive, it also exposes, in a most vivid manner, the physical nature of the estimate for increasing sample size, and so provides insight into a mechanism which is useful in many more general problems. Referring to the above equation, the previous estimate $\hat{a}_{k-1}$ is modified in proportion to the error between the observation of the random variable and the latest estimate of its mean value. Consequently, we will devote a complete chapter to this type of recursive algorithms, i.e. Chapter 7.

Iterative method. In the third computational technique, the whole data set is used sequentially to obtain a solution. Just as in the recursive method, a sequence of solutions is obtained that converges to the en bloc estimate:

$$\hat{a}_N^1, \hat{a}_N^2, \hat{a}_N^3, \dots, \hat{a}_N^m$$

In contrast to recursion where the sequence is due to the increasing data set, the sequence created in the iterative method originates differently. In the iterative case, all data are available before the computation starts, but the en bloc method cannot be applied directly and it has to be approximated by successive calculations on the whole data set. Only the estimate is updated at each iterative step, not the data set.

It is clear that the iterative method is "overkill" for the estimation of the mean. However, to illustrate the point, let us assume that the algorithm which will compute the mean is very crude in carrying out divisions. To calculate (6.1), it will divide 1 by N approximately and then multiply $\sum_{i=1}^{N} y_i$. En bloc computation will yield an approximate value: $\hat{a}_N = \frac{\alpha}{N} \sum_{i=1}^{N} y_i$. Iterative processing can now be used to increase the precision of the sample mean estimate. Consider the following function:

$$f(x) = xN - \sum_{i=1}^{N} y_i \tag{6.6}$$

By definition, for $x = \bar{y}_N$, $f(\bar{y}_N) = 0$. An iterative scheme for finding a zero of a function is the Newton-Raphson procedure:

$$x_{k+1} = x_k - \beta \frac{f(x_k)}{f'(x_k)} \quad \text{with } 0 < \beta < 2 \tag{6.7}$$

For the problem at hand,

$$
\begin{aligned}
\hat{a}_N^1 &= \tfrac{\alpha}{N} \sum_{i=1}^{N} y_i \\
\hat{a}_N^{k+1} &= \hat{a}_N^k - \tfrac{\alpha}{N} \left(\hat{a}_N^k \cdot N - \sum_{i=1}^{N} y_i \right)
\end{aligned}
\tag{6.8}
$$

where α/N represents the approximate division. The algorithm generates a sequence of approximations which converge to a precise value. Note that the complete data base is available and used all the time and that only the estimate is updated. For a proper use of iterative schemes a suitable stopping rule has to be provided, e.g.

$$
\bar{y}_N = a_N^m \text{ for } m \text{ such that } \left| a_N^m - a_N^{m-1} \right| < \varepsilon \text{ or } m = m_{max}
\tag{6.9}
$$

For the purpose of parameter estimation, all the computational procedures presented in this section will be useful. The recursive methods will be dealt with in Chapter 7, whereas the two other methods are covered below.

Estimating the mean value: interpretations. Here above, only the computational aspects of estimating the mean have been dealt with, without specifying why equation (6.1) was proposed. The choice follows from classic estimation theory, which will not be discussed in detail here. However, the basic background is provided here.

First, we follow a statistical interpretation. If the random variable is Gaussian and if the samples are independent observations, then it can be proven that (6.1) is an optimal estimator in the sense that it has no bias and that its variance is smaller than for any other computational procedure. Under the conditions mentioned, equation (6.1) can be found as the maximum likelihood estimate. The recursive estimate (6.2) embodies a Bayesian point of view in the sense that with a priori knowledge of $\hat{a}_{k-1}$, a new sample y_k brings us to an a posteriori estimate $\hat{a}_k$. The concepts of maximum likelihood and Bayesian statistics will be shortly introduced in Section 6.4.

Second, we can look at the problem as a minimisation problem. When we define an objective function:

$$
J(\bar{a}) = \sum_{i=1}^{N} (y_i - \bar{a})^2
\tag{6.10}
$$

it is easily understood that minimisation of J corresponds to the search for a value that is in the centre of the samples. Note that no statistical assumptions are necessary here: all samples contribute to the objective function. The gradient per sample, also termed the instantaneous gradient, is $2(y_i - \bar{a})$. Equation (6.2), the recursive estimator, can thus be seen as a "gradient algorithm" in which the estimate $\hat{a}_{k-1}$ is updated in a direction defined by the gradient of the instantaneous cost and with a magnitude of step size dictated by $1/k$, a weighting factor that is not constant but, in fact, is inversely proportional to the size of the data base at computation time. Thus, as the algorithm proceeds and confidence in the estimate increases, less and less notice is taken of the gradient measure, since it is more likely to arise from noise than from an error in the previous estimate of the mean value.

Third, reconsidering (6.10) we can obtain the en bloc solution of the mean by differentiating J with respect to $\bar{a}$, and setting it to zero:

$$\frac{\partial J(\bar{a})}{\partial \bar{a}} = 2 \sum_{i=1}^{N} (y_i - \bar{a}) = 2 \sum_{i=1}^{N} y_i - 2N\bar{a} = 0 \qquad (6.11)$$

yielding the minimum of the objective function, i.e.:

$$\bar{a} = \bar{y}_N = argmin J(\bar{a}) = \frac{1}{N} \sum_{i=1}^{N} y_i \qquad (6.12)$$

and, consequently, the en bloc solution (6.1).

6.2.2 Example 2: A Simple Nonlinear Parameter Estimation Problem

A nonlinear regression model is one in which the parameters appear nonlinearly, for example:

$$y = x^{\theta} \qquad (6.13)$$

where θ is the parameter to be estimated. The fact that the model is nonlinear in θ can easily be checked by taking the partial derivative of the right hand side with respect to θ and conclude that the result is still a function of θ.

Let us now estimate the parameter θ using the same objective function as in example 1 above, i.e. the least squares objective function

$$J(\theta) = \sum_{i=1}^{N} (y_i - \hat{y}_i)^2 = \sum_{i=1}^{N} (y_i - x_i^{\theta})^2 \qquad (6.14)$$

where $\hat{y}_i$ is the prediction of the model output at sampling time instant i. The minimum of this objective function can be obtained by differentiating (6.14) with respect to θ, setting the derivative to zero, as follows:

$$\frac{\partial J(\theta)}{\partial \theta} = -2 \sum_{i=1}^{N} (y_i - x_i^{\theta}) ln(x_i) x_i^{\theta} = 0 \qquad (6.15)$$

and attempting to solve for θ, the solution which is denoted $\hat{\theta}$. However, it appears impossible to obtain a nice explicit relationship as found in the en bloc method for the estimator of the mean, equation (6.1). Instead, the resulting rearranged equation:

$$\sum_{i=1}^{N} y_i ln(x_i) x_i^{\hat{\theta}} = \sum_{i=1}^{N} ln(x_i) x_i^{2\hat{\theta}} \tag{6.16}$$

can yield the estimate $\hat{\theta}$ only by an iterative procedure starting for some assumed initial guess of $\hat{\theta}$. At this point, it is important to realise that most parameter estimation problems we will tackle in wastewater treatment models will be nonlinear in the parameters and will therefore depend largely on iterative solution methods that may be computationally demanding. Parameter estimation is indeed a time consuming task.

6.3 Preliminary Steps in Parameter Estimation

Before we introduce the actual estimation of parameters as a solution of a minimisation problem, some preliminary activities that can or must be conducted at the start of a parameter estimation exercise are reviewed.

As mentioned in the introduction of this book, the quality of a model is assessed by a validation step. For this validation (or better, corroboration) model predictions are often compared to a data set that is not being used for identification. Some criteria that are useful to divide an overall data set in an estimation and validation subset are mentioned. Next, we will discuss the selection of parameters that will be estimated, leaving the others fixed to certain (assumed) values. Different methods that support this selection are reviewed. Since linearity of parameters makes their estimation considerably easier, much work has been devoted at finding linear parameters and trying to estimate them separately. Moreover, models have been rewritten (transformed, reparameterised) in such a way that originally nonlinear parameters become linear in the new model formulation. Some examples of this are given as well and commented upon. In most algorithms for parameter estimation initial guesses of the parameter values must be given to start the iterative search procedure. The proper choice of these initial guesses is dealt with as well because they may determine whether the parameter estimation is successful. On some parameters physical or user-defined (inequality) constraints apply. Although constrained optimisation algorithms are able to deal with such problems, they are not covered in this chapter. Rather, alternative ways to include constraints in parameter estimation are introduced.

6.3.1 *Selecting Data Subsets for Estimation and Validation*

One of the most important preparatory steps in a parameter estimation exercise is to split the available data in two subsets:

1. A first set of data for estimation of the parameters;
2. The remaining data for validation of the model.

The first set of data will be used to calculate the estimates of the parameters using an appropriate estimation algorithm. The second data set will be used to verify that the model with these parameters is able to describe or predict the dynamics of the process (other aims of the model may yield different definitions of the objectives of the validation step). The independence of these two data subsets is fundamental for reliable validation. We may indeed expect that a model identified on the basis of a particular data set is able to reproduce these data well, but it is much more important to know whether this model is able to reproduce quite different data that have not been used for calibration of the parameters.

The separation of the data must be done in such a way that the first data set (for calibration) is sufficiently informative and covers a sufficiently large spectrum of experimental conditions (using for instance optimal experiment design methodology). The remaining data, on the other hand, must still contain sufficient data to allow for a validation that is as credible as possible. Hence, one will prevent to split the data in subsets that are unbalanced in terms of number of data, although typically the first (calibration) data set is larger than the second (validation) data set.

6.3.2 Selection of Parameters to be Estimated

We shortly refer to the important distinctions that have to be made here between constants, parameters and variables (see Section 1.2.1). In this chapter we will only deal with parameter estimation, as variables will be calculated by the model or given as time series whereas constants are assumed to be given from prior knowledge as they apply, by definition, to any situation that can be modelled. Henceforth, only parameters must be estimated for the particular application at hand.

In the context of parameter estimation, it is important to realise that initial and boundary conditions of state variables and some inputs can also be formulated with the aid of parameters. Consequently, the set of parameters to be considered in the parameter estimation problem contains all of these and can be estimated simultaneously.

Selection of the parameters to be estimated is an important starting activity in a parameter estimation exercise. Below we introduce a number of criteria to select a certain subset of model parameters: structural and practical identifiability analysis, sensitivity analysis and numerical properties of the estimation algorithms.

Structural identifiability analysis. Structural identifiability analysis allows you to find out the possibly identifiable parameters or combinations thereof (provided the data are sufficiently rich in information, see Chapters 4 and 5). Hence, only the structurally identifiable subset will be contained in the parameter estimation problem. Note that for any identifiable parameter combination, e.g. in Section 4.6.2, $(1 - Y) * \mu * X / Y$, the following is allowed: Choose one of the three parameters to estimate (e.g. μ) and set values for the other two. When the parameter μ is now estimated, it means that its value found is conditioned by the choice of the values for the other two. Still, in case other values of the non-estimated parameters would be adopted, one does not need to repeat the estimation (of μ in the example) because one can always directly recalculate the corresponding value of the parameter that was estimated (in the example it was μ) by ensuring that the value of the parameter combination is maintained, i.e.:

$$(1 - \bar{Y})\frac{\bar{\mu}\bar{X}}{\bar{Y}} = (1 - \tilde{Y})\frac{\tilde{\mu}\tilde{X}}{\tilde{Y}} \tag{6.17}$$

Note that the above is something different than a model reparameterisation (see below) where the combination above would get a new "name" and associated meaning. This, for instance, has been done in [44] when the maximum growth rate parameter combination of the ASM1 model $\mu_A X_{BA}/Y_A$ was reparameterised into the maximum nitrification rate $r_{NH,max}$.

Practical identifiability analysis.   When practical identifiability problems are encountered, for instance, because the data are insufficiently informative to reliably estimate all parameters, a subselection of this parameter set can be made after an analysis of the parameter estimation error covariance matrix (or its inverse, the Fisher Information Matrix), see Chapter 5. Indeed, by eliminating the parameter that is causing the identifiability problem from the parameter set and giving it an assumed value, the estimation of the other parameters may be highly facilitated. Note that the estimates of these parameters will then be conditional on the assumed value of the non-identifiable parameters. Beck [23] proposes to adopt this scheme for any parameter estimation problem, i.e. first the estimation is performed for all parameters and an analysis is made of the parameter covariance matrix. Those parameters with the lowest confidence are then eliminated and a new parameter estimation exercise is launched. Despite its elegance, he also points to the contradiction of fixing the most uncertain parameters, since fixing its value gives the impression of having perfect knowledge on the parameter value. Detailed studies on the effect of this type of parameter space delimitation have been conducted in [268].

Sensitivity analysis.   Weijers and Vanrolleghem [282] and Reichert and Vanrolleghem [215] developed methods based on sensitivity analysis to preselect parameter subsets that ensure reliable estimation. However, these methods are computationally demanding due to the need to calculate the sensitivity functions (no attempt was made to do this analytically as in Chapter 5 in view of the complexity of the models to which the methods were applied). The applications made so far indicate that the two methods are leading to quite similar subsets of identifiable parameters. Furthermore, the fear existed that the identifiable parameter subsets would be highly dependent on the parameter values themselves (due to the nonlinear nature of the models this could be possible). If true this would mean that the selection of identifiable subsets should be done for each application of a model for which the parameter values were different. Luckily, for the cases studied by the authors the subsets of identifiable parameters turned out to be hardly dependent on the parameter values considered. Still, one should remain cautious in this matter.

Numerical properties of the estimation problem.   For numerical reasons it can be useful to restructure the overall parameter estimation problem in a stepwise procedure since a series of smaller parameter estimation problems are typically easier to solve than a single large one. For instance, one may first estimate the stoichiometric parameters before the kinetic parameter estimation is initiated. Another example may be that one first estimates the parameters related to the slow processes (decay, hydrolysis, nitrifier growth) and then the ones related to the fast processes (heterotrophic growth, aeration, ...). This type of stepwise parameter estimation typically induces a structuring of the data too. For instance, in case of the stoichiometric/kinetic parameter estimation split-up, typically long term averages of the data are taken for estimation of the stoichiometric parameters, whereas the more dynamic time series is used for estimation of the kinetic parameters.

In this stepwise approach it is important to realise the following. When the overall estimation of p parameters is started, we look for the best parameters in a p-dimensional parameter space. When we split the estimation problem into two sequential estimation problems, we will first estimate p_1 parameters and, given these p_1 parameter estimates, we will try to estimate the $p - p_1$ remaining parameters. However, it means that these $p - p_1$ other parameter estimates are conditional on the p_1 first estimates, possibly leading to biased estimation of the parameters. Combining the advantages of step-wise estimation with unbiased parameter estimation therefore leads to the following approach: first, the step-wise procedure is performed, but the estimates obtained are subsequently used to initiate an overall parameter estimation exercise where the search for the truly optimal estimates is done without any reduction in search space. Because the initial parameter estimates obtained from the stepwise procedure can be expected to be located relatively close to the overall best estimates, numerical problems are less likely to occur.

6.3.3   *Differentiating Linear and Nonlinear Parameters*

An important aspect of the models we try to identify is that many parameters appear in them in a nonlinear way. In Sections 1.2.2 and 2.6 we have introduced the definition and features of (non)linearity. Let us just recall that one of the easiest techniques to evaluate whether a function is nonlinear in a parameter is to take the partial derivative of the function with respect to that parameter and check whether the result is still a function of that parameter [218]. If the answer is affirmative the model is nonlinear in that parameter.

Below we will see that nonlinearity complicates the estimation problem considerably. It can therefore be advantageous to separate the linear from the nonlinear parameters and create subsets of these two types of parameters. Indeed, methods have been developed that can split the estimation problem in two, i.e. a sequence of a less complicated nonlinear and a fast/simple linear estimation procedure, rather than a single, more complicated, nonlinear parameter estimation. A nice example of this parameter set partitioning is the RAWN method for efficient neural network training [246].

In Section 2.6 another important aspect of nonlinear parameters was drawn to attention. We shortly repeat the example here. Consider the simple model (2.211-2.212):

$$\frac{dS}{dt} = DS_{in} - DS - \frac{1}{Y}\mu_{max}X \qquad (6.18)$$

$$\frac{dX}{dt} = -DX + \mu_{max}X \qquad (6.19)$$

This model is nonlinear in the parameters Y and μ_{max} (division by Y and "multiplication" μ_{max}/Y). However, it was shown that by transforming the parameters with

$$\theta_1 = \frac{1}{Y}\mu_{max}, \ \theta_2 = \mu_{max} \qquad (6.20)$$

the model can be rewritten in a form linear in the parameters θ_1 and θ_2. Such transformation is also termed reparameterisation [208], [209] and is a powerful method to reduce (or completely eliminate, as shown above) the estimation complications due to parameter nonlinearity. Its main drawback is that the physical meaning of parameters may be lost. Note, however, that in the example above θ_1 still has a well-known meaning in wastewater treatment modelling: it is the maximum specific substrate uptake rate.

Nonlinearity is not only a yes/no property of a model. Different levels of nonlinearity can be discerned and methods have been developed to determine the level of nonlinearity of, for instance, a parameter estimation problem. Pioneering work was done by Beale [18] and Bates and Watts [17] and a good introduction of nonlinearity assessment is given by Ratkowsky [208]. From a parameter estimation point of view it is important to minimise the level of nonlinearity so as to maximise the quality of a parameter estimation result. Below we will introduce several parameter estimation complications (bias on the parameter estimates and incorrect prediction of parameter confidence information) that are due to nonlinearity but which have not too large effects on the estimation results as long as the level of nonlinearity is small.

6.3.4   *Use and Misuse of Linearised Forms of Nonlinear Models*

It appears to be common practice to transform a model that is nonlinear in the parameters into a model with linear parameters. It is important to note that the "linearisation" discussed here is not based on a Taylor series approximation, but is really a transformation with no loss of accuracy. There are many models, though, that cannot be algebraically converted into linearised forms.

The best known example of such linearisation is probably the Lineweaver-Burk expression that transforms the Monod-type kinetic equation into a linear form:

$$\frac{1}{\mu} = \frac{K_S}{\mu_{max}} \cdot \frac{1}{S} + \frac{1}{\mu_{max}} \qquad (6.21)$$

The rationale for pursuing this type of transformation is that many people involved in dynamic modelling still lack good education of nonlinear parameter estimation but do have good notion of linear regression. This allows them to efficiently obtain estimates of the parameters (in fact, estimates of the parameters are calculated from new parameters appearing in the linearised version of the originally nonlinear model).

This ease of analysis is, unfortunately, accompanied by fundamental drawbacks. When data are transformed (e.g. in the Lineweaver-Burk expression we transform S into a new variable $1/S$ and μ into $1/\mu$), the measurement errors are transformed too. More particularly, although the actually measured variables may be characterised as iidN (*independent and identically distributed normally*), the transformed variables will typically not be. Moreover, error-free independent variables may no longer occur in the transformed equation, leading to an errors-in-variable problem (see the example below on the Eadie-Hofstee transformation). Tseng and Hsu [251] call this the destruction of the error structure.

As mentioned in the forthcoming sections on error characteristics and parameter estimation objectives, the type of errors determines which objective function should be applied. Consequently, if the wrong assumption is made on the error characteristics, biased parameter estimates can be expected. It was indeed found by many authors that different linearised forms of the same nonlinear model yield different estimates of the same parameters [218], [190].

Problems in addition to biased estimates are the difficulties one encounters when trying to get hold of the precision of the estimated parameters. Indeed, the parameters that are actually estimated are not the parameters one wants to estimate, but typically combinations thereof. For instance, in the Lineweaver-Burk approximation, the parameters that are estimated (and for which confidence intervals would become available, albeit probably wrong since the error structure is not as it should be for linear regression) are $\theta_1 = 1/\mu_{max}$ and $\theta_2 = K_S/\mu_{max}$. It is not trivial to calculate the confidence information on μ_{max} and K_S given confidence information on θ_1 and θ_2.

A final drawback of using linearised forms is that they typically require more data points to achieve the same accuracy of the parameter estimates. Also, the spacing of the data becomes important. For instance, when applying the Lineweaver-Burk approximation, many data points are located at low values of the variable

$(1/S)$ and only few are found at large values of $(1/S)$, leading to a very high sensitivity of the parameters (especially the slope of the linear regression) to the latter values.

Related to the above example, we want to point out that transformation into a linear form can allow proper, unbiased estimation of the parameters using linear regression when the proper weighting is given to the different data points. For instance, Cornish-Bowden [62] reminds us of the original approach of Lineweaver and Burk in 1934 where the data of μ were weighted with μ^4 in the weighted regression performed. It appears that this aspect is not passed on when the Lineweaver-Burk method is taught.

In some cases a transformation into a linear form may happen to transform the error structure into one that is closer to the so desired iidN that would allow OLS estimation. This will be a lucky coincidence though. However, it must be stressed that it may be wrong to generalise that transformation into a linear form is helpful or harmful. It may produce either effect. Which one is applicable depends on the error structure of the data and not on the model.

A final appropriate and even recommended use of the transformation into a linearised form is the use of it as an easy means of obtaining the initial guesses for parameters. These guesses are needed to start a nonlinear parameter estimation algorithm (see below).

6.3.5 *Reparameterisation of Models with Nonlinear Parameters*

Using a multitude of examples Ratkowsky [208] has pointed out to what extent the nonlinearity of parameters in a model can cause significant errors in the estimates, even if the measurement error structure is considered adequately and the proper objective function is selected (see above). These errors are caused by the fact that the estimators used in nonlinear regression can be badly biased, non-normally distributed and have variances greatly in excess of the minimum possible variance. However, as the number of data increases toward infinity, the bias diminishes, the distribution of the estimator becomes more normal and the excess variance decreases, thereby approaching more and more closely the condition for a linear model. Some nonlinear regression models approach the large-sample behaviour even in small samples and Ratkowsky [208] termed such models "close-to-linear" and advocated searching for such models for practical use. This search for new models can be done using what is called reparameterisation, i.e. the analytical reformulation of a model to obtain certain properties of the model.

The level of nonlinearity of a model[9], expressed for instance by the curvature measures of Bates and Watts [17], is one of the properties one wants to control.

---

[9]Note that the level of nonlinearity will also be a measure of the difference to expected between confidence regions of parameters calculated using a linear approximation versus the regions obtained with the method of Beale [18], see the section Evaluation of Parameter Estimation Quality.

The level of nonlinearity is composed of two terms, the intrinsic and the parameter-effects nonlinearity. It is the latter component one can influence by reparameterisation.

An example may illustrate the basic idea. Consider again the Monod kinetic model and its reparameterised version:

$$\mu = \frac{\mu_{max} S}{K_S + S} \quad \mu = \frac{S}{\theta_1 S + \theta_2} \tag{6.22}$$

It is noteworthy that the two parameters θ_1 and θ_2 are exactly the parameters that are estimated through linear regression in the Lineweaver-Burk linearisation method. It was pointed out by Ratkwosky (1986) that putting the parameters of this model in the denominator was the way to obtain a close-to-linear model. Note the difference between this approach and the Lineweaver-Burk method. Here the parameters are obtained using nonlinear parameter estimation methods, but with less nonlinearity induced bias, whereas in the Lineweaver-Burk method linear regression is used, but with quite major drawbacks in terms of bias and error properties. Ratkoswky (1986) also provides the ways in which the confidence information on the original parameters μ_{max} and K_S can be calculated given the standard errors on θ_1 and θ_2. Furthermore, he extends the approach of parameters-in-denominator reparameterisations for a class of kinetic expressions and also discusses the necessary weighting functions needed to deal with measurement errors that are proportional to the magnitude of the variable.

In general, no set of simple rules is available to find the proper reparameterisation of a highly nonlinear model. However, from histograms of the estimation results for each parameter separately, hints can be found on the proper reparameterisation. For instance, a histogram with a long right-hand tail (characteristic of a log-normal distribution) suggests replacement of the parameter in the model by the exponential of the parameter.

In summary, reparameterisation is useful to get closer to a linear estimation problem which has an undeniable beneficial effect on the statistical properties of the estimation problem but it is accompanied with a loss in "meaningfulness" of the new parameters. However, in this respect Ratkowsky [208] rightfully states that pH is a reparameterisation of the proton concentration that we got really used to.

6.3.6   *Initial Estimates of the Parameters*

Except for the (lucky) situation in which parameters can be readily calculated analytically from the objective function (e.g. in linear parameter estimation problems, see below), the minimisation algorithms used need initial guesses for the parameter values. The choice of a good initial guess can spell the difference between success and failure in locating the minimum or between rapid and slow convergence towards it [13]. Unfortunately, while we can prescribe algorithms for proceeding from the initial estimates, we must rely heavily on intuition and prior knowledge in selecting the initial guess.

Let us first of all warn that we should not exaggerate the importance of this initial guess either. It may be quite useful to just start with the first value that comes to mind and see where the algorithms bring us. A second and most obvious option is the use of prior knowledge, i.e. previous work done by yourself or others.

When a number of parameters are linear in the model, it is not necessary to select initial guesses for them too, since their optimal initial values can be readily calculated given the initial guesses of the nonlinear parameters [13]. For instance, suppose the nonlinear model to fit to a data set is $y(t) = \theta_1 exp(-\theta_2 t)$. If we have the initial guess $\theta_2 = 6$, the initial guess of θ_1 can readily be found by solving the linear ordinary least squares problem of minimising $J = \Sigma[y_i - \theta_1 exp(-6t_i)]^2$. This can be done in one calculation step (see below).

However, the most fruitful means of obtaining initial guesses is to substitute the original estimation problem by a simpler one. The solution to this simpler problem can then serve as good initial guesses for the original one. For instance, Ratkowsky [208] devotes a complete chapter of his book on nonlinear regression modelling to specify initial guess calculations for nine nonlinear models. Clearly the current chapter is not the place to introduce such dedicated methods, but should shed some light on some general principles for obtaining initial guesses. Below we introduce some approaches that have been applied successfully [13].

Reparameterisation in a linear form.   In this approach we try by means of a transformation of the variables, to come to a model formulation that is linear in the parameters (examples of this were given above). Although we know that the parameter estimation problem is no longer properly defined due to the loss of the error structure (see above), the parameter estimates obtained by, for instance, multiple linear regression (that does not need initial guesses), can be considered to be good initial guesses for the original parameter estimation problem.

Multistage estimation.   This type of estimation splits the set of parameters in groups of auxiliary parameters that are estimated in sequence on the basis of different subsets of data. The original parameters are then calculated by combining the separately estimated auxiliary parameters and their values are used as initial guesses for the original parameter estimation problem. Let us illustrate this approach with a simple example in which we try to determine the parameters describing the temperature dependency of a conversion process rate:

$$r = kSe^{-\frac{E}{T}} \tag{6.23}$$

where r is the rate, T is the temperature, S the substrate concentration, and k and E are the kinetic parameters to be estimated. Suppose we have measurements r_i

for q different temperatures T_i. We can now use the data taken at T_i to estimate an auxiliary parameter K_i that combines the parameters k and E:

$$r_i = K_i S \qquad (i = 1, 2, \cdots q) \qquad (6.24)$$

This estimation is, again, a simple linear regression problem. The estimated K_I can then be used (as "data") to estimate k and E in the linearised model (with its known statistical limitations)

$$log(K_i) = log(k) - ET_i \qquad (i = 1, 2, \cdots q) \qquad (6.25)$$

The values obtained for k and E can then be used as initial guesses for the original estimation problem.

Model simplification. In many instances it is possible to see the model under study as constructed from several submodels in which various subprocesses are considered. For instance, we may regard the Activated Sludge Model No. 1 as composed of different (interacting) submodels such as decay processes, nitrification, hydrolysis. We can try to approach the final model through a sequence of simpler ones, in which various effects are neglected and the corresponding parameters suppressed. After parameter estimation with the simpler model, these are used as initial guesses for a next parameter estimation of a more complex model. Basically this is the approach used in many activated sludge modelling protocols where, for instance, first the decay, hydrolysis and nitrification parameters are estimated before the overall model calibration is tackled with the parameter estimates found in these separate parameter estimations as initial guesses [193] [195].

6.3.7 *Inequality Constraints on the Parameters*

In many instances we know from prior knowledge that parameters are bound within a certain interval. For instance, by definition, all Monod half-saturation constants are non-negative, $K_S \geq 0$. Setting inequality constraints allows the domain of parameter values within which the estimate is to be found to be limited. The presence of inequality constraints often exerts a beneficial influence on the convergence of a minimisation algorithm since the algorithms are no longer "lost" in irrelevant regions of parameter space. It could therefore be argued [13] that imposing generous, though not unreasonable, bounds should be recommended in all parameter estimation problems.

The minimisation algorithms we will be discussing in the next section are the more traditional unconstrained minimisation algorithms. However, they are quite powerful and we would like to use them to solve constrained problems too. To that end we will have to modify our definition of the minimisation problem. Alternatively, we may turn to constrained optimisation methods, but we will not introduce these here as it would lead us too far (for the interested reader, reference is made to, for instance, [87].

Basically two approaches have been used to modify the minimisation problem. In the first one the objective function is extended by a term that penalises leaving the feasible parameter domain. In the other approach the interval of allowed parameters is projected onto the complete real axis via an appropriate transformation, ensuring that unconstrained minimisation in this new parameter domain guarantees constraining the original parameters.

Penalty functions. In this approach we modify the objective function in such a way that it remains almost unchanged in the interior of the feasible domain, but increases drastically when one of the parameters approaches the constraints. The approach of penalty functions is simple: for each inequality constraint $g_j(\theta) \geq 0$ we just add a term to the objective function to be minimised:

$$J_{constrained}(\theta) = J_{unconstrained}(\theta) + \sum_j J_{penalty,j}(\theta) \qquad (6.26)$$

This penalty function is typically nearly zero when the constraint function g is strongly positive, but increases sharply as the constraint function approaches zero from above. A typical penalty function proposed by Carroll [48] is:

$$J_{penalty,j}(\theta) = \frac{\alpha_j}{g_j(\theta)} \qquad (6.27)$$

where α_j is a small positive constant ensuring $J_{penalty,j}$ to be small compared to $J_{unconstrained}$ when in the feasible domain. To find the parameter estimates that minimise the original objective function $J_{unconstrained}$ and at the same time fulfill the constraints, an iteration over the value of α_j is necessary. Each time a minimum is found in $J_{constraint}$ with a certain α_j the estimates are used as initial guesses for a minimisation with an objective function with decreased value of α_j. This iteration is continued until the parameter estimates no longer changes significantly upon reduction of α_j.

The most pronounced penalty function is also called "barrier" function. It boils down to setting $J_{penalty,j} = \infty$ (or, practically, a very large number) whenever a parameter gets a value outside the feasible domain. The problem with this "barrier" approach is that a search algorithm that gets into the unfeasible domain is unable to find a direction back into the feasible domain as no gradient down can be found. To overcome this problem, it is suggested to linearly increase the "high cost function value" in function of the distance that a constrained parameter drifts away outside its boundaries. In other words, the high cost function plateau is replaced by a high cost function steep hill with a positive slope. The latter method was implemented by Van Vooren [269] and was found to perform without any problems.

The penalty function is easy to program and has been found to work well when the solution is known to be in the interior of the feasible domain. However, when the solution is likely to be on the boundary, other methods should be used (see for instance [13] or [87]).

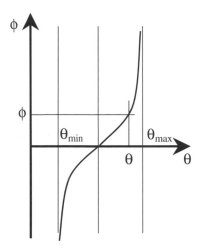

FIG. 6.1. Scaled tan-function used to transform the bounded interval $[\theta_{min}, \theta_{max}]$ of the parameter θ into the whole real axis for the parameter ϕ.

Transformation of parameters.   A constrained minimisation problem can be transformed into an equivalent unconstrained one by transforming the parameters bounded in a feasible domain to the real axis by transforming the parameters. For instance, to minimise $J(\theta)$ with θ to be positive is equivalent to minimising $J(\phi^2)$ with ϕ free to assume any value. Bard [13] also mentions the transformation of a minimisation problem where we wish $\beta + \theta \geq \alpha$ into a minimisation of $J([\alpha + \beta]/2 + [(\alpha - \beta)/2]sin[\phi])$ where ϕ can again assume any value.

Another example of a useful parameter transformation is the mapping by a scaled *tan* function (corrected from [213]): The parameter θ within the interval between θ_{min} and θ_{max} is mapped to the whole real axis by the transformation (Figure 6.1):

$$\phi = tan\left(\frac{\pi}{2} \frac{2\theta - \theta_{max} - \theta_{min}}{\theta_{max} - \theta_{min}}\right) \qquad (6.28)$$

The minimisation with the unconstrained algorithm can now be performed in the coordinates ϕ and the solution in the original coordinates is obtained by the inverse transformation:

$$\theta = \frac{1}{2}(\theta_{max} + \theta_{min}) + (\theta_{max} + \theta_{min})\frac{arctan(\phi)}{\pi} \qquad (6.29)$$

which maps the real axis to the interval between θ_{min} and θ_{max}.

Schuetze [226] complained about this approach by stating that the minimisation algorithms he used were apparently not able to find a direction of improvement when searching close to the boundaries.

6.4   Error Characteristics

Although parameter estimation appears to be rather straightforward at first sight, i.e. we want to obtain values of the parameters that give the "best" fit to a given set of data, the definition of what is best requires considerable attention. To define "best" it is essential to characterise the errors one is confronted with during para-meter estimation, i.e. measurement errors and model fitting residuals (the devia-tions between model predictions and data). After characterisation of these errors, a decision can be made on the objective function to be used in the parameter esti-mation exercise (Section 6.5) and only then the steps can be undertaken to prepare all inputs for the parameter estimation itself (Section 6.6).

6.4.1   *Measurement Errors and Residuals*

Assuming the selected model is a perfect representation of the system under study, the residuals between the model predictions $\hat{y}$ and the experimental data y are only due to measurement error. This error ε can be made explicitly visible in the standard model representation given in Section 1.2.1 (equations 1.7-8) as:

$$\frac{dx}{dt} = f(x, t, u, \theta), \qquad\qquad x(t = 0) = x_0$$
$$y = h(x, t, u, \theta) + \varepsilon(t) \qquad\qquad\qquad (6.30)$$

Hence, in parameter estimation we aim to find the parameters θ in such a way that the predicted residuals $\hat{\varepsilon} = y - \hat{y}$ possess properties that are similar to the prop-erties one may expect of plain measurement errors, characterised e.g. via repeat measurements.

It is worth noting that this objective can typically only be reached when the model is adequate since otherwise model error is confounded with measurement error. Consequently, we have seen the development of model selection and valida-tion criteria that focus on a thorough analysis of the residuals (see Chapter 3).

What are now typical properties of measurement errors? In many cases the errors are assumed to be "iidN", independent and identically distributed normally. In other words the errors are assumed to be random variables that are normally distributed with zero mean and constant variance (homoscedasticity) equal to one.

An example of a homoscedastic residual sequence is given in Figure 6.2(left). The distribution of the data in this illustration is not taken to be normal, but uni-form. In the same figure on the right side a series of heteroscedastic (relative) errors is shown, i.e. the variance of these particular data increases along the X-axis.

6.4.2   *Autocorrelated Residuals*

Another important characteristic of errors concerns their serial dependency or au-tocorrelation. Figure 6.2 shows two residual series that are serially independent. Figure 6.3 on the other hand shows an experimental residuals series where auto-correlation is clearly apparent. The autocorrelation with time lag τ quantifies the dependency of a variable at any time t_k and the variable at time t_k-τ:

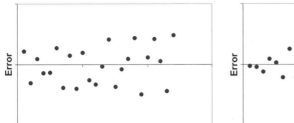 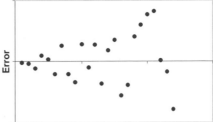

FIG. 6.2. Series of uniform homoscedastic (left) and heteroscedastic (right) errors.

$$r_\varepsilon(\tau) = \frac{1}{r_\varepsilon(0)} \sum_{k=\tau}^{N-\tau} \varepsilon(t_k - \tau) \cdot \varepsilon(t_k)$$

where $r_\varepsilon(0) = \sum_{k=\tau}^{N-\tau} \varepsilon^2(t_k)$

In the lower part of Figure 6.3 the autocorrelation for this data series is indeed found significant for 4 of the 20 first time lags (not considering time lag 0 because this always gives autocorrelation =1).

In Chapter 3 autocorrelation based tests for model selection were given. They allow to evaluate whether the residual sequence has the properties one may expect if the model is adequate, e.g. they should be independent if the measurement errors are. Another method introduced there concerned the runs test that also evaluates the dependency of residuals.

In this framework, an important feature to consider is that the N residuals obtained after fitting a model are in principle never uncorrelated (independent) because there are only N-p degrees of freedom left among them after estimation of p parameters [208]. Nevertheless, this violation of the assumptions is not really important and is therefore hardly ever focused upon.

In case autocorrelation is really significant, an approach that allows to still apply standard parameter estimation approaches is to correct for the correlated residual errors by including and identifying an autocorrelation model as part of the overall model [160]. Note that this approach complicates the modelling tasks because an autocorrelation model must be selected and its parameters estimated as well.

$$y(t_k) = h(x(t_k), t, \theta, u(t_k)) + z(t_k)$$
$$z(t_k) = \varepsilon(t_k) + \Phi(\varepsilon(t_{k-1}))$$

(6.31)

Another approach consists in eliminating correlation from the data set by subsampling, i.e. by dropping data points from the raw data set. Evidently, information is lost in this way, but the principal estimation inaccuracies associated with correlated errors are eliminated. As an example, the autocorrelated data series given in Figure 6.3 are subsampled (retaining 1 in 6 values) and yield the non-correlated series of Figure 6.4

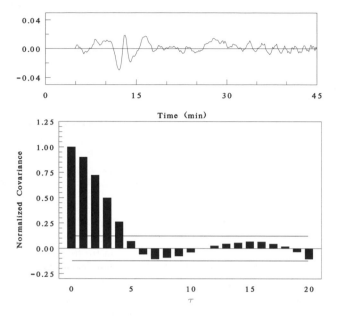

FIG. 6.3. Autocorrelated residual sequence (top) and autocorrelation for lags 0 to 20.

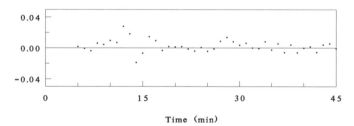

FIG. 6.4. Subsampled non-correlated data series of Figure 6.3.

6.4.3 *Estimation of the Measurement Error Covariance Matrix*

As will be more clear from the developments below, the determination of the measurement error covariance V is an important activity prior to many parameter estimation exercises. In the following example it will be described how the measurement error covariance matrix is estimated for a two variable data set collected in a combined respirometric-titrimetric set-up [193] [195]. The data series under study containing r_O (respiration rates) and Hp (protons produced) data are given in Figure 6.5.

Note that for neither of the two data series, repeat measurements are available for assessment of the measurement errors. Hence, another approach is required.

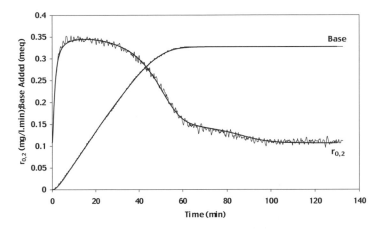

FIG. 6.5. Combined respirometric/titrimetric experimental data set [193] [195].

It is based on the selection of a period in which it can be assumed that the error remaining after fitting a (simple) model is pure measurement error and not containing modelling error.

For r_O data the measurement variance σ_O^2 is estimated based on a data series obtained during endogenous respiration (typically before or after the substrate addition). In the example (Figure 6.5) the measurement errors were estimated from t = 100-120 min where the model $r_O = r_{O,end} = constant$ was fitted. This selected data series is blown up in Figure 6.6 together with the average value and the residuals ($\varepsilon(t) = r_{O,end} - r_O(t)$). $\hat{\sigma}_O^2$ is estimated via equation (6.32) where N is the number of considered measurements and p is the number of adjusted parameters (here just 1).

$$\hat{\sigma}^2 = \frac{\displaystyle\sum_{i=1}^{N} \varepsilon_i^2}{N - p} \tag{6.32}$$

For the Hp data, however, an unrealistically optimistic picture would be obtained when the measurement variance were estimated at the point where substrate degradation is terminated. Indeed, the Hp profile in the example is a completely horizontal line with nearly no error due to the way the sensor operates [98]. As a consequence the σ_H^2 is estimated based on the data series from t = 15-35 minutes where, this time, the slope is assumed to be constant. Thus the data is not compared to an average value but to a model of the simple form $Hp = a*t + b$. The data series, model and residuals are illustrated in Figure 6.7. σ_H^2 is also estimated via equation (6.32), only now is p=2 (note that whether p is 1 or 2 does not really matter in these cases since $N >> p$).

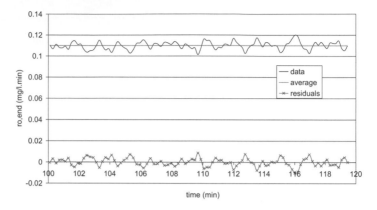

FIG. 6.6. Estimation of measurement variance on r_O data [193] [195].

To calculate the covariance between Hp and r_O data, σ^2_{OH}, the following calculation is made:

$$\hat{\sigma}^2_{OH} = \frac{\sum\limits_{i=1}^{N} \varepsilon_i^{Hp} \varepsilon_i^{ro}}{N - p} \tag{6.33}$$

The covariance among the residuals was estimated to be:

$$V = \begin{bmatrix} \sigma^2_O & \sigma^2_{OH} \\ \sigma^2_{OH} & \sigma^2_H \end{bmatrix} = \begin{bmatrix} 1.385 \cdot 10^{-5} & 2.541 \cdot 10^{-7} \\ 2.541 \cdot 10^{-7} & 1.617 \cdot 10^{-6} \end{bmatrix}$$

And the correlation matrix was calculated to be:

$$R = \begin{bmatrix} \dfrac{\sigma^2_O}{\sigma^2_O} & \dfrac{\sigma^2_{OH}}{\sigma_O \sigma_H} \\ \dfrac{\sigma^2_{OH}}{\sigma_O \sigma_H} & \dfrac{\sigma^2_H}{\sigma^2_H} \end{bmatrix} = \begin{bmatrix} 1 & 5.367 \cdot 10^{-2} \\ 5.367 \cdot 10^{-2} & 1 \end{bmatrix}$$

It was tested, via a test for correlation (t-test), that the correlations between the measurement errors of the two data sets were insignificant at test level 5%. Thus, the measurement error covariance matrix V can finally be determined to be:

$$V = \begin{bmatrix} \sigma^2_O & \sigma^2_{OH} \\ \sigma^2_{OH} & \sigma^2_H \end{bmatrix} = \begin{bmatrix} 1.385 \cdot 10^{-5} & 0 \\ 0 & 1.617 \cdot 10^{-6} \end{bmatrix}$$

The variance of r_O data is about 10 times higher than the variance on Hp data, which will be important to consider to obtain reliable parameter estimation results.

6.4.4   Errors-in-Variables Problems

One of the assumptions used in nearly all parameter estimation approaches is that the independent variable (typically time, but it can be another experimental setting

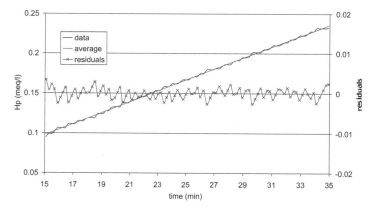

FIG. 6.7. Estimation of measurement variance on Hp data [193] [195].

such as the substrate concentration) for which measurements are made, is free of error. However, it may be that this assumption does not hold.

Occurrence of errors-in-variables problems.   When the independent variable itself is also a measurement, e.g. when it represents an experimental condition, the independent variable is not error-free. It is noteworthy that even the time measurement may contain an error if insufficient attention is given to its assessment!

   Another frequent origin of errors in the independent variable is due to a transformation of the model, often made to facilitate parameter estimation (i.e. to transform it into a linear regression form), e.g. the Eadie & Hofstee transformation of the Monod function:

$$\mu = -K_S \frac{\mu}{S} + \mu_{max}$$

where, since μ contains measurement error, the independent variable (μ/S) in this simple linear regression is not error free (it may be that the original independent variable S can be considered error free as it is an experimental setting).

   The more popular Lineweaver-Burk linearisation of the Monod-function

$$\frac{1}{\mu} = \frac{K_S}{\mu_{max}} \cdot \frac{1}{S} + \frac{1}{\mu_{max}}$$

on the other hand, is not violating this assumption since the independent variable (1/S) in this transformed equation is still free of error. We have seen above, however, that this transformation has other deficiencies and should be looked at carefully too when parameters are to be estimated with it.

   Errors in the independent variable may go up to 10% of the errors in the dependent variable without major effect on parameter estimates, i.e. the assumption of error free independent variables is then sufficiently met [218].

Parameter estimation in an errors-in-variables setting. The question we address here is what we should do in case we have to deal with an errors-in-variables problem. Basically, the parameter estimation problem just becomes a special nonlinear estimation problem, even for a model that is linear in the parameters [218]. In the case of a least squares objective function (see below), two kinds of residuals must be considered simultaneously and adjoined in a single residual vector. Calling Y_k the (vector of) dependent variables and X_k the (vector of) independent variables at time t_k, the residual vector becomes:

$$\varepsilon_k(\theta) = \begin{bmatrix} Y_k - \hat{Y}_k(\theta) \\ X_k - \hat{X}_k(\theta) \end{bmatrix} \tag{6.34}$$

With this residuals vector the least squares problem can be addressed if the covariance matrix V_k between these residuals is known entirely, i.e. the objective function to minimise becomes

$$J(\theta) = \sum_{k=1}^{N} \varepsilon_k^T(\theta) V_k^{-1} \varepsilon_k(\theta) \tag{6.35}$$

It is assumed in this development that the predicted values do not deviate too much from the real values.

The example below will illustrate how the problem can be tackled even in case a correlation exists between the "dependent" and "independent" variables.

Example of an errors-in-variables problem: Interpretation of a respirogram. As mentioned above, errors-in-variables problems may be created (sometimes unintentionally) when a parameter estimation problem is reformulated in a supposedly easier form. This was purposefully done to facilitate the introduction of the estimation of biokinetic parameters from respirograms in Kong *et al.* [149]. We repeat the example here.

Plotting a typical OUR_{ex} versus time data set typically results in a profile as in Figure 6.8. This oxygen uptake rate curve contains the similar information as the Monod growth curve in defining the relationship between growth rate μ and substrate concentration S.

The substrate degradation rate r_s and S are related to the measured OUR_{ex} in the following way (see also Section 3.2):

$$r_S = \frac{OUR_{ex}}{1 - Y}$$
$$S(t) = \frac{1}{1 - Y}\left(\int_0^{t_{fin}} OUR_{ex}(t)\,dt - \int_0^t OUR_{ex}(t)\,dt\right) \tag{6.36}$$

Where: Y = yield coefficient (mg COD biomass/mg COD substrate consumed)
t_{fin} = time at which OUR_{ex} returns to zero (min.)

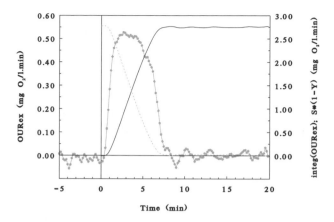

FIG. 6.8. OUR data collected in a respirometer (symbols); integral of the respiration rate, i.e. the consumed oxygen (up-curve) and remaining substrate in the reactor (down-curve).

r_s = substrate degradation rate (mg O_2/l. min)

For illustrative purposes, the cumulative OUR_{ex} and corresponding substrate concentrations in function of time have been combined in Figure 6.8.

Kong *et al.* [149] illustrate how the well-known Monod hyperbolic curve appears when one plots the "growth rate" *(1-Y)*r_s* as a function of the "substrate concentration" *(1-Y)*S* (Figure 6.9).

This curve is similar to the one obtained from continuous culture experiments in which growth rates are measured for different substrate concentrations. Note, however, that the substrate concentration in such chemostat experiments is measured very accurately and that the independent variable can therefore be considered to be essentially free of error. Hence, standard nonlinear parameter estimation can be applied to estimate the biokinetic parameters.

However, the assumption of absence of error in the substrate concentrations plotted in Figure 6.9 certainly does not hold. Indeed, it is calculated from the original OUR_{ex} data that contain considerable measurement error (see Figure 6.8).

Spanjers and Keesman [236] tackled this errors-in-variables problem in the proper way, i.e. they established the objective function (6.35) with residuals vector:

$$\varepsilon_k(\theta) = \begin{bmatrix} r_k - \hat{r}_k(\theta) \\ S_k - \hat{S}_k(\theta) \end{bmatrix}$$

where, for notational convenience, r_k is the respiration rate (i.e. *OURex* in Figures 6.8) and S_k the short term BOD (i.e. *(1-Y)S* in Figures 6.8 and 6.9) at time t_k.

The covariance matrix V could be determined quite nicely because the functional relationship between the "dependent" and the "independent" variable can be

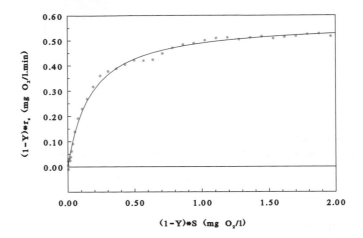

FIG. 6.9. Plot of the respiration rate versus the remaining substrate in the reactor as given in Figure 6.8.

formulated (by applying the trapezium integration rule to equation (6.36)):

$$S_k = \frac{1}{2} \sum_{i=k}^{N-1} \frac{r_{i+1} + r_i}{t_{i+1} - t_i}$$

The expectation of S_k for constant sampling interval and integration interval h then follows from:

$$E\left[S_k\right] = E\left[\frac{1}{2} \sum_{i=k}^{N-1} \frac{r_{i+1} + r_i}{h}\right] = \frac{1}{2h} \sum_{i=k}^{N-1} (\bar{r}_{i+1} + \bar{r}_i)$$

where E is the expectation operator and $\bar{r}$ indicates expected respiration rates. Also, the variance of S_k at each time instant can be expressed in terms of the characteristics of the measurement error ξ_k of r_k, i.e. an average of zero and a constant variance λ,

$$var\left[S_k\right] = var\left[\frac{h}{2} \sum_{i=k}^{N-1} (r_{i+1} + r_i)\right]$$

$$= \frac{h^2}{4} var\, (r_k + 2r_{k+1} + 2r_{k+2} + \ldots + 2r_{N-1} + r_N)$$

$$\cong h^2\, (N - k)\, \lambda$$

where it is assumed that $var[S_N]=h^2\lambda$. Consequently, for the covariance between S_k and r_k we can derive

$$var\left[S_k, r_k\right] = E\left[(r_k - E\left[r_k\right])\, (S_k - E\left[S_k\right])\right]$$

$$= E\left[\quad (\xi_k) \quad \left(\frac{h}{2} \sum_{i=k}^{N-1} (r_{i+1} + r_i) - \frac{h}{2} \sum_{i=k}^{N-1} (\bar{r}_{i+1} + \bar{r}_i) \right) \right]$$

$$= \frac{h}{2} E\left[\quad (\xi_k) \quad (\xi_k + 2\xi_{k+1} + 2\xi_{k+2} + \ldots + 2\xi_{N-1} + \xi_N) \right]$$

$$\cong \frac{h}{2} \lambda$$

This is also the covariance between the error in S_k and the measurement error in r_k. Hence, the measurement error covariance matrix is written as:

$$V_k = \lambda \begin{bmatrix} h^2 (N - k) & \frac{h}{2} \\ \frac{h}{2} & 1 \end{bmatrix}$$

6.5   Objectives in Parameter Estimation: Estimators

In order to objectify the estimation of parameters (subjective "guessing" parameters by visual inspection of model predictions and data, is still often applied in practice), functions have to be defined that represent the wish to fit a model to the data. These functions are conventionally arranged such that small values represent close agreement between model and data. The model parameters are then adjusted to achieve a minimum in these functions, yielding best-fit parameters. Consequently, this adjustment process is a minimisation problem in many dimensions (i.e. as many as there are parameters to be estimated).

These functions are termed loss, merit, cost or objective function. The choice of such function is indeed one of the first problems to be solved when model parameters are to be estimated. The best known objective function for parameter estimation is the sum of squared errors function

$$J(\theta) = \sum_{i=1}^{N} \left(y_i - \hat{y}_i(\theta) \right)^2 \tag{6.37}$$

where y_i are the observations (in total N observations are available) and $\hat{y}_i(\theta)$ are the model predictions for a given parameter set θ. The objective function J represents all of the information contained in the observations that is not explained by fitting the model to the data [218].

Although this objective function is very well known, its origin is less known to the users. Basically all classic objective functions we will introduce below originate from maximum likelihood (ML) estimators and can be seen as simplifications of the maximum likelihood objective function under given assumptions.

6.5.1   Maximum Likelihood Estimation

The starting point in maximum likelihood estimation is the following [203]. If we consider the possible values that parameters can have, we have the intuitive feeling that some values are more likely than others for which the model predictions look nothing like the data. The problem, however, is how to quantify that intuition. The approach taken in ML estimation starts from the fact that the experimental data are a random sample drawn from the universe of data sets. The question can therefore be asked what the probability is that this data set could have occurred given a particular set of parameters (assuming the model is correct). If the probability of occurrence for a certain parameter set is very small, it can be concluded that the parameters under consideration are unlikely to be correct. From this, the probability of the data given the parameters is identified as the likelihood of the parameters given the data. This identification is based entirely on intuition and has no mathematical basis. If we accept this, then it is, from this point on, only a small step to propose ML estimation as the procedure in which parameters get the values that maximise the likelihood of occurrence of the observations.

Of course, this more philosophical discussion must be concretised in operational functions. Below we will introduce the likelihood functions in case the measurement errors are assumed independent and normally distributed, either homoscedastic (constant variance) or heteroscedastic (non-constant variance). We will see that maximising the resulting likelihood functions leads to reasonably simple objective functions.

6.5.2   Weighted Least Squares (WLS) or χ^2 Estimation

When the assumption is made that the measurement errors are independent (uncorrelated) and originating from normal distributions, the likelihood function can easily be constructed as the product of these normal distributions [213]

$$L\left(\bar{y}\,|\theta\right) = \prod_{i=1}^{N} \frac{1}{\sqrt{2\pi}} \frac{1}{\sigma_i} exp\left(-\frac{1}{2}\sum_{i=1}^{N}\left(\frac{y_i - \hat{y}_i\left(\theta\right)}{\sigma_i}\right)^2\right) \qquad (6.38)$$

in which $\bar{y}$ is the set of N observations, σ_i is the (estimated) standard deviation of the measurements y_i. For a given data set $\bar{y}$ the maximum likelihood estimates $\theta(\bar{y})$ of the parameters are those values for which the above equation has its maximum. This is equivalent to finding the minimum of the function (due to the negative sign)

$$J\left(\theta\right) = \chi^2\left(\theta\right) = \sum_{i=1}^{N} \frac{1}{\sigma_i^2}\left(y_i - \hat{y}_i\left(\theta\right)\right)^2 \qquad (6.39)$$

since all other terms are composed of constants. Minimising this objective function is termed weighted least squares – for obvious reasons – and the estimates are termed weighted least squares estimates.

Alternatively this minimisation is also termed chi-squared fitting. The origin of the latter term warrants some explanation [203]. The χ^2 statistic is a sum of N squares of normally distributed quantities, each normalised to unit variance. However, because we have used the above equation 6.39 to estimate the best parameter set, the terms in the sum are no longer statistically independent. Still, the probability density function for different values of χ^2 can be derived analytically at its minimum, and is the χ^2-distribution with N-p degrees of freedom (with p the number of estimated parameters). Comparing the computed χ^2 with tabulated values of the χ^2-distribution for N-p degrees of freedom gives a quantitative measure of the goodness-of-fit of the model. If the result fails the statistical test, it means either that

- the residuals are unlikely due to chance fluctuations or, more probably, that
- the model is wrong, or that
- the measurement standard deviations σ_i were underestimated, or that
- the measurement errors were not normally distributed (however, luckily, the test is not very sensitive to this type of deviation).

It may also occur that the χ^2 test is too good to be true. This problem is not likely to be due to a non-normal distribution of the measurement errors, but frequently is caused by an overestimation of the measurement standard deviations. As a rule of thumb it can be stated that a good fit yields a typical value of χ^2 equal to the number of degrees of freedom. Asymptotically, the statistic χ^2 becomes normally distributed with mean N-p and variance $2(N$-$p)$.

In case the measurement errors cannot be estimated, weights can be assigned according to engineering judgement [23]

$$J(\theta) = \sum_{i=1}^{N} w_i (y_i - \hat{y}_i(\theta))^2 \tag{6.40}$$

For instance, one may have the insight that the errors are proportional to the measured value (for instance from an understanding of the measurement principle) [68]. It is then appropriate to use the squared of the inverse of the measured value $(1/y_i)^2$ as the weight w_i. In case transformation of variables is performed (see section 6.3), the weights may often be deduced analytically from the measurement error distributions of the original data [218].

6.5.3   Ordinary Least Squares (OLS) Estimation

In case the standard deviations of the measurements are (assumed to be) constant (homoscedastic), the common factor σ_i in equation 6.39 can be dropped from the sum, leading to the objective function

$$J(\theta) = \sum_{i=1}^{N} (y_i - \hat{y}_i(\theta))^2 \tag{6.41}$$

to be minimised to yield the (ordinary) least squares estimates.

6.5.4 *Bayesian Estimation*

As mentioned above, maximum likelihood estimation assumes the parameters to be constants whereas the data are considered to be sampled from a universe of data sets. Bayesian estimation takes this a step further and treats both parameters and measurements as random variables. Bayesian estimation basically updates prior knowledge by considering experimental evidence [23], [213]. To obtain the conditional probability density function (PDF) of occurrence of the parameter set θ given the measurement set $\bar{y}$, Bayes' rule

$$p\left(\theta \,|\, \bar{y}\right) = \frac{p\left(\bar{y} \,|\, \theta\right)}{p\left(\bar{y}\right)} p\left(\theta\right)$$

is applied to the a priori PDF of the parameters $p(\theta)$ and measurements $p(\bar{y})$ and the conditional PDF that a data set $\bar{y}$ occurs given the parameters θ. The posterior PDF as a function of θ is thus proportional to the likelihood function $p(\bar{y}, \theta)$ multiplied by the prior PDF. The three probability density functions on the right hand side have to be specified by the user on the basis of prior knowledge (and the collected experimental data) which makes this approach quite demanding.

It is important to note that this result is not giving a parameter estimate, but rather the complete distribution of the parameter values for the given experimental data set and prior knowledge. If a particular parameter estimate is required for further work with the model, the posterior distribution can be analysed in various ways to yield a parameter point estimate. For instance, taking the mode of the posterior distribution results in the maximum a posteriori (MAP) estimator [19].

It is also noteworthy that if the data do not contain any information on a parameter, this will mean that the posterior PDF of that parameter will not be updated and will remain the same as the prior PDF. This is an important result in case an experiment is non-informative. The parameter non-identifiability that would occur in classic least squares estimation does not cause major problems in obtaining a Bayesian estimate of the parameters [218], [214].

6.5.5 *Robust Estimation*

Problems in nonlinear regression with ordinary or weighted least squares are due to three phenomena:

1. In contrast to what is often assumed, the residual errors are not necessarily normally distributed;
2. In order to weight the residual errors properly, it is necessary that the variance of the error at each measurement point is known. This requires the availability of repeat measurements which are often not made. Extrapolation from a restricted number of repeat measurements (and modelling the evolution of the variance along the time series) may be adequate but may also lead to incorrect weighting, leading to biased estimates;

3. Data points that do not belong to the distribution type to which the majority of the data adhere, so-called outliers, may occur and lead to inaccurate parameter estimates and biased confidence regions.

Rousseeuw and Leroy [221] provide a good introduction into the extensive field of robust estimation and specifically focus on how to detect and deal with outliers.

The robust non-parametric method (i.e. not relying on an assumption concerning a distribution of the errors) applied by Atkins [9] for the estimation of biodegradation model parameters is used here as an illustration of these approaches. It works as follows: For a given parameter set of size p, form p equations by taking p data points from the n available data,

$$
\begin{aligned}
h\left(\theta, x_1\right) - y_1 &= 0 \\
h\left(\theta, x_2\right) - y_2 &= 0 \\
\vdots \quad - \quad \vdots \ &= 0 \\
h\left(\theta, x_p\right) - y_p &= 0
\end{aligned}
$$

Seek a solution for this set of (nonlinear) equations and repeat this process until all possible combinations of p equations from the N data have been used. List and sort the values for each parameter. Take the median value of that sorted list as the estimate for the parameter considered.

It should be noted that this method heavily relies on the adequacy of the solution method for the many sets of (nonlinear) equations. Atkins [9] indeed reports failure of the [45] method he applied in his work and also complains that the amount of computer time needed to obtain an estimate with this non-parametric method could be prohibitive. On the other hand he also illustrates that bias in the estimates due to the inappropriate use of a WLS objective function could be minimised.

Another robust procedure, applied by Hardwick *et al.* (1991), seeks to maximise the number of runs (also termed zero crossings, see Section 3.3.2) of the residuals sequence obtained with a parameter set. Own experience on respirometric data, however, has found that this method is not very successful due to many local minima problems, i.e. the search algorithm gets stuck at a given number of runs and is not progressing further as no gradient in the number of runs objective function can be discerned. Modification of the method by adding a term that reflects the quality of fit gave a slight but not sufficient improvement in the search efficiency.

6.5.6 *Alternative Objective Functions*

Many other, less statistically underpinned objective functions have been applied, all of them reflecting a certain interpretation given by the users of an optimal parameter estimate.

- Absolute deviations (1-norm) (minimise the sum of absolute deviations),
- Min-max objective function (infinite norm) (minimise the maximum absolute error),
- Maximise the number of sign changes in the sequence of residuals. Note the similarity with one of the structure characterisation methods (run test),
- Minimise first lag autocorrelation. Here too, note the link with structure characterisation methods.

6.5.7 Selecting a Set of Feasible Parameters

In some approaches the aim of parameter estimation is not to obtain a parameter point estimate but one is content with a set of feasible parameters that result in model predictions with certain properties, for instance, they give predictions for which the maximum deviation from the data points is less than a certain value. In this respect the development and use of the so-called HSY approach by Hornberger and Spear [129] and Young [288], the set-membership method by Keesman and van Straten [143] and the GLUE-methodology by Beven and Binley [28] are noteworthy. Let us go a bit more in detail on the set-membership and GLUE approaches.

Set-membership.   In case detailed characterisation of errors is not possible due to a limited length of data records, or if the residuals have non-random components as a result of model inadequacy or systematic measurement errors, a statistical approach will give unreliable results. Under these circumstances, a deterministic error characterisation in terms of lower and upper bounds only will be a good alternative. This reasoning has led to the development of the so-called set-membership methods [174].

Basically, the parameter space is divided into a behaviour and a non-behaviour space, where the former space contains all parameter sets that give rise to model predictions that are completely contained within an error band around the experimental data (called the behaviour set). The MCSM (Monte Carlo Set-Membership) algorithm [141], [144] can handle this division. The key idea is that randomly selected parameter vectors which result in a model response consistent with the behaviour set belong to the feasible set. Note that by choosing an appropriate error bound, the feasible parameter set can be reduced to a singleton and in this way a unique parameter estimate is obtained from this feasible parameter set method.

As an illustration the work done by Vanrolleghem and Keesman [264] is given. The data set given in Figure 6.10 depicts a respiration rate time series collected by an on-line respirometer at a Dutch wastewater treatment plant. The outer full lines delimit the behaviour set corresponding to the data depicted as symbols and a certain error bound. The feasible model output set is given by the two lines closer to the data points. This set contains all simulation results corresponding to

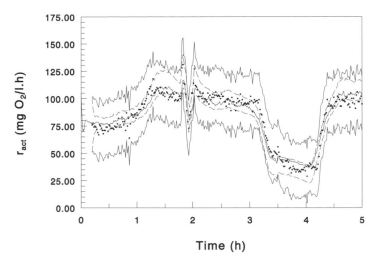

FIG. 6.10. Set-membership method applied to a respiration rate time series collected at a full-scale WWTP: Measurement set (symbols), behaviour set (outer full lines), feasible model output data set (middle dashed lines) and singleton trajectory (centre full line).

parameters whose simulation results were contained completely in the behaviour set. Finally, the single line closest to the data points is the trajectory corresponding with the unique parameter set that is obtained when the error on the measurement set is reduced to such an extent that only a single parameter set is left that gives simulation results that are all contained in the behaviour set.

In another example Vanrolleghem and Keesman [264] also compared the MCSM feasible parameter set with the 95% confidence region of parameters obtained with other parameter estimation methods. The result of a three-parameter estimation problem in which a Monod-model was fitted to a batch respirometric data set, is given in Figure 6.11. It shows that the approximate confidence region is contained completely in the MCSM parameter set, albeit that it is lying on the lower end of the three parameters estimated. Note that the range of the MCSM feasible parameter set is considerably larger than the 95% confidence region. This is of course caused by the large error band that was chosen in this study around the respiration rate data. Theoretically, the size of the feasible parameter set can be reduced to a singleton by reducing this error band adequately.

GLUE. Similarly to the Set-membership approach, the basis of the GLUE approach of Beven and Binley [28] is that any parameter set combination that predicts output variables reasonably well is considered equally likely. It is based upon making a large number of runs of a given model with different sets of parameter values, chosen randomly from specified distributions. Each set of parameters evaluated is assigned a goodness-of-fit value of being the "true" system simulator. The

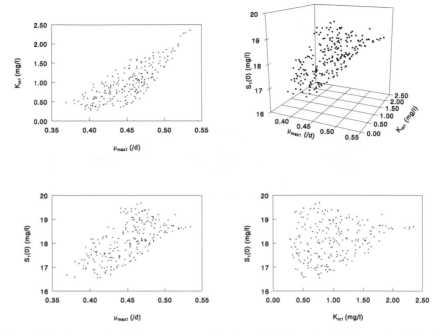

FIG. 6.11. Set-membership parameter estimation of a Monod-model for a set of batch respiration rate data using an error band of 0.1 mg O_2/l.min.

GLUE approach is divided into the following steps:

1. Define a goodness-of-fit function for output data. The choice of function can be crucial to the results of the procedure. Further, a criterion based on the goodness-of-fit function for accepting or rejecting a parameter set must be determined.

2. Define initial ranges or distributions of parameter values to be considered.

3. Sample the parameter space to obtain realisations or simulations of the model. It is most common to use Monte Carlo simulation with uniform parameter distributions.

Going through the steps above yields empirical joint distributions for model parameters. The scatter plots in Figure 6.12 exemplify the approach.

6.5.8 *Multi-Objective Functions*

For some applications, a model must be able to simulate different aspects of the system. For instance, one aim of a model may be to allow a good representation of the average behaviour of the system, whereas the focus of another application may be in the accurate prediction of peak behaviour.

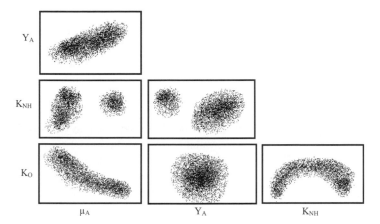

FIG. 6.12. Feasible parameter set obtained by applying the GLUE approach to a nitrification model (4-dimensional) parameter estimation problem.

Many studies have been devoted to find the most adequate objective functions for particular applications. Although in many cases it may be quite straightforward to select an appropriate objective function (see discussion above), no clear-cut answer or selection procedure is currently available. Hence, one is left with the concept that a modeller can choose among different parameter sets obtained with different objective functions that seem equally plausible [235]. It is noteworthy that this multi-objective equivalence of several parameter sets is quite different from the rationale behind the above mentioned set membership, GLUE or HSY approaches in which a single objective can be fulfilled by a range of possible parameter sets.

To deal with the multiple objectives, conveniently written as a set of objective functions

$$J(\theta) = \{J_1(\theta), J_2(\theta), \cdots, J_m(\theta)\}$$

a Pareto set P of solutions is pursued corresponding to various trade-offs among the objectives. This Pareto set is defined such that any member parameter set θ_i has the following properties [235]:

1. For all non-members θ_j at least one member θ_i exists such that the objective $J(\theta_i)$ is strictly less than $J(\theta_j)$. This allows to partition the parameter sets into "good" solutions (Pareto solutions) and "bad" solutions;

2. It is not possible to find θ_j within the Pareto set such that $J(\theta_j)$ is strictly less than $J(\theta_i)$. By "strictly less than" it is meant that $J_k(\theta_j) < J_k(\theta_i)$ for all k=1,...,m. In the absence of additional information, it is not possible to distinguish any of the "good" (Pareto) solutions as being objectively better than any of the other good solutions (i.e. there is no uniquely "best" solution).

The resulting "good" (Pareto) parameter sets therefore are each able to fulfil one of the objectives better than any other member of the Pareto set, but the trade-off will be that some other characteristics of the system's behaviour are less-well described/predicted.

6.5.9 Multivariate Estimation

So far all objective functions were only dealing with a single variable for which a number of N observations were available to fit to the model. In practice we of course have many experimental set-ups in which more than one variable is measured, for instance substrate and biomass measurements. It is then quite logical to try to fit a model to each of the outputs it can predict. However, it is also evident that not all of the variables are as trustworthy and that, in view of a certain purpose of the model, it may be more important to predict some variables better than others. To deal with this, it is logical to adopt the Weighted Least Squares (6.39) or ML objective function to express the "optimality" of a parameter set and use the weighting factors to reflect the importance or reliability of the different variables.

Recall that the definition of the Fisher Information Matrix (see Section 5.3) was already developed on the basis of a general WLS objective function written in matrix format:

$$J(\theta) = \sum_{i=1}^{N} \left(y_i\left(\hat{\theta}\right) - y_i \right)^T Q_i \left(y_i\left(\hat{\theta}\right) - y_i \right) \tag{6.42}$$

We recognise the vector of different output variables y_i available for each time instant i and the weights matrix Q_i in which each output variable and each combination of variables is given a weight in the calculation of the objective function value. Note that this weights matrix can be different for each time instant i, for instance to reflect a time-varying quality of the measurements or process conditions. Typically the weights are chosen as the inverse of the measurement error covariance matrix, just as can be seen in the definition of the WLS-objective functional for a single output variable (6.39).

6.5.10 Multiresponse Estimation

The above describes how to deal with multiple output variables for which data are available. In some cases multiple experiments are conducted to estimate a single set of parameters. The simplest example is the availability of repeat experiments. However, in some cases it can be advantageous to perform multiple experiments under slightly different conditions but with the aim to estimate a common set of parameters. De heyder et al. [68] took advantage of such experiment design to solve an identifiability problem: experiments were conducted at different mass transfer intensities $K_L a$. The fact that multiple experimental data sets are available for the same output variable is known under the term multiple response data sets. The estimation of parameters with such data sets is called multiresponse estimation. An early example of this is given in Johnson and Berthouex [135]. Mathematically, multiple responses are considered in the same way as different output variables. However, each of the responses is typically weighted equally although one may want to express in the weights that one experiment was more reliable than another.

As an example below a quite general weighted least squares criterion is given in which multivariable (number of variables *Nvar*) data sets are available for *Nresponse* experiments.

$$J(\theta) = \sum_{k=1}^{Nresponse} w_k \sum_{j=1}^{Nvar_k} w_{jk} \sum_{i=1}^{Ndata_{jk}} w_{ijk} \left(y_{ijk}\left(t_{ijk}\right) - \hat{y}_{ijk}\left(\theta, t_{ijk}\right)\right)^2$$

6.6   Minimisation Approach

As mentioned above finding the best parameter estimates typically involves minimising the deviation of the model's predictions from the data points using one of the objective functions *J* given above. Depending on whether the parameters are linear in the model or nonlinear, the solution methods of this minimisation problem are quite different. For linear parameters a one step calculation gives the best estimates, whereas for nonlinear parameters (as an illustration, see the second introductory example of this chapter) we have to resort to numerical methods that search the parameter space in a systematic way.

6.6.1   *Linear Parameter Estimation*

For problems where the parameters are linear in the model (e.g. linear regression using least squares), the parameter estimates are easily found by differentiating the objective function *J* with respect to each of the parameters, set these derivatives to 0, and solve the resulting system of equations for the unknown parameters.

In general, a model linear in the parameters can be rewritten in the following form:

$$y_i = \phi_i^T \theta \tag{6.43}$$

where θ is the parameter vector, ϕ_i is termed the regressor, y_i contains the terms that are independent of the parameters and the index i refers to time. The choice of a discrete time representation of the model is natural considering that estimation will be done on the basis of experimental data that are available at sampled time instants.

As an example we will consider a mass balance for a substrate that takes part in two conversions: a growth reaction (with a first order dependency of the conversion rate to the substrate concentration) and a maintenance reaction:

$$\frac{dS}{dt} = DS_{in} - DS - \frac{1}{Y}\alpha S X - m_S X \tag{6.44}$$

Let us assume that α (growth rate constant) and m_S (maintenance coefficient) are unknown and that we have data on substrate and biomass concentrations in the reactor, the dilution rate D and the inlet substrate concentration S_{in} with sampling period T. We further assume that the yield coefficient Y is known. The parameter vector θ therefore equals:

$$\theta = \begin{bmatrix} \alpha \\ m_S \end{bmatrix} \tag{6.45}$$

By approximating the derivative dS/dt by a finite difference $(S_{i+1}\text{-}S_i)/T$ (where i corresponds to the time instant $t = i.T$), we can rewrite the mass balance (6.44) as follows:

$$S_{i+1} - S_i - T D_i S_{in,i} + T D_i S_i = -\frac{1}{Y} S_i X_i T \alpha + X_i T m_S \tag{6.46}$$

In other words, in the formalism of equation (6.30), y_i and ϕ_i correspond to:

$$y_i = S_{i+1} - S_i - T D_i S_{in,i} + T D_i S_i \quad , \quad \phi_i = \begin{bmatrix} -\frac{1}{Y} S_i X_i T \\ X_i T \end{bmatrix}$$

In the simple case of a linear equation with unknown parameters θ, linear regression can be applied, for instance a weighted least squares objective function:

$$J(\theta) = \sum_{i=1}^{N} \beta_i \left(y_i\left(\hat{\theta}\right) - y_i \right)^2$$

The least squares estimator that minimises this criterion, can easily be deduced to be:

$$\hat{\theta} = \left[\sum_{i=1}^{N} \beta_i \phi_i \phi_i^T \right]^{-1} \sum_{i=1}^{N} \beta_i \phi_i \phi_i^T$$

As mentioned before, a possible choice for the weights β_i are the inverses of the variance σ_i^2 of the measurement y_i.

6.6.2 Nonlinear Parameter Estimation

Not all parameter estimation problems can be solved as easily as the problems in which parameters appear linearly in the model. In some exceptional cases nonlinear parameters can be determined analytically by solving (a set of) nonlinear equations. The problems, however, turn very quickly mathematically intractable and one needs to try to find the minimum of a nonlinear objective function. Luckily, finding the minimum of a multivariate function f is a common problem in many research fields and the available expertise is substantial.

Overall, the purpose is to find as efficiently as possible values of θ that make $J(\theta)$ minimal. However, typical for nonlinear functions is that the minimum can

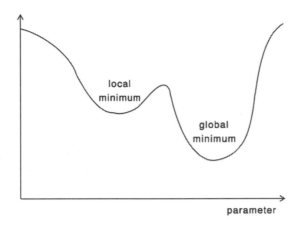

FIG. 6.13. Local and global minima in an objective function.

either be global (the lowest function value in the whole parameter space) or local (the lowest function value in a finite neighbourhood), see Figure 6.13. Despite
extensive efforts, no perfect minimisation algorithm for nonlinear objective functions exists (so far), and consequently, finding the global minimum for nonlinear
problems cannot be guaranteed [203]. Rather we must accept that additional efforts will be needed and care should be taken to maximise the confidence that one
is not ending up with a bad local minimum of the objective function, i.e. ending up with sub-optimal parameter estimates. Consequently, one of the important
characteristics of minimisation algorithms will clearly be their sensitivity to local
minima.

A helpful visualisation of a nonlinear objective function is a landscape with
hills and valleys. The minimisation algorithm should search for the lowest point
in this landscape, but can eventually end up in a local minimum instead of the
global minimum. Minimisation algorithms typically need a set of initial values
where they start from on their quest for the parameter set that gives the lowest J.
A property of nonlinear function minimisation is that the minimum found by the
algorithm (global or local) can be influenced by the choice of the starting values for
the parameters $\theta(0)$. Indeed, a search algorithm can get "stuck" in a local minimum
when it comes down one way, whereas it may never come near that local minimum
when it comes another way. For instance, in Figure 6.13 one could imagine getting
stuck in the local minimum when coming from the left, whereas one would end up
in the global minimum when $\theta(0)$ was put on the right side of the figure.

The overall procedure of nonlinear parameter estimation is schematised in Figure 6.14. Initially, the model structure, of which we want to estimate certain selected parameters, and the experimental data need to be specified. To start the
algorithm, first guesses of the parameters have to be given. The minimisation algo-

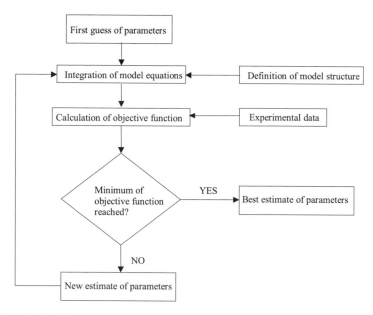

FIG. 6.14. Illustration of parameter estimation routine (modified from [280]).

rithm will then request for model predictions corresponding to this first parameter set. These model predictions are obtained by solving the set of model equations and are passed on to the routine where the objective function is calculated by confronting the predictions with the data. On the basis of rules that are different for each minimisation algorithm either a new proposal for parameters is made and sent to the model solver or, if certain criteria are met, the parameter values are passed on to the user as best estimates. Stopping criteria may be that the maximum number of iterations is reached or that no improvement in objective function is found in recent iterations.

It is important to note here that most calculation time needed to find the best estimates is spent in the box "integration of model equations". Therefore, any approach that can minimise the time spent in this box is very important. For instance, one may aim for the fastest model solution methods. The main gain is, however, obtained by selecting a minimisation algorithm that finds the minimum with the smallest number of iterations through that box.

Schuetze [226] attempted to classify the different minimisation methods in two main groups, local and global minimisation methods. However, he indicated that it may not be possible at all to come up with a strict taxonomy since there are many interrelations between the approaches. For example, most global search methods include at some stage a local procedure in order to refine an approximation of a solution which has been found by the global procedure.

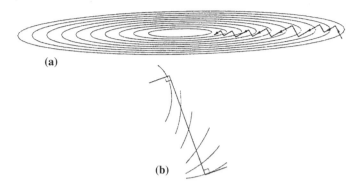

(a)

(b)

FIG. 6.15. Inefficient steepest decent search along valley.

6.6.3 *Local Minimisation Algorithms Using Derivative Information*

A wide range of minimisation procedures has been developed for locating a local minimum, and many among them make use of information about the gradient of the objective function with respect to the parameters to be estimated. This information is either assumed to be directly available or is computed by numerical approximation.

Steepest descent. The basic idea of these gradient based methods is that, first, a direction is sought in parameter space along which a minimisation of the objective function is pursued. Once the direction is decided upon the step size with which the parameters will be changed is to be determined.

For the steepest descent method, the path of steepest descent is followed as long as J decreases. When the minimum along this direction is reached, a new steepest descent direction is searched for and the parameters are changed according to this new direction. While, theoretically, the method will converge, it may do so in practice with agonising slowness [80] after some rapid initial progress. It happens particularly when the path of steepest descent zigzags slowly down a narrow valley, each iteration bringing only a slight reduction in J (see Figure 6.15).

Gauss-Newton method. A well-known alternative, the Gauss-Newton method, approximates the objective function J locally (around a parameter set θ_i reached at a certain point in the minimisation procedure) by a Taylor series expansion:

$$J_i'(\theta) = J(\theta_i) + \left(\left.\frac{\partial J}{\partial \theta}\right|_{\theta_i}\right)^T (\theta - \theta_i) + \tfrac{1}{2}(\theta - \theta_i)^T \left.\frac{\partial^2 J}{\partial \theta^2}\right|_{\theta_i} (\theta - \theta_i)$$

$$= c_i + b_i^T \cdot \delta\theta + \tfrac{1}{2}\delta\theta^T \cdot A_i \cdot \delta\theta$$

where $c_i \equiv J(\theta_i)$ $b_i \equiv \nabla J|_{\theta_i}$ $[A_i]_{kl} \equiv \left.\frac{\partial^2 J}{\partial \theta_k \partial \theta_l}\right|_{\theta_i}$ $\delta\theta = \theta - \theta_i$

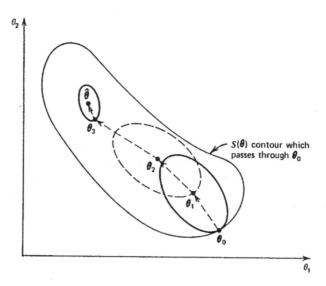

FIG. 6.16. Newton method for parameter estimation by consecutive linearisations of the objective function.

For this approximate linear objective function the minimum can be calculated analytically, i.e. by setting the gradient of J' to zero

$$\frac{\partial J_i'}{\partial \theta} = b_i + A_i \cdot \delta\theta = 0$$

we obtain that the parameter set minimising this approximation of the objective function is found at:

$$\theta_{i+1} = \theta_i - A_i^{-1} b_i$$

Evidently, for a nonlinear problem the minimum obtained in this way is not the true minimum (due to the Taylor series approximation) and the linearisation must be carried out at this point again to make a move to the next minimum, eventually ending in the true minimum. Figure 6.16 shows the procedure for a two-parameter example. Note the nonlinear contour line of J and consecutive linear approximations of J' (ellipsoidal contour lines). The main drawback of the method is that it may converge very slowly (or even diverge) and oscillate around the solution. Our own experience is that the method is not very useful for the type of estimation problems one is encountering in wastewater treatment modelling.

Moreover, the Newton method requires the evaluation of the second derivative matrix A (Hessian) of the objective function at any θ. This is analytically feasible in case the objective function can be written explicitly, but this is often not the case.

Rather, A will need to be approximated numerically, with all the corresponding problems.

When we consider a weighted least squares problem, we can write the elements $[A]_{kl}$ to be:

$$[A]_{kl} \equiv \frac{\partial^2 J}{\partial \theta_k \partial \theta_l} = 2 \sum_{i=1}^{Ndata} \frac{1}{\sigma_i^2} \left[\frac{\partial y_i}{\partial \theta_k} \frac{\partial y_i}{\partial \theta_l} - \left(y_i - \hat{y}_i\left(\theta\right)\right) \frac{\partial^2 y_i}{\partial \theta_k \partial \theta_l} \right]$$

In the Gauss modification of Newton's method, the second term is neglected because the error $[y_i\text{-}y_i(\theta)]$ multiplying the second derivative should be random and can have either sign. Hence, they tend to cancel out when summed over i [203]. Therefore, we no longer need to evaluate the second derivatives and obtain for the Gauss-Newton methods that

$$\theta_{i+1} = \theta_i + \left(Y_{\theta i} Y_{\theta i}^T\right)^{-1} Y_{\theta i} \varepsilon_i$$

in which $Y\theta_i$ is the matrix of output sensitivities to the parameters and ε_i is the vector of residuals for the parameter set θ_i.

It is to be noted that some "fiddling" with the A-matrix has no effect at all on the final result of the minimisation but will only affect the downhill path that is taken to get there. Indeed, the condition at the minimum that b should be zero is independent of how A is defined.

Levenberg-Marquardt.   Probably the best-known modification ("fiddling") of the above two basic methods is the Levenberg-Marquardt algorithm [168] where a compromise is sought between the above inverse Hessian and the steepest-descent method. The idea of Marquardt can be explained briefly as follows. First, both linearisation and steepest-descent methods are "asked" for their optimal direction for a next parameter update step. The Marquardt algorithm then provides a method for interpolating between these two directions and for obtaining a suitable step size as well. We shall not go into any further detail here. This algorithm is good in the sense that it almost always converges and does not slow down as the steepest-descent method often does [80].

Quasi-Newton methods.   The basic idea of the quasi-Newton or variable metric methods [203] is to build up, iteratively, a good approximation to the inverse Hessian matrix A^{-1}. Hence, the term "quasi" points to the fact that we only use the current approximation of the inverse Hessian to move forward in the Newton parameter update formula. As explained in Press *et al.* [203] it is found that this approximation often works better than the true Hessian, because it always guarantees in the beginning to move downhill, whereas this guarantee is not given in the Newton method. Of course, close to the minimum the approximation converges to the true Hessian and we can benefit from the quadratic convergence of

Newton's method. The two best known examples of quasi-Newton methods are the BFGS (Broyden, Fletcher, Goldfarb and Shanno) and DFP (Davidon-Fletcher-Powell) methods [87]. These algorithms only differ in details but it has generally become recognised that, empirically, BFGS is superior to DFP. Vanrolleghem and Keesman [264] reported, however, that while these methods converge very quickly, they appear quite sensitive to local minima too.

6.6.4 *Derivative-Free Local Minimisation Algorithms*

In the parameter estimation problems we have discussed in this book on dynamic models, the objective function is defined by a system of ordinary or partial differential equations that are mostly not linear in the system variables and parameters. Consequently, the objective function is evaluated by numerical integration of the dynamic system. Parametric derivatives generally must be found by further integration of a large, derived system (of sensitivity functions), one for each parameter. In situations such as these, derivative-free algorithms are particularly attractive - especially ones that make efficient use of previously computed function values [207]. It is clear that these methods need less preparation for implementation since no calculation of the derivatives must be supported. Also, numerical problems associated with calculating the derivatives may induce convergence problems. In contrast to, for instance, [13], we do not consider methods that use finite differences to approximate the derivative as derivative-free methods. According to us they belong to the first group of local minimisation algorithms.

Rosenbrock method.   One of the oldest methods for derivative-free minimisation was introduced by Rosenbrock in 1960 [220]. It was specifically developed for problems in which the objective function is characterised by ridges and valleys and deals with these by rotating the search axes at the end of every stage in the direction of the valley. Riefler *et al.* [217] found this method very insensitive to the local minima which other methods such as BFGS were suffering from.

Brent's algorithm.   The direction set method of Powell [202] with refinements proposed by Brent [42] is one of the best derivative-free local minimisation methods [203]. It is based on a repeated combination of one-dimensional searches along a set of various directions [96]. Direction-set methods consist of prescriptions for updating the set of directions as the method proceeds, attempting to come up with a set which either

- includes some very good directions that will take us far along narrow valleys, or else

- includes some number of "non-interfering" directions with the special property that minimisation along one is not "spoiled" by subsequent minimisation along another, so that interminable cycling through the set of directions can be avoided.

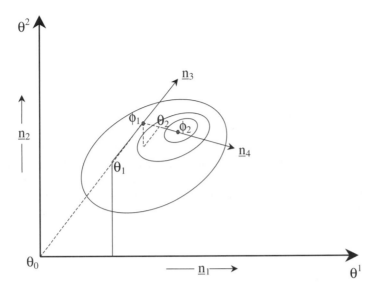

FIG. 6.17. Elementary steps in a direction set algorithm with direction set containing conjugate directions adapting according to Brent [42].

Interesting to mention is that Brent also incorporated some random "jumps" into the procedure to avoid some "local minima" problems that Powell's original algorithm still suffered from.

In Figure 6.17 a typical sequence of searches is presented to illustrate the basic algorithm. We start in an initial parameter guess θ_0 and search sequentially for the minimum along each of the directions in the direction set $\{\underline{n}_1,\underline{n}_2\}$. When all directions are passed through (leading to θ_1), the oldest direction in the set is replaced by a new direction constructed as $\underline{n}_3 = \theta_1$-$\theta_0$. This first iteration is ended by searching along this new direction for the minimum ϕ_1. Then a new iteration is started along directions $\{\underline{n}_2,\underline{n}_3\}$ leading to θ_2 and the new direction $\underline{n}_4$ which leads to the minimum ϕ_2. This sequence of steps continues until a stop criterion is reached (e.g. maximum number of allowed iterations or a lack of further decrease in objective function).

Vanrolleghem and Keesman [264] confirmed the statement by Press et $al.$ [203] that this method is probably one of the best ones in terms of optimally compromising convergence rate and insensitivity to local minima. Schuetze [226], using the original Powell method, indeed complained about its sensitivity to local minima.

Simplex method.   Another well-known local minimisation method, which does not require derivative information, was proposed by Nelder and Mead [184], better known as the simplex method (not to be mixed up with the linear programming simplex method !).

The main concept used by this method is the geometrical concept of a simplex.

A simplex is the geometrical figure consisting, in p dimensions, of $p+1$ points (or vertices) and all their interconnecting line segments, polygonal faces etc. Examples are a triangle ($p=2$) or a tetrahedron ($p=3$). The simplex method only makes use of nondegenerate simplexes, i.e. if any point of a nondegenerate simplex is taken at the origin, then the other p points define vector directions that span the p-dimensional parameter space.

Starting from an initial simplex, elementary steps (Figure 6.18) are developed to find a minimum by evaluating the objective function value at the vertices of the simplex and replacing the vertex with the highest value by another point in p-dimensional space. These steps make sure that the objective function value of each new point of the new simplex is closer to the optimum than the old one. Furthermore, it is ensured that the elementary steps maintain the simplex nondegeneracy property.

The typical progress of iterations is illustrated in Figure 6.19 using a two-dimensional example. Vertices 1,2 and 3 form the initial simplex and increasing numbers indicate the new vertices added at each iteration. Note that vertex 7 has the largest function value for the simplex {4,6,7} but is not reflected immediately since it is the newest vertex in that simplex. When simplex {6,9,10} is reached, vertex 6 has been in the current simplex for four iterations leading to a contraction to the new simplex {6,11,12}. The iteration continues from this simplex and the algorithm will continue to reflect and contract the simplex until the stop criterion has been achieved.

The simplex method is well appreciated for its robustness to local minima, ease of implementation and reasonable convergence rate [203], [264], [226].

Secant or DUD algorithm. The name under which this algorithm was proposed in Ralston and Jennrich [207], DUD (doesn't use derivatives), is clearly indicative of what the authors found to be a main feature of the algorithm. Essentially, the DUD algorithm can be considered a derivative-free Gauss-Newton algorithm. For instance, in the weighted least squares case,

$$J\left(\theta\right) = \sum_{i=1}^{Ndata} \frac{1}{\sigma_i^2} \left(y_i - \hat{y}_i\left(\theta\right)\right)^2$$

the Gauss-Newton algorithm approximates $\hat{y}_i\left(\theta\right)$ by a first order Taylor series expansion about the current value of the parameters θ, whereas the DUD algorithm approximates $\hat{y}_i\left(\theta\right)$ by a linear function $F(\theta)$ that exactly agrees with $\hat{y}_i\left(\theta\right)$ at $(p+1)$ points. This function describes a secant to the nonlinear function $\hat{y}_i\left(\theta\right)$.

The $(p+1)$ points are initially selected as in the Simplex method and an updating mechanism is used to replace one of the $(p+1)$ points. In the weighted least squares case, the objective function calculated with the secant has exactly one minimum which can be easily located. Usually, the point with the highest value of $J(\theta)$ is replaced by this minimum except when it is too close to the best solution found

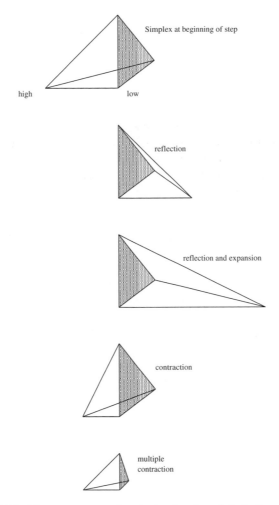

FIG. 6.18. Elementary operations in the simplex minimisation method.

so far. Then, the point with the highest value of J is contracted by a factor of 10 towards the best point to improve the approximation of J by its secant function at the next iteration. If, on the other hand, the minimum found from the secant approximation is larger than any value of the current $(p \mid 1)$ points, convergence is improved by a step reduction mechanism [213].

Reichert [213] extended the original algorithm to take into account individually weighted data and simple inequality constraints. He reported that the parameter transformation approach mentioned in the section on inequality constraints in this chapter could not be used for DUD because the unlimited step size of the

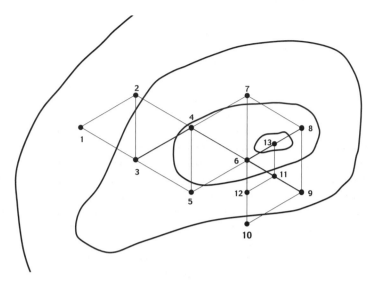

FIG. 6.19. The Simplex algorithm for a two-dimensional optimisation problem.

secant method would lead to rapid divergence of parameters lying on constraints. Rather, the alternative approach to deal with inequality constraints that is based on its translation into an equality constrained problem and minimising an objective function augmented with these constraints and Lagrange multipliers (e.g. [13]) was advocated.

6.6.5 *Global Minimisation Algorithms*

All local minimisation methods suffer from the problem that finding a local minimum does not ensure that a global minimum has been found. Consequently, global minimisation methods have been proposed which do not make use of local properties of the objective function. Thus, the problem of getting stuck in a local minimum is circumvented. However, this advantage of global methods is usually paid for by a higher number of function evaluations necessary for obtaining a solution.

Global optimisation methods can be grouped roughly into two main groups [226], the first of which consists of (purely) deterministic methods, such as gridding. The gridding methods consist of evaluating the objective function at a large number of points predefined on a grid laid in the parameter space. If the number of function evaluations is sufficiently high, there might be some chance to come at least close to minima. This method is clearly very inefficient because it does not allow for any scope of learning from the function evaluations. An obvious refinement is of course to refine the grid after a series of evaluations around the best value found so far.

The second group of global optimisation methods can be termed randomised methods since random decisions are involved in the search[10]. The degree of randomness varies significantly in different approaches: Uniform random search can be considered as being entirely probabilistic since the objective function is evaluated at every point of a random sample of points in parameter space. Again, as in the gridding approach, no use is being made of information gained in previous evaluations. Methods that do make use of such information are called adaptive random search methods. The main idea here is to perform evaluations of the objective function at points centred around promising points.

Among the first realisations of adaptive random search techniques is the Simulated Annealing technique [203]. The main idea is that the search is not always going towards a candidate solution (eventually ending up in a local minimum) but to allow – occasionally – also a step in a different direction. The probability of such random jumps is slowly decreased during the minimisation process. This method can be regarded as the predecessor of such currently popular methods as the genetic algorithms (GA) and controlled random search (CRS) techniques. These algorithms, which can also be termed "evolutionary" [225], start with an initial population of candidate solutions (a bit similar to the vertices in the simplex method) sampled randomly from the parameter space. The function values of the individuals of this population influence the process of generating new candidate solutions. Various techniques have been proposed for this process. In genetic algorithms [101] new generations of candidate solutions are obtained by imitation of the biological evolution processes of cross-over, mutation and selection of appropriately encoded representations (e.g. bit-strings) of populations of parameters. The definition of the parameters of the algorithm itself is crucial to the success of its application.

In the CRS methods [204], new candidate solutions are generated by reflections on the centre of gravity of the current set of candidate solutions. Duan *et al.* [81] developed the shuffled complex evolution (SCE) algorithm which constitutes an extension of the CRS procedure, to which elements of the simplex method of Nelder and Mead [184] and of competitive evolution were added. In SCE the initial set of candidate solutions is split into different communities ("complexes") which are allowed to develop separately by a combination of CRS and simplex search. From time to time, these complexes are mixed ("shuffled"), thereby passing on information about the parameter space gained independently by each community. It is generally accepted that the SCE method is robust and more efficient than a genetic algorithm procedure [226].

---

[10]It should be noted that many local minimisation algorithms include random steps to "jump" out of local minima, in this way allowing for increased robustness against this major problem of nonlinear minimisation.

6.6.6 *Comparison of Different Minimisation Algorithms*

Although many studies are certainly available in the literature, only a few are mentioned here for illustrative purposes. In a study by Tseng and Hsu [251] the estimates obtained from three linear transformations of a nonlinear model (Lineweaver & Burk, Eadie & Hofstee and Eisenthal & Cornish-Bowden) were compared with the estimates obtained with their own random search method. They evaluated the quality of fit (expressed as the sum of squared errors) and the computation time for the different methods. A remarkable result was that estimate bias became larger for the transformation-based methods when more data points were included for parameter estimation, again stressing the problems caused by the loss of the required error structure due to these transformations.

In 1992, Fuente and co-workers [94] evaluated a large range of minimisation algorithms for calibration of an activated sludge treatment plant model. They compared steepest descent and the Davidon-Fletcher-Powell methods as examples of unconstrained minimisation and compared different methods for constrained minimisation based on the penalty function method including infinite barriers in addition to log and inverse penalties. The conclusion of the work was that the minimisation algorithms had no difficulties in finding parameter estimates that fitted the data. However, although the fits are similar, the parameter estimates differ substantially which points to identifiability problems.

The study of Vanrolleghem and Keesman [264] compared a series of methods, including BFGS, Simplex, Brent, Random Search and Monte Carlo Set-Membership on identifying Monod-based models with three to six parameters on respirometric data. The results confirmed the relatively high rate of convergence of the BFGS and Brent methods and the robustness to local minima of the Brent, Simplex and the Random Search and Set-Membership methods.

Riefler *et al.* [217] also evaluated the BFGS, steepest descent and Rosenbrock methods on the estimation of biofilm kinetics. The steepest descent method converged too slowly to be practically useful, whereas the BFGS algorithm suffered from sensitivity to local minima as deduced from the sensitivity of its results on the initial parameter guesses. The authors concluded that the Rosenbrock method was most fit for their estimation problem which was characterised by a narrow objective function valley. The steep walls prevented derivative approximations with sufficient accuracy to determine a descent direction on the direction of the valley.

Finally, Schuetze [226] did a comprehensive evaluation of minimisation algorithms, comparing

- the Controlled Random Search (CRS) method,
- a genetic algorithm (GA) according to Caroll and a modification by Krishnakumar, called micro genetic algorithm (μGA), that uses very small population sizes and does not include mutation,

- the original Powell [202] method as presented in [203] without the modifications by [42] since they were considered too computationally demanding,
- a gridding of the parameter space where four values were evaluated in each parameter dimension,
- an approach in which simple substitute models were created for the objective function, i.e. a function was fitted to the objective function values obtained and, given an appropriate choice of the function, its minimum was obtained in an analytical way,
- the Response Surface Methodology (RSM) which requires the building of a quadratic substitute model f over the entire feasible parameter space. The $(p+1)(p+2)/2$ coefficients α in p variables (here the parameter values for which we look for a minimum in the objective function)

$$f\left(\theta_1, \cdots \theta_p\right) = \sum_{\substack{i, j = 1 \\ i \geq j}}^{p} \alpha_{ij}\theta_i\theta_j + \sum_{i=1}^{p} \alpha_{i0}\theta_i + \alpha_{00} \qquad (6.47)$$

are determined by least squares regression over all 3^p combinations of the smallest, mean and largest value of the range of each of the p parameters. The parameter estimates are then easily found by minimising f.

The best performance was obtained with the CRS method (giving the lowest objective function within a given maximum number of function evaluations). In this study, the Powell method was found to be quite sensitive to local minima and the methods based on substitute models for the objective function were found to work unsatisfactorily. As expected, allowing a larger number of function evaluations turned out more favourable for the genetic algorithms. As a side remark it was stated that the tangent transformation used to still work with constraints when algorithms for unconstrained minimisation are used, was not very successful in several minimisation runs. Apparently, when the search approaches a region close to the constraint, the tangent transformation (expanding even small differences in the parameters in such region) prevents convergence or further successful search steps of the procedure.

6.7 Evaluation of Parameter Estimation Quality

When a model is calibrated to a data set, it is essential to evaluate the success of this step, even before actual validation of the model is performed. Two aspects need to be evaluated. First, we need to know whether the fitting process itself went well (i.e. did we reach an adequate model fit ?). Second, we want to find out how accurate we have estimated the parameters and conclude whether this accuracy is sufficient for our purpose or whether additional data need to be collected (or a different model built). The first task will, for instance, be based on residuals analysis whereas the second task will typically focus on the parameter estimation confidence regions, which in most cases will be based on the parameter estimation error covariance matrix. We will, however, also draw attention to alternative methods for parameter accuracy evaluation because the covariance matrix can only be considered an approximation of the true parameter error distribution due to – again – the nonlinearity of the model. These alternative methods will, however, be characterised – again – by high computational demands making their use less popular than the approximate covariance matrix based analysis.

6.7.1 *Residuals Analysis*

Extensive methodology has been developed for investigating whether a calibrated model provides a good description of the data. The methods usually involve examination of the residuals, i.e. the differences between the observed data and the model predictions. It is quite evident that these methods have therefore also been used for structure characterisation (see Chapter 3). Basically, the whole methodology of residuals analysis is based on the feature that an adequate model leaves only measurement error in the residuals. Hence, if the model is appropriate and the model calibration process has been successful, we may expect that the residuals are characterised in the same way as the measurement errors, for instance randomness, homoscedasticity (constant variance), normal distribution. A comparison of their characteristics therefore allows us to conclude whether the calibration was successful.

Analysis of randomness can be based on autocorrelation assessment methods as introduced in Chapter 3 for model selection purposes, i.e. the autocorrelation tests, the runs test, ... A note should be made, however, when correlation analysis is performed on the basis of small data sets. The modeller can indeed be misled when his examination is based on the residuals' (in)dependency since there are only $N_{data}\text{-}p$ degrees of freedom left among them after the model identification [208]. On the other hand, the presence of autocorrelation may have statistically serious consequences, more particularly it will lead to an underestimation of the size of the confidence regions, or in other words, we will get too much confidence in our parameter estimates [19]. Thus, investigators should not delude themselves into thinking that more data points are better for least squares estimation of parameters, unless they know that the measurement errors are uncorrelated [218].

Other residuals analysis methods evaluate whether the other assumptions made when choosing the objective function J are fulfilled. For instance, the residuals' homoscedasticity may be evaluated by plotting the residuals against the independent variable (mostly time) or against the measured variable (see for instance Figure 6.2). Trends in the residuals sequence may suggest a switch from an ordinary least squares to a weighted least squares objective function. Testing whether the residuals are indeed normally distributed can be done by performing adequate distribution analysis tests found in statistical handbooks.

6.7.2 *Parameter Estimation Error Covariance Matrix Determination*

Even if the nonlinearity of the model parameters makes it only an approximation and other methods are in principle more correct (see below), the parameter estimation error covariance matrix is the corner stone of parameter estimation accuracy evaluation. Indeed, once this covariance matrix is available, one may use several statistical techniques to evaluate the quality of the estimated parameters. In the next sections some of these techniques are discussed, namely:

- Computation of confidence intervals.
- Determine and draw confidence region ellipses for two parameters.
- Determine whether the zero vector lies within the confidence region.
- Computation and evaluation of correlation values.

Because of its importance, we will therefore review the different methods that are being used to obtain the parameter estimation error covariance matrix.

Linear model.   In case we are dealing with a linear parameter estimation problem (see the example in Section 6.6.1), $Y = X\theta$ where Y is the vector of output measurements and X the regressor vector, we obtain that the estimation error covariance matrix can be readily calculated as (under iidN assumption of the residuals):

$$V = \left(X^T X\right)^{-1} \sigma^2$$

where σ^2 is the measurement error, typically estimated as $s^2 = J_{opt}/(N_{data}-p)$. In case modelling errors exist, one needs to have an independent measure of the measurement error s_e^2, which can substitute for σ^2. It must be stressed that the matrix inversion that is needed may be troublesome. Indeed, these matrices are often characterised by poor condition numbers due to highly correlated parameters (remember the Modified E criterion for optimal experiment design that could allow this inversion problem to be solved, see Chapter 5).

In case the measurement errors are not equal (but still random and distributed normal), and the weighting of the errors in the objective function J is achieved via the measurement error covariance matrix Q (which is a diagonal matrix if the measurements are uncorrelated but with different variance σ_i^2):

$$Q = \begin{bmatrix} \frac{1}{\sigma_1^2} & 0 & \cdots & 0 \\ 0 & \frac{1}{\sigma_2^2} & \cdots & 0 \\ \vdots & \vdots & \ddots & \vdots \\ 0 & 0 & \cdots & \frac{1}{\sigma_{Ndata}^2} \end{bmatrix}$$

the parameter estimation error covariance matrix is again readily calculated, i.e.

$$V = \left(X^T Q X\right)^{-1}$$

In some cases this value of V is multiplied with a term $\chi^2/(N_{data}\text{-}p)$ where χ^2 is the minimum value obtained for the objective function J_{opt} in this weighted least squares parameter estimation. Indeed, whereas the above equation gives the parameter variance as a function of measurement errors only, the multiplication with χ^2 leads to more reasonable (larger) estimates of errors in cases in which the standard deviations of the measurements do not take all sources of error into account (e.g. some modelling error). The danger involved in employing this multiplication with χ^2 is that systematic errors may be treated as statistical errors [213]. The multiplication with χ^2 is, however, useful when the σ_i^2 are not given in absolute terms, but merely as relative magnitudes, something which typically happens in case no detailed analysis of the measurement errors is conducted (as for instance in the example of Section 6.4.3.). In that case the χ^2 value obtained as a result of the parameter estimation corrects for this.

From the above, it is clear that calculating the covariance matrix does not involve a lot of calculations as it only requires some matrix manipulations with the available data.

Nonlinear model. For nonlinear parameter estimation problems an approximate parameter error covariance matrix can be calculated by replacing the X-terms in the above by the output sensitivity functions with respect to the parameters, i.e.

$$V = \left(\frac{\partial y}{\partial \theta}^T Q \frac{\partial y}{\partial \theta}\right)^{-1}$$

where $\partial y/\partial \theta$ is a vector of the output sensitivities at each of the N_{data} measurement points. Note that again we see the link between the parameter estimation error covariance matrix and the Fisher Information Matrix indicated in Chapter 5.

We immediately can see that the actual evaluation of this covariance matrix will involve many more calculations. Indeed, the evaluation of the sensitivity function either requires a simultaneous solution of a considerable set of differential equations (see the simple example given in Chapter 5). Alternatively, a numerical approximation of the sensitivity functions can be made by performing p additional simulations around the nominal trajectory where each parameter is perturbed with a small perturbation $\delta\theta$. The adequate choice of this perturbation parameter $\delta\theta$ is the Achilles' heel of this method (see also the similar Nelder and Mead [184] method below).

Luckily, some minimisation algorithms such as the Levenberg-Marquardt and Newton algorithms need or approximate (quasi-Newton and Gauss-Newton methods) the Hessian, which is the inverse of the covariance matrix. Only some minor modification to these algorithms is therefore needed to get access to this Hessian matrix which after (careful) inversion leads to the parameter estimation error co-variance matrix V. Some other algorithms also build up information that leads to an approximation of the Hessian, e.g. Brent's algorithm. However, Van Vooren [269] learnt that this approximation is too crude in some parameter estimations since the method only asymptotically converges to the Hessian.

The method proposed by Spendley *et al.* [237], and extended by Nelder and Mead [184], can also be adopted. It allows the covariance matrix around the minimum for any parameter estimation method to be calculated as it is basically an add-on exercise. The technique is based on the construction of a quadratic surface around the minimum of the cost function f. If $(p+1)$ points in p dimensions are given by P_0, P_1, P_p, then "half-way points" $P_{ij}=(P_i+P_j)/2$, $i \neq j$ are calculated, and a quadratic surface is fitted to this combined set of $(p+1)(p+2)/2$ points. The points P_i may be taken as:

$$P_0 = \left(\hat{\theta}_1, \hat{\theta}_2, \cdots, \hat{\theta}_p \right)$$
$$P_1 = \left(\hat{\theta}_1 + \delta_1, \hat{\theta}_2, \cdots, \hat{\theta}_p \right)$$
$$P_2 = \left(\hat{\theta}_1, \hat{\theta}_2 + \delta_2, \cdots, \hat{\theta}_p \right)$$
$$\vdots$$
$$P_p = \left(\hat{\theta}_1, \hat{\theta}_2, \cdots, \hat{\theta}_p + \delta_p \right)$$

in which $\hat{\theta}_i$ is the estimated optimum parameter value and δ_i is the step size, a user-defined small value or a value automatically chosen as function of the machine precision. In order not to exceed the parameter boundaries (e.g. under constrained optimisation), the step may be chosen positive or negative.

The matrix with the step sizes δ is called the direction matrix D.

$$D = \begin{bmatrix} \delta_1 & 0 & \cdots & 0 \\ 0 & \delta_2 & \cdots & 0 \\ \vdots & \vdots & \ddots & \vdots \\ 0 & 0 & \cdots & \delta_p \end{bmatrix}$$

A quadratic approximation to the objective function J in the neighbourhood of the minimum can be obtained using Taylor series expansion

$$J'(\theta) = J(P) + \sum_i \left. \frac{\partial J}{\partial \theta_i} \right|_P \theta_i + \frac{1}{2} \sum_{i,j} \left. \frac{\partial^2 J}{\partial \theta_i \partial \theta_j} \right|_P \theta_i \theta_j + \cdots \quad \cong c - b \cdot \theta + \frac{1}{2} \theta \cdot A \cdot \theta$$

where $c \equiv J(P)$ $b \equiv -\nabla J|_P$ $[A]_{ij} \equiv \left. \frac{\partial^2 J}{\partial \theta_i \partial \theta_j} \right|_P$

The coefficients of the Hessian matrix A can be estimated as:

$$a_{ii} = 2\left(y_i + y_0 - 2y_{0i}\right) \qquad i = 1, \cdots, p$$
$$a_{ij} = 2\left(y_{ij} + y_0 - y_{0i} - y_{0j}\right) \qquad i = 1, \cdots, p \; ; \; j = 1, \cdots, p \; ; \; i \neq j$$

where y_i are the values of the objective function J at P_i and y_{ij} those at P_{ij}. The Hessian or Fisher Information Matrix in the original coordinate system is given by[11]

$$2\left(D^{-1}\right)^T A D^{-1}$$

so that the parameter estimation error covariance matrix is given by

$$\frac{1}{2} D A^{-1} D^T$$

In many cases, the sum of squares of residuals is minimised, and normal equal-variance independent errors are assumed, thus this matrix must be multiplied by $2\sigma^2$ [184]. As usual σ^2 is estimated by $J_{opt}/(N_{data}-p)$.

Van Vooren [269] implemented this technique and investigated the effect of the step sizes δ_i. It was found that the final results were not much influenced by different choices of the δ_i's. However, the choice of the step size will depend on the rounding errors, and it is advised in Nelder and Mead [184]) that the step size should be at least 10^3 times that rounding error. A too large step size should be avoided too because then the linear approximation of the objective function may no longer hold (depending on its level of nonlinearity). Hence, a compromise must be sought. A pragmatic approach is to perform the calculation with halved step size and compare the results. If they are sufficiently close, they can be accepted.

Clearly, one of the drawbacks of this method is that it requires a considerable number of additional function evaluations [213]. However, in many parameter estimation cases, the extra calculation time necessary to run this extra algorithm is negligible compared to the overall parameter estimation time.

6.7.3 Confidence Regions

In order to give a meaningful value for the estimated parameters, they should be supplied together with a confidence region. A confidence region is a p-dimensional interval in which, with a certain probability (e.g. 68.3 %), the true parameter (a p-dimensional vector) lies. The most common confidence region is the one-dimensional region, the confidence interval. A visualisation of a two-dimensional confidence region and a test whether the (n-dimensional) 0 lies within the n-dimensional confidence region are also given below.

---

[11]Note that in the original paper Nelder and Mead [184], the factor 2 was erroneously omitted as was corrected in an erratum to this paper in *Comput. J.*, 8, 27, 1965.

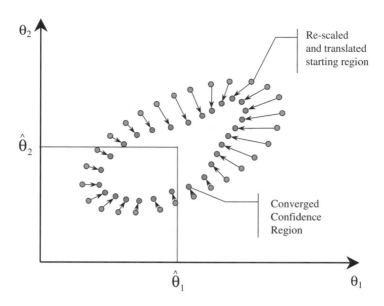

FIG. 6.20. Exact confidence region determination for two parameters θ_1 and θ_2 using the [162] approach.

First, however, we will discuss the actual nature of the confidence region in a nonlinear parameter estimation problem. Let us start from a linear parameter estimation problem in which a weighted least squares objective function was minimised and in which the residuals can be assumed to be independent distributed normal. Under those conditions, the confidence region of the parameter estimates can be exactly calculated from the covariance matrix. As soon as the model is nonlinear in the parameters, this covariance matrix that we have constructed with the methods given above, only allows to approximate the true confidence region. The exact boundaries of a $100(1\text{-}\alpha)\%$ confidence[12] region for the parameter estimate $\hat{\theta}$ are defined as those parameters for which the following equality holds [18]:

$$J_{crit} = J_{opt} \cdot \left(1 + \frac{p}{N_{data} - p} F_{\alpha;\, p, N_{data} - p} \right) \qquad (6.48)$$

where J_{opt} is the minimum value found for the objective function and $F_{\alpha;\, p, N_{data} - p}$ is the value of the F-distribution with p and $N_{data} - p$ degrees of freedom and a confidence level α.

The $100(1\text{-}\alpha)\%$ confidence region for $\hat{\theta}$ is then the locus of values for θ which result in a value of J below this critical value.

---

[12]In case we do not know the true residuals' distribution, we are unable to obtain a specified probability level [80]. For residuals with normal distribution, we do have that certainty.

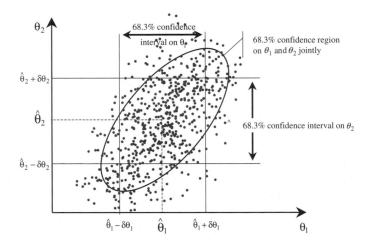

FIG. 6.21. Terminology of confidence intervals in the two-dimensional case.

The problem of finding this boundary may, however, be computation demanding as no easy analytical calculation is feasible. Rather, we have to search for those parameter sets fulfilling the above (in)equalities, i.e. we have to run a considerable number of additional simulations. Typical approaches are to evaluate the parameter space around the parameter estimate $\hat{\theta}$ over a grid or via random sampling using Monte Carlo techniques. However, this method quickly becomes impractical as the number of parameters increases or if the desired resolution increases. Consequently, Lobry *et al.* [162] proposed an elegant, pragmatic method for finding the confidence region boundary with a minimum of additional simulations. The method is introduced in Figure 6.20. First, an initial trial confidence region is proposed that completely encloses the true confidence region, i.e. all $J(\theta)$ should be larger than the value given by the Beale-formula above. Then an algorithm is initiated that makes each of the points converge towards the centre, i.e. where the parameter estimate $\hat{\theta}$ is found. This convergence is stopped when a $J(\theta)$ is found that corresponds to the critical J value given by Beale and a certain α.

Confidence intervals.   The difference between a confidence interval in one and two dimensions is illustrated in Figure 6.21. The same fraction of evaluated points lies (i) between the vertical lines, (ii) between the two horizontal lines and (iii) within the ellipse. Overall, a confidence interval of a parameter θ_i is given by $[\theta_i - \delta\theta_i, \theta_i + \delta\theta_i]$ where

$$\delta\theta_i = \sqrt{\Delta\chi_1^2}\sqrt{V_{ii}}$$

V_{ii} is the diagonal element of the covariance matrix corresponding to the i-th parameter. $\Delta\chi_1^2$ will be obtained from distribution data for a given $100(1-\alpha)\%$ confidence and $(N_{data}-p)$ degrees of freedom. For instance, when the 68, 90 and 99%

confidence intervals are to be calculated for one parameter, the $\Delta \chi_1^2$ values to be used are 1.00, 2.71 and 6.63 respectively [203]. For two parameters these $\Delta \chi_1^2$ values would be 2.30, 4.61 and 9.21.

Draper and Smith [80] rightly point to the danger of regarding the square in Figure 6.21 given by the two horizontal and two vertical interval lines, as the joint confidence region. In that case the correlation between the two parameters is completely ignored. The "joint" message of individual confidence intervals should be regarded with caution.

Finally, the marginal confidence intervals of a parameter are given by the projection of the confidence region on the corresponding parameter axis and are therefore bigger than the normal confidence intervals specified above [160].

Confidence region ellipsoids. A confidence region ellipsoid (Fig. 6.22) (two-dimensional confidence region) can be drawn on the basis of the parameter estimation error covariance matrix using the following equation

$$\lambda_i \left(\left[w_{ii} w_{ij} \right] \delta\theta \right)^2 + \lambda_j \left(\left[w_{ji} w_{jj} \right] \delta\theta \right)^2 = \Delta \chi_2^2$$

where $\delta\theta = [\delta\theta_i \quad \delta\theta_j]^T$ and $\Delta \chi_2^2$ are taken for a given $100(1-\alpha)\%$ confidence region. The ellipsoids will be obtained using

$$\delta\theta_i = \frac{\Delta \chi_2^2}{\lambda_i} w_{ii} \cos(\phi) - \frac{\Delta \chi_2^2}{\lambda_j} w_{ij} \sin(\phi)$$

$$\delta\theta_j = \frac{\Delta \chi_2^2}{\lambda_i} w_{ij} \cos(\phi) - \frac{\Delta \chi_2^2}{\lambda_j} w_{jj} \sin(\phi)$$

where ϕ varies between 0 and 2π. The w_{ij} and the λ_i are the elements of the eigenvector and eigenvalue decomposition of the covariance matrix V:

$$V = \left[V_{ij} \right] = W \cdot D^{-1} \cdot W^T \qquad W = \left[w_{ij} \right] = \left[\underline{w}_i \right] \qquad D = diag\,(\lambda_i)$$

The centre of the ellipse will be the vector of estimated parameter values.

6.7.4 *Significance of a Single Parameter*

To test whether a parameter estimate is significant, can be done easily by calculating the following t-value [208]

$$t = \frac{\hat{\theta}_i}{\sqrt{V_{ii}}}$$

i.e. the ratio of the parameter estimate $\hat{\theta}_i$ to its standard error, estimated by the square root of the variance of the estimate V_{ii}. This value can then be tested by reference to a Student's t-distribution with $(N_{data}-p)$ degrees of freedom and a user-defined confidence level $100(1-\alpha)\%$. A high t-value associated with a parameter estimate tends to indicate that the estimate is well determined in the model; conversely, a low t-value tends to indicate that the estimate is poorly determined, although sometimes in multiparameter models, a t-value may be low because of high correlation of the parameter with other parameters in the model.

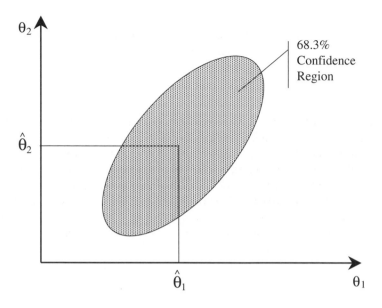

FIG. 6.22. A confidence region ellipsoid for the parameters θ_1 and θ_2.

6.7.5   *Significance of the Parameter Set*

Testing whether *0* lies within the *p*-dimensional *100(1-α)%* confidence region is relevant as it allows assurance that the parameter set found is significantly different from the zero vector. This test can be performed by evaluating whether

$$\Delta\chi_v^2 > \sum_{i=1}^{p} \lambda_i \left(\underline{w}_i \cdot (-\theta)\right)^2$$

Here $\underline{w}_i$ (the columns of W) and λ_i are the eigenvectors and the corresponding eigenvalue of V. The value $\Delta\chi_v^2$ is obtained for a user-supplied *100(1-α)%* confidence level and a $F_{\alpha;p,N_{data}-p}$ distribution. If this inequality holds, *0* will lie within the specified confidence region.

6.7.6   *Correlation Among Parameters*

The correlation coefficient r_{ij} between two parameters θ_i and θ_j gives a measure for the correlation between these two parameters. If r_{ij} is close to *-1* or *1* the parameters are said to be highly correlated. An r_{ij} close to *0* implies a low correlation. The correlation coefficient is given by

$$r_{ij} = \frac{V_{ij}}{\sqrt{V_{ii}V_{jj}}}$$

When a high correlation is encountered this may point to practical identifiability problems that may, for instance, be solved by collecting additional data from an optimally designed experiment (see Chapter 5).

6.8 Conclusions

The purpose of the present chapter was to introduce quite different aspects of parameter estimation from a complete data set. In the next chapter we will focus on how this work can also be done on-line, i.e. with a data set that is continuously growing. To clearly make the difference between both types of estimation, the chapter was introduced with simple, illustrative examples. Next necessary preliminary steps for a parameter estimation problem were introduced, such as the selection of parameters to be estimated, reparametrisation of the model, the guessing of initial values for the parameters. An important section also dealt with the way one deals with constraints on parameter values when using unconstrained minimisation algorithms. Quite some part of this chapter was dedicated to the different aspects of measurement errors since the method that is to be used for parameter estimation depends on the error characteristics. From this and other features, the objective function of the parameter estimation can be selected among the different ones presented in Section 5.5. The real calculation work is then focused upon in Section 6.6, where some linear and nonlinear parameter estimation algorithms are introduced in a rather qualitative manner. Note again that it is important to realise that most of the algorithms presented in this chapter are basically optimisation algorithms that can also be useful in other tasks than parameter estimation, e.g. optimisation of an experiment design. Last but not least, the chapter ends with methods for the assessment of the quality of the parameter estimation performed. It also adresses the important question of the quality of the parameter estimates themselves and the generation of confidence information.

7

Recursive State and Parameter Estimation

7.1 Introduction

This final chapter is dedicated to a question closely related to those already addressed in the preceding chapters. Up to now, we have built models, or we have selected one of them for our specific applications, and we have shown how to calibrate its parameters. In terms of modelling, we can consider that the task is completed. We can use our model(s) for simulating numerically the dynamical behaviour of our processes, possibly for predicting and/or analysing their behaviour. But we can also use the models for monitoring, i.e. to predict the time evolution of the process variables and parameters on-line. The developed monitoring tools can be used to follow the time evolution of variables and/or parameters that are not accessible from on-line measurements; they can also be used for diagnosis about the operation of the plant and help the operator or a supervision system to take appropriate actions to maintain the process in good (ideally, "optimal") operating conditions, diagnose possible process, sensor or actuator failures, or prevent accidents. The supervision can (should) be connected to control loops.

In the following, we shall call these monitoring tools *software sensors*. A software sensor can be defined as an algorithm built from a dynamical model of (part of) the process to estimate on-line unmeasured variables (typically process component concentrations like biomass, substrates or products) and/or unknown (or badly known) parameters (e.g. specific reaction rates, or some other kinetic or yield co-efficients) from the (few) measurements that are available on-line (typically liquid flow rates, nutrient concentrations (N, P), respiration rate, turbidity, pH, ORP, or gaseous outflow rates). In that sense, these tools can be viewed as "sensors" based on an algorithm (software), i.e. in our nomenclature, as *software sensors*.

In this chapter, we shall introduce two types of software sensors.

1. State observers for reconstructing on-line the time evolution of unmeasured process component concentrations.

2. On-line parameter estimators for unknown or badly known parameters.

The need for monitoring systems and automatic control in order to optimise process operation or detect disturbances in wastewater treatment processes is obvious. Generally speaking, the problems arising from the implementation of these processes are similar to those of more classical industrial processes. Nevertheless, monitoring and automatic control of wastewater treatment processes is clearly developing very slowly. There are at least two main reasons for this:

a) The internal working and dynamics of these processes are as yet badly grasped and many problems of methodology in modelling remain to be solved. It is difficult to develop models taking into account the numerous factors which can influence the specific bacterial growth rate and the yield coefficients which characterise microorganism growth. The modelling effort is often tedious and requires a great number of experiments before producing a reliable model. Reproducibility of experiments is often uncertain due to the difficulty in obtaining the same environmental conditions. Moreover, as these processes involve living organisms, their dynamic behaviour is strongly nonlinear and non-stationary, as we already noted in Chapter 2. Model parameters may not remain constant over a long period[13]. They will vary e.g. due to metabolic variations of biomass or to random and unobservable physiological or genetic modifications. It should also be noted that the lack of accuracy of the measurements often leads to identifiability problems (see Chapter 5).

b) Another essential difficulty lies in the absence, in most cases, of cheap and reliable instrumentation suited to real-time monitoring. To date, the market offers very few sensors capable of providing reliable on-line measurements of the biological and biochemical parameters required to implement high performance automatic control strategies. The main variables, i.e. biomass, substrate and product concentrations, generally need determination through

---

[13]This may appear ambiguous with the terminology used in this monograph where we consider parameters as being "constant" for all times. However due to the simplicity of the considered models with regard to the complexity of the processes, it appears that different values of the parameter are sometimes necessary to describe satisfactorily the process dynamics in different conditions.

laboratory analyses. The cost and duration of the analyses obviously limit the frequency of the measurements and their timely availability.

To reconstruct the state of the system from the only on-line available measurements and to control [14] [192] biological variables such as the biomass, the substrates or the products, appropriate algorithms have to be developed. The efficiency of any monitoring system highly depends on the design of the monitoring techniques and the care taken in their design. Indeed, monitoring algorithms will prove to be efficient if they are able to incorporate the important well-known information on the process while being able to deal with the missing information (lack of on-line measurements, uncertainty on the process dynamics,...) in a "robust" way, i.e. such that the missing information will not significantly deteriorate the performance of the monitoring system.

In this chapter, we shall show how to incorporate well-known knowledge about the dynamics of wastewater treatment processes (basically, the reaction network and the material balances) in monitoring algorithms which may moreover be capable of dealing with the process uncertainty (in particular on the reaction kinetics) via on-line estimation schemes for the uncertain kinetic parameters.

The chapter is organised as follows. Section 7.2 will concentrate on the notion of state observability as applied to WWTP. The following sections will deal with state observation, i.e. the on-line reconstruction of process components, via different approaches:

1. "classical observers" (extended Luenberger observers, and extended Kalman filter) (Section 7.3);
2. "asymptotic" observers (which are observers independent of the process kinetics) (Section 7.4);
3. an intermediate class of observers, in between classical observers and asymptotic observers, applicable when the reaction rate model structure is known but the model parameters are badly known (Section 7.5).

Section 7.6 will introduce two types of algorithms for recursive estimation of uncertain parameters: a least squares estimation algorithm, and an observer-based estimator. Practical issues and real-life results will be presented in both sections.

7.2   State Observability

Simply speaking, the notion of observability can be defined as the possibility to connect the state variables of a dynamical system to the measured variables via the dynamical model of the system. Essentially, a system is observable if every state variable of the system affects some of the process variables. An important consequence of the observability of a system is the ability to reconstruct the time evolution of the state variables from measured variables in an arbitrary finite time from any initial conditions (see e.g. [151]). This notion is very important in practical applications because it implies that if a system is observable, then it is formally

possible to design an "observer" that is theoretically capable of correctly reconstructing the time evolution of the unmeasured variables, theoretically after a finite
arbitrarily chosen time.

Observability is clearly a critical issue in dynamical systems in general and in
chemical and biochemical systems in particular. For instance, the implementation
of Kalman or Luenberger observers for an application to bioreactors is based on
the a priori knowledge of the observability of the process. Because of the nonlinear
aspects of their dynamics, the observability analysis is rather complex in biochemical process applications; and the usually large uncertainty in the kinetics of the
biochemical reactions and analytical expressions used to describe them makes the
approach even more difficult. As a matter of fact, very few works deal with the observability of nonlinear biochemical processes (e.g. [3], [7]) and they are usually
concerned with particular process applications.

Let us see how the concept of observability can be formalised mathematically,
or more precisely how we can test if a process is observable or not.

For simplicity, we shall consider the notion of observability for *linear* systems
(in the state), and for which the matrices (A and B) have constant parameters (i.e.
time invariant systems). Let us consider the following dynamical equation system:

$$\frac{dx}{dt} = Ax + Bu \tag{7.1}$$

with $\dim(x) = N$. Consider that the measured variables (the *output*) are denoted y
($\dim(y) = p$) and are related to the state variables x by the following relationship:

$$y = Cx \tag{7.2}$$

where C is a matrix (The simplest obvious case is when the measured variables y
are some of the process components (states); then C is a matrix whose entries are
"1" or "0"). The linear system (7.1) (7.2) is said to be observable if the following
matrix O

$$O = \begin{bmatrix} C \\ CA \\ CA^2 \\ \vdots \\ CA^{N-1} \end{bmatrix} \tag{7.3}$$

is full rank. It is well known (e.g. [151]) that the linear stationary system (7.1),
(7.2), is observable if and only if

$$rank(O) = N \tag{7.4}$$

Let us now consider linear approximations of nonlinear models, e.g. of the General
Dynamical Model introduced in Chapter 2:

$$\frac{d\xi}{dt} = -D\xi + Y\rho(\xi) + F(\xi) - Q(\xi) \tag{7.5}$$

If we define $f(\xi, D, F, Q)$ as follows:

$$f(\xi, D, F, Q) = -D\xi + Y\rho(\xi) + F(\xi) - Q(\xi) \tag{7.6}$$

the linear approximation of the General Dynamical Model (7.5) around some operating point obtained from the Taylor series expansion of the model equations (also called the *linearised tangent model*, see also Section 2.7.3 in Chapter 2) is written as follows:

$$\frac{dx}{dt} = A(\bar{\xi})x + B(\bar{\xi})u \tag{7.7}$$

where A, B, x are equal to:

$$A(\bar{\xi}) = [\frac{\partial f}{\partial \xi}]_{\xi=\bar{\xi}} = -\bar{D}I_N + Y[\frac{\partial \rho}{\partial \xi}]_{\xi=\bar{\xi}} + [\frac{\partial (F-Q)}{\partial \xi}]_{\xi=\bar{\xi}} \tag{7.8}$$

$$B(\bar{\xi}) = [\frac{\partial f}{\partial u}]_{\xi=\bar{\xi}} \tag{7.9}$$

$$x = \xi - \bar{\xi} \tag{7.10}$$

and u is the input vector which may be constituted by the values varying around the solution $\xi = \bar{\xi}$ of some of the feedrates F_i and/or the dilution rate D.

The observability test can still be applied to the linear model (7.7), but it is worth noting that then the observability test (7.4) is only a sufficient observability condition for the nonlinear system (7.5) (e.g. [152]): if the linearised tangent model (7.7)-(7.10) is observable at $\xi = \bar{\xi}$, then the nonlinear system (7.5) is observable around this point.

7.2.1 *Example #1: Two Step Nitrification Process*

Let us consider the two step nitrification model introduced in Section 2.3.4:

$$S_{NH} \longrightarrow X_1 + S_{NO_2} \tag{7.11}$$

$$S_{NO_2} \longrightarrow X_2 + S_{NO_3} \tag{7.12}$$

characterised by the following matrices and vectors in the General Dynamical Model format:

$$\xi = \begin{bmatrix} S_{NH} \\ S_{NO_2} \\ S_{NO_3} \\ X_1 \\ X_2 \end{bmatrix}, \quad Y = \begin{bmatrix} -\frac{1}{Y_1} & 0 \\ Y_3 & -\frac{1}{Y_2} \\ 0 & Y_4 \\ 1 & 0 \\ 0 & 1 \end{bmatrix}, \quad F = \begin{bmatrix} DS_{NH,in} \\ 0 \\ 0 \\ 0 \\ 0 \end{bmatrix} \tag{7.13}$$

$$\rho = \begin{bmatrix} \rho_1 \\ \rho_2 \end{bmatrix}, \quad Q = 0 \tag{7.14}$$

If we assume that the reaction rates are only a function of the components intervening in the reaction (i.e. $\rho_1(S_{NH}, X_1, S_{NO_2})$ and $\rho_2(S_{NO_2}, X_2, S_{NO_3})$), the matrix A of the linearised tangent model is equal to:

$$A(\bar{\xi}) =$$

$$
\begin{bmatrix}
-D - \frac{1}{Y_1}\frac{\partial \rho_1}{\partial S_{NH}} & -\frac{1}{Y_1}\frac{\partial \rho_1}{\partial S_{NO_2}} & 0 & -\frac{1}{Y_1}\frac{\partial \rho_1}{\partial X_1} & -\frac{1}{Y_1}\frac{\partial \rho_2}{\partial X_2} \\
Y_3\frac{\partial \rho_1}{\partial S_{NH}} & -D + Y_3\frac{\partial \rho_1}{\partial S_{NO_2}} - \frac{1}{Y_2}\frac{\partial \rho_2}{\partial S_{NO_2}} & -\frac{1}{Y_2}\frac{\partial \rho_2}{\partial S_{NO_3}} & Y_3\frac{\partial \rho_1}{\partial X_1} & -\frac{1}{Y_2}\frac{\partial \rho_2}{\partial X_2} \\
0 & Y_4\frac{\partial \rho_2}{\partial S_{NO_2}} & -D + Y_4\frac{\partial \rho_2}{\partial S_{NO_3}} & 0 & \frac{\partial \rho_2}{\partial X_2} \\
\frac{\partial \rho_1}{\partial S_{NH}} & \frac{\partial \rho_1}{\partial S_{NO_2}} & 0 & \frac{\partial \rho_1}{\partial X_1} & 0 \\
0 & \frac{\partial \rho_2}{\partial S_{NO_2}} & \frac{\partial \rho_2}{\partial S_{NO_3}} & 0 & \frac{\partial \rho_2}{\partial X_2}
\end{bmatrix}
$$

It can be checked that all the states of the linearised tangent model with the above matrix A is observable if S_{NH}, S_{NO_2} and S_{NO_3} are measured, and is not observable if the measured variables are S_{NO_2} and S_{NO_3}, or if these are S_{NH}, X_1 and S_{NO_2} when the reaction rate is independent of S_{NO_3}.

7.2.2 Simple Local Observability Tests

Let us now see how to derive a very simple necessary observability condition for the system (7.7), (7.8), (7.9), (7.10). For simplicity, we shall consider that F is independent of ξ and that there is no gaseous outflow rate ($Q = 0$). Let us first define the matrix $\tilde{A}(\bar{\xi})$:

$$\tilde{A}(\bar{\xi}) = Y[\frac{\partial \rho}{\partial \xi}]_{\xi=\bar{\xi}} = Y\rho_{\xi} \tag{7.15}$$

where ρ_{ξ} is a more compact notation for $[\frac{\partial \rho}{\partial \xi}]_{\xi=\bar{\xi}}$. Let us denote by $\tilde{O}$, the "observability" matrix computed from $\tilde{A}(\bar{\xi})$. Then we have the following results for the observability of the system (7.7), (7.8), (7.9), (7.10).

Theorem 1: $\operatorname{rank}(O) \leq \min(N, p + R)$
with $R = \operatorname{rank}(Y)$, $N = \dim(x)$, $p = \dim(y)$

Proof: see [71]

The following important consequence can be derived from Theorem 1.

Corollary 1: *A necessary condition for the observability of (7.7), (7.8), (7.9), (7.10) is that:*

$$p + R \geq N \tag{7.16}$$

i.e. the number of measured components + the rank of the yield coefficient matrix must be larger than or equal to the number of process components.

Comment #1: Theorem 1 gives an upper bound for the rank of the observability matrix of system (7.7), (7.8), (7.9), (7.10). This result is obviously local since it applies to the linearised tangent model of (7.5). But it is worth noting that the result is generic in the sense that it is independent of the mathematical structure of the reaction rates $\rho(\xi)$.

Comment #2: the above condition is a necessary condition but not a sufficient one. If it can be useful to detect in a very simple manner the possible lack of observability of the process, the fulfillment of condition $p + R \geq N$ does not guarantee its observability. This will be illustrated in Example #2 here below.

7.2.3 *Example #1: Two Step Nitrification Process (Continued)*

From Theorem 1, we know that system (7.15) will be unobservable if two (or less) components are measured on-line ($R + p \leq 4 < 5$).

7.2.4 *Example #2: Simple Microbial Growth Process*

Let us consider a simple microbial growth process involving one substrate S and one population of microorganisms X, and synthesising a reaction product P. The process can be described by the following reaction scheme:

$$S \longrightarrow X + P \tag{7.17}$$

The dynamical behaviour of the process in a stirred tank reactor is described by the following equations:

$$\frac{d}{dt} \begin{bmatrix} S \\ X \\ P \end{bmatrix} = -D \begin{bmatrix} S \\ X \\ P \end{bmatrix} + \begin{bmatrix} -\frac{1}{Y_1} \\ 1 \\ Y_2 \end{bmatrix} \rho + \begin{bmatrix} DS_{in} \\ 0 \\ 0 \end{bmatrix} \tag{7.18}$$

where Y_1 and Y_2 are the yield coefficients and $\rho = \mu X$. The matrix $\tilde{A}$ of the linearised tangent model is here equal to:

$$\tilde{A}(\bar{S}, \bar{X}, \bar{P}) = \begin{bmatrix} -\frac{1}{Y_1}\rho_S & -\frac{1}{Y_1}\rho_X & -\frac{1}{Y_1}\rho_P \\ \rho_S & \rho_X & \rho_P \\ Y_2\rho_S & Y_2\rho_X & Y_2\rho_P \end{bmatrix} \tag{7.19}$$

with:

$$\rho_i = [\frac{\partial \rho}{\partial i}]_{\xi = \bar{\xi}} \qquad\qquad i = S, X, P \qquad\qquad (7.20)$$

From Theorem 1, we know that the linearised system of (7.18) will be unobservable if only one component is measured on-line (R + p = 2 < 3). As a matter of fact, if, for instance, P is available for on-line measurement, the "observability" matrix $\tilde{O}$ is equal to:

$$\tilde{O} = \begin{bmatrix} 0 & 0 & 1 \\ Y_2\rho_S & Y_2\rho_X & Y_2\rho_P \\ Y_2\rho_S\bar{\rho} & Y_2\rho_X\bar{\rho} & Y_2\rho_P\bar{\rho} \end{bmatrix} \qquad\qquad (7.21)$$

with $\bar{\rho} = -\frac{1}{Y_1}\rho_S + \rho_X + Y_2\rho_P$.

It is obvious that $\tilde{O}$ is not full rank (the last two rows are proportional to each other). Now, if two components are available for on-line measurement (e.g. S and P), the "observability" matrix becomes:

$$\tilde{O} = \begin{bmatrix} 1 & 0 & 0 \\ 0 & 0 & 1 \\ -\frac{1}{Y_1}\rho_S & -\frac{1}{Y_1}\rho_X & -\frac{1}{Y_1}\rho_P \\ Y_2\rho_S & Y_2\rho_X & Y_2\rho_P \\ -\frac{1}{Y_1}\rho_S\bar{\rho} & -\frac{1}{Y_1}\rho_X\bar{\rho} & -\frac{1}{Y_1}\rho_P\bar{\rho} \\ Y_2\rho_S\bar{\rho} & Y_2\rho_X\bar{\rho} & Y_2\rho_P\bar{\rho} \end{bmatrix} \qquad\qquad (7.22)$$

It will be full rank as long as ρ_X is different from zero. In the particular (but widely encountered) situation when the specific growth rate μ is only a function of the limiting substrate S, we can expect the process to be observable whenever $\mu(S)$ is different from zero (which only happens when $S = 0$ with most kinetic models available in the literature (see [14])).

Finally, not any choice of two components will guarantee the observability of the process. Assume for instance in line with many practical situations that the reaction rate ρ is independent of the product concentration P, i.e. $\rho_P = 0$. Consider now that S and X are measured on-line. Then the "observability" matrix $\tilde{O}$ is equal to:

$$\tilde{O} = \begin{bmatrix} 1 & 0 & 0 \\ 0 & 1 & 0 \\ -\frac{1}{Y_1}\rho_S & -\frac{1}{Y_1}\rho_X & 0 \\ Y_2\rho_S & Y_2\rho_X & 0 \\ -\frac{1}{Y_1}\rho_S\bar{\rho} & -\frac{1}{Y_1}\rho_X\bar{\rho} & 0 \\ Y_2\rho_S\bar{\rho} & Y_2\rho_X\bar{\rho} & 0 \end{bmatrix} \qquad\qquad (7.23)$$

with:

$$\tilde{\rho} = -\frac{1}{Y_1}\rho s + \rho x \tag{7.24}$$

It is obvious that since the third column is equal to zero, rank($\tilde{O}$) < 3 and therefore, the linearised system is not observable. However the result is straightforward to obtain by simply looking at the linearised tangent model equations of S, X, P for ρ_P equal to zero and by recalling that an important condition for the observability of the process is that the dynamical equations of the measured variables must incorporate connections with the unmeasured variables. In the example here, the dynamical equations of S and X no longer contain any term related to P if ρ is taken independent of P.

7.3   Classical Observers

7.3.1   *The Basic Structure of a State Observer*

An interesting alternative to on-line measurement for monitoring biomass, reactant and product concentrations that circumvents and exploits the use of a model in conjunction with a limited set of measurements is the use of Luenberger or Kalman observers. In these techniques, a model which includes states that are measured as well as states that are not measured is used in parallel with the process. The model states may then possibly be used for feedback. This configuration may be used to reduce the effect of noise on measurements as well as to reconstruct the states that are not measured. An introduction to these ideas can be found in e.g. Kwakernaak and Sivan[151].

Let us derive the general structure of state observers. Let us consider the following nonlinear state space model:

$$\frac{dx}{dt} = f(x, u) \tag{7.25}$$

The measured variables, denoted y, are related to the process states by the following relation:

$$y = h(x) \tag{7.26}$$

The general structure of a state observer is then written as follows:

$$\frac{d\hat{x}}{dt} = f(\hat{x}, u) + K(\hat{x})(y - \hat{y}) \tag{7.27}$$

where $\hat{x}$ and $\hat{y}$ are the on-line estimations of x and y given by the state observer (7.27) and:

$$\hat{y} = h(\hat{x}) \tag{7.28}$$

and where $K(\hat{x})$ is the "gain" of the observer. The above observer equation can be interpreted as a copy of the process model plus a correction term $K(y - \hat{y})$ proportional to the output observation error $y - \hat{y}$. This term is indeed the driving term of the observer; it disappears in presence of perfect estimation (obviously, under the assumption that the process model is perfect too). The design of the state observer consists of choosing an appropriate gain $K(\hat{x})$.

The above state observer was originally developed for linear problems. Because of the nonlinear characteristics of bioprocess dynamics, it is of interest to extend these concepts and exploit particular structures for biochemical engineering application problems. The design of the gain matrix K is based on a linearised version (the linearised tangent model) of the process dynamics observation error (computed from a Taylor's series expansion of a state space model around some operating point. If we define the observation error e as follows:

$$e = x - \hat{x} \tag{7.29}$$

the dynamics of the observation error are readily derived from equations (7.25) and (7.27):

$$\frac{de}{dt} = f(\hat{x} + e, u) - f(\hat{x}, u) - K(\hat{x})(h(\hat{x} + e) - h(\hat{x})) \tag{7.30}$$

If we consider the linearisation of the above equation around the observation error $e = 0$, we obtain:

$$\frac{de}{dt} = (A(\hat{x}) - K(\hat{x})C(\hat{x}))e \tag{7.31}$$

where $A(\hat{x})$ and $C(\hat{x})$ are respectively equal to:

$$A(\hat{x}) = [\frac{\partial f}{\partial x}]_{x=\hat{x}}, \ C(\hat{x}) = [\frac{\partial h}{\partial x}]_{x=\hat{x}} \tag{7.32}$$

Thus the design problem can be formulated as the choice of the matrix $K(\hat{x})$ such that the linearised error dynamics (7.31) has desired properties. This has resulted in two typical state observation designs: the extended Luenberger observer, and the extended Kalman observer. The word "extended" obviously emphasises that these observers are extensions of the original linear versions to nonlinear systems. These modified observers, particularly the extended Kalman filter (EKF), have found applications in some biochemical processes (e.g. [21], [22], [24] [242], [153], [287], [47]).

7.3.2 *Extended Luenberger Observer*

In the extended Luenberger observer, the objective is to select $K(\hat{x})$ such that the linearised error dynamics (7.31) are asymptotically stable (i.e. the error converges to zero). This is achieved by choosing $K(\hat{x})$ such that (see e.g. [14]):

1. the matrix $A(\hat{x}) - K(\hat{x})C(\hat{x})$ and its time derivative are bounded:

$$\|A(\hat{x}) - K(\hat{x})C(\hat{x})\| \le C_1, \qquad \forall \hat{x} \tag{7.33}$$

$$\|\frac{d}{dt}(A(\hat{x}) - K(\hat{x})C(\hat{x}))\| \le C_2, \qquad \forall \hat{x} \tag{7.34}$$

2. the eigenvalues of $A(\hat{x}) - K(\hat{x})C(\hat{x})$ have strictly negative real parts:

$$Re(\lambda_i[A(\hat{x}) - K(\hat{x})C(\hat{x})]) \leq C_3 < 0, \quad \forall \hat{x} \text{ and } i = 1 \text{ to } N \qquad (7.35)$$

The importance of the requirement of state observability of the linearised system becomes clearer at this point: if the linearised system is not observable, it is then not possible to assign freely the dynamics of the observation error (or in other words, to have perfect estimation after some defined time), i.e. to select arbitrary eigenvalues λ_i. An example will be given in Section 7.3.4 here below.

7.3.3 Extended Kalman Observer

Although the Kalman filter has been originally introduced in a stochastic framework, it can also interpreted as the solution of a (deterministic) optimisation problem (see e.g. [14], [66]). Indeed the design of the extended Kalman observer consists of finding the gain matrix $K(\hat{x})$ that minimises the mean square observation error:

$$E = \int_0^t e^T W e \, d\tau \qquad (7.36)$$

with the dynamical model (7.31) ("under the constraints of the dynamical model (7.31)", in the usual notations of optimisation theory). W is a weighting matrix that allows different weightings to the different terms of the error e with a view to standardise the error norm, e.g. when the different components of e are not of the same dimension.

The gain matrix $K(\hat{x})$ can be shown to be equal to:

$$K(\hat{x}) = R(\hat{x})C^T \qquad (7.37)$$

where the N × N symmetric matrix $R(\hat{x})$ is a solution of the following dynamical Riccati matrix equation:

$$\frac{dR}{dt} = -RC^T W C R + R A^T(\hat{x}) + A(\hat{x})R, \quad R = R^T, \ R(0) = R_0 = R_0^T \quad (7.38)$$

As we shall see here below, the choice of the intial values of the entries of the matrix R plays an important role in the convergence properties of the extended Kalman observer.

7.3.4 Application to the General Dynamical Model

Let us see how to design extended Luenberger and Kalman observers to the General Dynamical Model (7.5). Let us consider that some (exactly, p) process components are accessible for on-line measurement, i.e. the output vector is equal to:

$$y = C\xi \qquad (7.39)$$

with C an elementary matrix (i.e. with only "0" and "1" as entries). The matrix $A(\hat{\xi})$ of the linearised tangent model associated to the General Dynamical Model (7.5) is equal to:

$$A(\hat{\xi}) = -DI_N + Y[\frac{\partial \rho}{\partial \xi}]_{\xi=\hat{\xi}} + [\frac{\partial(F - Q)}{\partial \xi}]_{\xi=\hat{\xi}} \qquad (7.40)$$

The observer equations specialise as follows:

$$\frac{d\hat{\xi}}{dt} = -D\hat{\xi} + Y\rho(\hat{\xi}) + F - Q(\hat{\xi}) + K(\hat{x})(y - C\hat{\xi}) \qquad (7.41)$$

Then the design equations presented precedingly apply.

Example: simple microbial growth process. Let us consider the following example, i.e. a simple microbial growth process in a continuous stirred tank reactor (CSTR)[14]:

$$S \longrightarrow X \qquad (7.42)$$

Recall that the model dynamics in a CSTR are given by the following equations:

$$\frac{dS}{dt} = -\frac{1}{Y_1}\mu X + DS_{in} - DS \qquad (7.43)$$

$$\frac{dX}{dt} = \mu X - DX \qquad (7.44)$$

With $\xi = [S, X]^T$, the matrix A is now equal to:

$$A = \begin{bmatrix} -\frac{1}{Y_1}\hat{\rho}_S - D & -\frac{1}{Y_1}\hat{\rho}_X \\ \hat{\rho}_S & \hat{\rho}_X - D \end{bmatrix} \qquad (7.45)$$

Let us consider as a matter of illustration one particular kinetic model for the specific growth rate μ, e.g. the Haldane model:

$$\mu = \frac{\mu_0 S}{K_S + S + \frac{S^2}{K_I}} \qquad (7.46)$$

Assume that the substrate concentration S is available for on-line measurement while the biomass concentration X is not. The state observer specialises as follows. The output matrix C and the gain matrix K of the observer are equal to:

$$C = [1 \ \ 0], \ \ K = \begin{bmatrix} k_1 \\ k_2 \end{bmatrix} \qquad (7.47)$$

---

[14]The choice of the bioreactor example is arbitrary, yet it will be used throughout the following sections as the illustration support.

And the matrix $A - KC$ is equal to:

$$A - KC = \begin{bmatrix} -\frac{1}{Y_1}\hat{\rho}_S - D - k_1 & -\frac{1}{Y_1}\hat{\rho}_X \\ \hat{\rho}_S - k_2 & \hat{\rho}_X - D \end{bmatrix} \qquad (7.48)$$

The observability matrix is equal to:

$$O = \begin{bmatrix} 1 & 0 \\ -\frac{1}{Y_1}\hat{\rho}_S - D & -\frac{1}{Y_1}\hat{\rho}_X \end{bmatrix} \qquad (7.49)$$

By computing the determinant of the matrix O, we can conclude that the system is observable if $\hat{\rho}_X$ is different from zero ($\hat{\rho}_X \neq 0$). If $\rho = \mu X$, this means that $\mu + \frac{\partial \mu}{\partial X} X$ must be different from zero. For the Haldane model, this means that μ must be different from zero, which may happen only if $S = 0$. Hence the system is observable from S.

The observer equations are then equal to:

$$\frac{d\hat{S}}{dt} = -\frac{1}{Y_1}\mu_{max}\frac{\hat{S}}{K_S + \hat{S} + \frac{S^2}{K_I}}\hat{X} + DS_{in} - D\hat{S} + k_1(S - \hat{S}) \qquad (7.50)$$

$$\frac{d\hat{X}}{dt} = \mu_{max}\frac{\hat{S}}{K_S + \hat{S} + \frac{S^2}{K_I}}\hat{X} - D\hat{X} + k_2(S - \hat{S}) \qquad (7.51)$$

The extended Luenberger observer design consists of choosing the eigenvalues of the matrix $A - KC$. In practice, this can be achieved by computing the characteristic polynomial of $A - KC$ (i.e. $\det(\lambda I - A + KC)$) and comparing it with the characteristic polynomial corresponding to the desired observer dynamics. If you select a desired error dynamics with stable real poles, this latter polynomial is here equal to $(\lambda + \lambda_1)(\lambda + \lambda_2)$ with λ_1 and λ_2 strictly positive (we have two poles because the observer is of order two). The above computation formalises as follows:

$$det(\lambda I - A + KC) = \lambda^2 + \lambda(k_1 + 2D + \frac{1}{Y_1}\hat{\rho}_S - \hat{\rho}_X)$$

$$+(\frac{1}{Y_1}\hat{\rho}_S + D + k_1)(D - \hat{\rho}_X) + \frac{1}{Y_1}\hat{\rho}_X(\hat{\rho}_S - k_2) \quad (7.52)$$

$$= (\lambda + \lambda_1)(\lambda + \lambda_2) \qquad (7.53)$$

By identifying the different terms of the two characteristic polynomials (7.52) (7.53), we obtain the following relationships:

$$k_1 + 2D + \frac{1}{Y_1}\hat{\rho}_S - \hat{\rho}_X = \lambda_1 + \lambda_2 \qquad (7.54)$$

$$(\frac{1}{Y_1}\rho_S + D + k_1)(D - \hat{\rho}_X) + \frac{1}{Y_1}\hat{\rho}_X(\hat{\rho}_S - k_2) = \lambda_1\lambda_2 \qquad (7.55)$$

This leads to the following values for the gains k_1 and k_2 from a chosen set of poles $-\lambda_1$ and $-\lambda_2$:

$$k_1(\hat{S}, \hat{X}) = -2D - \frac{1}{Y_1}\hat{\rho}_S + \hat{\rho}_X + \lambda_1 + \lambda_2 \qquad (7.56)$$

$$k_2(\hat{S}, \hat{X}) = \hat{\rho}_S - \frac{Y_1}{\hat{\rho}_X}[\lambda_1\lambda_2 - (\frac{1}{Y_1}\hat{\rho}_S + D + k_1)(D - \hat{\rho}_X)] \qquad (7.57)$$

The extended Luenberger observer is then given by equations (7.50)(7.51) with the values of k_1 and k_2 computed as above. The implementation of the observer requires explicit values for $\hat{\rho}_S$ and $\hat{\rho}_X$ in the above expression. For the Haldane model, these are equal to:

$$\hat{\rho}_S = \frac{\mu_0\hat{X}(K_S - \hat{S}^2/K_I)}{(K_S + \hat{S} + \frac{\hat{S}^2}{K_I})^2}, \quad \hat{\rho}_X = \frac{\mu_0\hat{S}}{K_S + \hat{S} + \frac{\hat{S}^2}{K_I}} \qquad (7.58)$$

Note that the values of the observer gains k_1 and k_2 are changing with time as a function of the estimates $\hat{S}$ and $\hat{X}$.

Let us now consider the extended Kalman observer. The matrix R has the following form here:

$$R = \begin{bmatrix} r_1 & r_3 \\ r_3 & r_2 \end{bmatrix} \qquad (7.59)$$

Then the extended Kalman observer is given by equations (7.50)(7.51) in which the gains k_1 and k_2 are calcuated according to the following equations:

$$k_1 = r_1 \qquad (7.60)$$

$$k_2 = r_3 \qquad (7.61)$$

$$\frac{dr_1}{dt} = -r_1^2 - 2(\frac{1}{Y_1}\hat{\rho}_S + D)r_1 - 2\frac{1}{Y_1}\hat{\rho}_X r_3 \qquad (7.62)$$

$$\frac{dr_2}{dt} = -r_3^2 + 2(\hat{\rho}_X - D)r_2 + 2\hat{\rho}_S r_3 \qquad (7.63)$$

$$\frac{dr_3}{dt} = -r_1 r_3 + \hat{\rho}_S r_1 - \frac{1}{Y_1}\hat{\rho}_X r_2 + (\hat{\rho}_X - \frac{1}{Y_1}\hat{\rho}_S - 2D)r_3 \qquad (7.64)$$

The last three equations (7.62)(7.63)(7.64) are the Riccati equations.

Recall that in the preceding section we have mentioned that the lack of observability will not allow the observer dynamics to be assigned. Let us illustrate this.

Let us consider the simple microbial growth process with a product P (Example #2 in Section 7.2). We have seen that the model is not observable with one

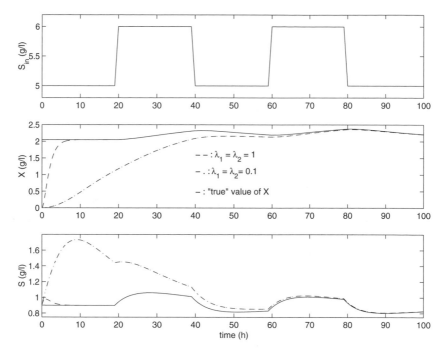

FIG. 7.1. Simulation results of the extended Luenberger observer.

measured component. As in Section 7.2, let us consider that P is the measured component. The matrix $A - KC$ is then equal to:

$$A - KC = \begin{bmatrix} -\frac{1}{Y_1}\hat{\rho}_S - D & -\frac{1}{Y_1}\hat{\rho}_X & -\frac{1}{Y_1}\hat{\rho}_P - k_1 \\ \hat{\rho}_S & \hat{\rho}_X - D & \hat{\rho}_P - k_2 \\ Y_2\hat{\rho}_S & Y_2\hat{\rho}_X & Y_2\hat{\rho}_P - D - k_3 \end{bmatrix} \quad (7.65)$$

It is then routine to check that one of the eigenvalues of the matrix $A - KC$ is equal to $-D$, whatever the values of the gains k_1, k_2 and k_3.

This can be straightforwardly deduced from the characteristic polynomial:

$$det(\lambda I - A + KC) = (\lambda + D)(\lambda^2 + \alpha\lambda + \beta) \quad (7.66)$$

with:

$$\alpha = k_3 + 2D + \frac{1}{Y_1}\hat{\rho}_S - \hat{\rho}_X - Y_2\hat{\rho}_P \quad (7.67)$$

$$\beta = (k_3 + D)(D - \hat{\rho}_X + \frac{1}{Y_1}\hat{\rho}_S) + Y_2(k_1\hat{\rho}_S + k_2\hat{\rho}_X - D\hat{\rho}_P) \quad (7.68)$$

The time response of the observation error is then typically equal to the weighted sum of exponential terms with negative exponents, where one is equal to e^{-Dt}. It

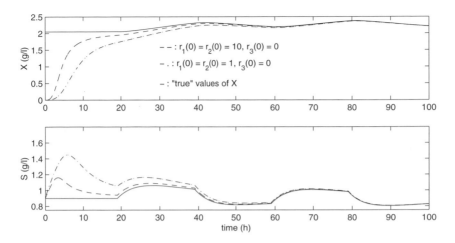

FIG. 7.2. Simulation results of the extended Kalman observer.

is then hopeless to try to look for a convergence rate faster than $1/D$: whatever the values of the observer gains, one pole of the observer will remain equal to $-D$! In the automatic control terminology, the process is not observable (because it is not possible to assign freely the dynamics of the observer) but is detectable (because the dynamics of the observer error is asymptotically stable due to the presence of three poles $(-D, -\alpha/2 - \sqrt{\alpha^2 - 4\beta}/2$ and $-\alpha/2 + \sqrt{\alpha^2 - 4\beta}/2)$ whose real part is negative).

Simulation results.   Let us illustrate the performances of the extended Luenberger observer and of the extended Kalman observer on the above simple microbial growth process with Haldane kinetics, and on the simple microbial growth process with one product P.

The parameters and initial conditions of the simulation model are based on experimental values from a degradation process of lactoserum by *Rhodopseudomonas capsulata* ([275]); they have been set to the following values:

$$\mu_0 = 0.33\ h^{-1},\ K_S = 5\ g/l,\ K_I = 25\ g/l,\ Y_1 = 0.5,\ Y_2 = 0.6$$
$$S(0) = 0.9\ g/l,\ X(0) = 2.05\ g/l,\ P(0) = 1.22\ g/l$$
$$S_{in}(0) = 5\ g/l,\ D(0) = 0.05\ h^{-1}$$

Figures 7.1 and 7.2 illustrate the performances of the extended Luenberger observer (ELO) and of the extended Kalman observer (EKO), respectively with a square variation of the influent substrate S_{in} for different observer gain values. The higher the values of the poles λ_1 and λ_2 of the ELO (from 0.1 to 1, here), the quicker the convergence of the estimated values. Similarly, the higher the initial values of the entries of the matrix R (from 1 to 10), the better the convergence of

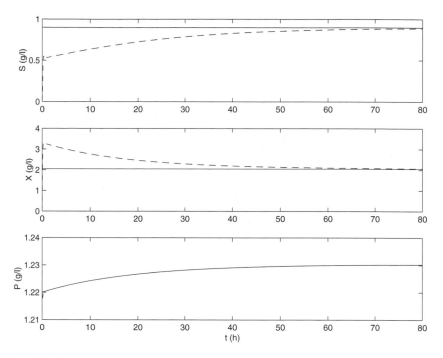

FIG. 7.3. Simulation results of the extended Luenberger observer for an unobservable process.

the EKO. The initial values of the estimates in the observers have been set to the following values: $\hat{X}(0) = 0 \ g/l$, $\hat{S}(0) = 0.9 \ g/l$ (ELO & EKO) = 0 g/l (ELO for the unobservable process), $\hat{P}(0) = 1.22 \ g/l$ (ELO for the unobservable process).

Figure 7.3 illustrates the inability to assign arbitrarily the dynamics of a classical observer (here, the ELO) when the system is unobservable, as is the case for the simple microbial growth with only one on-line measurement (here, P). The chosen poles λ_1 and λ_2 correspond to much faster dynamics than the residence time $1/D$ ($\lambda_1 = \lambda_2 = 10 >> D = 0.05$). We note on Figure 7.3 that after a fast transient, the dynamics of the observer are dominated by the residence time, i.e. $1/D$ (it takes about three times the residence (60 hours) for the observer to converge).

7.3.5   *Performance of Classical Observers in Presence of Model Uncertainties*

Let us now illustrate the performance of a classical observer when some of the model parameters are badly known. Let us consider here the ELO applied to the estimation of the biomass concentration X from measurements of the substrate concentration S in a simple microbial growth process[15]. Let us consider Monod kinetics[16]:

$$\mu = \mu_{max} \frac{S}{K_S + S} \qquad (7.69)$$

with μ_{max} the maximum specific growth rate (h^{-1}) and K_S the saturation constant (g/l). Assume that the yield coefficient Y_1 is known while the kinetic parameters μ_{max} and K_S may be such that their values may have been determined with some uncertainty. The observer equations are then equal to:

$$\frac{d\hat{S}}{dt} = -\frac{1}{Y_1}\tilde{\mu}_{max}\frac{\hat{S}}{\tilde{K}_S + \hat{S}}\hat{X} + DS_{in} - D\hat{S} + k_1(S - \hat{S}) \qquad (7.70)$$

$$\frac{d\hat{X}}{dt} = \tilde{\mu}_{max}\frac{\hat{S}}{\tilde{K}_S + \hat{S}}\hat{X} - D\hat{X} + k_2(S - \hat{S}) \qquad (7.71)$$

$\hat{S}$ and $\hat{X}$ represent the estimations of S and X given by the observer. The parameters $\tilde{\mu}_{max}$ and $\tilde{K}_S$ used in the observer computation may be different from their "true" values μ_{max} and K_S.

Recall that in the ELO for this system, the gains k_1 and k_2 are selected as follows:

$$k_1 = \lambda_1 + \lambda_2 - \frac{1}{Y_1}\frac{\tilde{K}_S\tilde{\mu}_{max}\hat{X}}{(K_S + \hat{S})^2} + \frac{\tilde{\mu}_{max}^2\hat{S}}{\tilde{K}_S + \hat{S}} - 2D \qquad (7.72)$$

$$k_2 = Y_1\frac{\tilde{K}_S + \hat{S}}{\tilde{\mu}_{max}\hat{S}}[-\lambda_1\lambda_2 + (\lambda_1 + \lambda_2)(D - \frac{\tilde{\mu}_{max}\hat{S}}{\tilde{K}_S + \hat{S}})$$
$$-(D - \frac{\tilde{\mu}_{max}\hat{S}}{\tilde{K}_S + \hat{S}})^2 + \frac{1}{Y_1}\frac{\tilde{\mu}_{max}\tilde{K}_S\hat{S}\hat{X}}{(\tilde{K}_S + \hat{S})^3}] \qquad (7.73)$$

in order to have observer dynamics assigned to desired values λ_1 and λ_2 (see Section 7.3.1). Let us test the performance of the observer with a wrong value of one of the kinetic parameters. The numerical simulation conditions are the following:

$$Y = 0.5, \quad \mu_{max} = 0.33\,h^{-1}, \quad K_S = 5\,g/l, \quad S_{in} = 5\,g/l$$
$$D = 0.05\,h^{-1}, \quad X(0) = 1\,g/l, \quad S(0) = 0.5\,g/l$$

The observer has been initialised with $\hat{S}(0) = S(0)$ and $\hat{X}(0) = 0$. Figure 7.4 shows the estimation results with 10 % error on K_S ($\tilde{K}_S = 4.5$ g/l instead of 5) for two sets of design parameters k_1 and k_2 (one corresponding to $\lambda_1 = \lambda_2 = -0.1$ ("slow" dynamics), the other for $\lambda_1 = \lambda_2 = -10$ ("fast" dynamics))[17]. The results are quite

[15]The choice of the extended Luenberger is obviously arbitrary and is not based on any priori idea about its advantages and drawbacks.

[16]The choice of the Monod model is obviously also arbitrary. For instance, we could have as well chosen the Haldane model in the preceding section.

[17]Similar results are obtained with an error on μ_{max}.

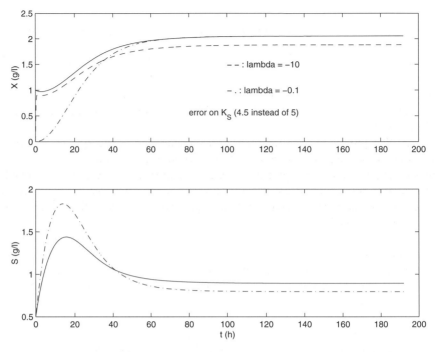

FIG. 7.4. Extended Luenberger observer with a wrong value of K_S (initial transient)(-: "true" simulated values).

typical of classical observers but also of any "high gain" observers for nonlinear systems: the higher the gain, the worse the state estimation results. Indeed high gain (fast dynamics) will result in a good estimation of the measured variable (here S)(it is impossible to distinguish on the figure between S and $\hat{S}$!) while rejecting the parameter uncertainty on the estimated variable (here X). This simple example shows that there is a need to develop observers that can handle parameter uncertainty, a typical situation encountered in (bio)chemical process applications.

7.4 Asymptotic Observers

One of the reasons for the popularity of the EKO/ELO is that it is easy to implement since the algorithm can be derived directly from the state space model. However, since (as the ELO) it is based on a linearised model of the process, the stability and convergence properties are essentially local and valid only around some operating point, and it is rather difficult to guarantee its stability over wide ranges of operation. Ljung [158] shows that the EKO for state and parameter estimation of linear systems may give biased estimates or even diverge if it is not carefully initialised.

One reason for the problem of convergence of EKO/ELO is that, in order to guarantee the (arbitrarily chosen) exponential convergence of the observer, the process must be locally observable, i.e. the linearised tangent model must be observable and fulfill the classical observability rank condition. This condition, as it turns out, is restrictive in many practical situations (as illustrated in Figure 7.3) and may account for the failure of EKO/ELO to find widespread application (e.g. [14], [16], [71]).

Another problem is that the theory for the extended Luenberger and Kalman observers is developed using a perfect knowledge of the system model and parameters, in particular of the process kinetics: it is difficult to develop error bounds and there is often a large uncertainty on these parameters. The performance of ELO/EKO in the presence of badly known parameters has been illustrated in the preceding Section 7.3.5.

It appears from the above remarks that there is a clear incentive to develop alternative methodologies for the on-line estimation of the unmeasured concentration variables in wastewater treatment processes that do not rely on parts of the dynamical model that may be largely uncertain, and that do not require in particular the explicit use of kinetic models. Indeed, the objective of this section is to propose an alternative to EKO/ELO and use process mechanisms in a more direct manner to develop a nonlinear observer applicable to the estimation problem of wastewater treatment processes. The proposed observer is based on the well-known nonlinear model of the process without the knowledge of the process kinetics being necessary. In order to advance the application of this method, we discuss its stability and convergence properties. We would like to emphasise that the presented results are global (i.e. independent of the initial conditions) as opposed to the local properties for EKO/ELO (see e.g.[14]).

This section is organised as follows. We shall first present the general methodology for single tank bioprocesses and discuss its theoretical convergence properties and the practical implementation aspects. Then we shall present a real-life application on an anaerobic digestion process. Finally we shall introduce the extension to fixed bed reactors.

7.4.1   *Asymptotic Observers for Single Tank Bioprocesses*

The derivation of the asymptotic observer equations are based on the Key State Transformation introduced in Section 2.8 and on the following assumptions:

1. p ($\geq M$ (the number of reactions)) components are measured on-line.
2. The feedrates F, the gaseous outflow rates Q and the dilution rate D are known either by measurement or by choice of the user.
3. The yield coefficient matrix Y is known.
4. The reaction rate vector ρ is unknown.
5. The M reactions are irreversible and independent, i.e. rank$(Y) = R = M$

From assumption 1, we can define the following state partition:

$$\xi = \begin{bmatrix} \xi_1 \\ \xi_2 \end{bmatrix} \tag{7.74}$$

where ξ_1 and ξ_2 hold for the measured component concentrations and the unmeasured ones, respectively.

Let us recall the state transformation ζ introduced in Chapter 2 (Section 2.8):

$$\zeta = C_a \xi_a + C_b \xi_b \tag{7.75}$$

where C_a and C_b are solutions of the matrix equation:

$$C_a Y_a + C_b Y_b = 0 \tag{7.76}$$

and with the following dynamical equations (independent of the reaction rate ρ!):

$$\frac{d\zeta}{dt} = -D\zeta + C_a(F_a - Q_a) + C_b(F_b - Q_b) \tag{7.77}$$

Let us consider one (arbitrarily chosen) transformation ζ defined by (7.75, 7.76). The variable ζ can be rewritten as a linear combination of the measured and unmeasured states ξ_1 and ξ_2, i.e.:

$$\zeta = A_1 \xi_1 + A_2 \xi_2 \tag{7.78}$$

The equations (7.78)(7.77) are the basis for the derivation of the asymptotic observer. The dynamical equations of ζ are used to calculate an estimate of ζ on-line, which is used, via equation (7.78) and the on-line data of ξ_1, to derive an estimate of the unmeasured component ξ_2. Let us further assume that the matrix A_2 is (left) invertible. The following two cases can be differentiated.

Case #1: $p = M$

In this case, the asymptotic observer is written as follows:

$$\frac{d\hat{\zeta}}{dt} = -D\hat{\zeta} + C_a(F_a - Q_a) + C_b(F_b - Q_b) \tag{7.79}$$

$$\hat{\xi}_2 = A_2^{-1}[\hat{\zeta} - A_1 \xi_1] \tag{7.80}$$

Comment: if we consider the most simple and straightforward choice for the state transformation ζ, i.e. with $\xi_1 = \xi_a$ and $\xi_2 = \xi_b$, and with $C_b = I_{N-p}$, then we have:

$$A_1 = -Y_2 Y_1^{-1}, \quad A_2 = I_{N-p} \tag{7.81}$$

And therefore the condition on the invertibility of A_2 is in fact a condition on the invertibility of the submatrix Y_1 (i.e. Y_1 is full rank or rank(Y_1) = M). This condition is indeed fulfilled from assumption 5 (independent reactions) and if the measured variables are independent.

Case #2: p > M

Assume that the number of measured components is larger than the number which is strictly necessary than the one needed, i.e. $p > M$. Then the asymptotic observer is modified as follows:

$$\frac{d\hat{\zeta}}{dt} = -D\hat{\zeta} + C_a(F_a - Q_a) + C_b(F_b - Q_b) \tag{7.82}$$

$$\hat{\xi}_2 = A_2^+[\hat{\zeta} - A_1\xi_1] \tag{7.83}$$

where A_2^+ is a left inverse of A_2 (see the example (case #3) here below).

The observer (7.79)(7.80) or (7.82)(7.83) is completely independent of the process kinetics and can be implemented without the knowledge of the reaction rates $\rho(\xi)$ being required.

Theoretical convergence of the asymptotic observer.   The convergence properties can be summarised in the following theorem.

Theorem 2: *If the dilution rate D is a persistently exciting signal, i.e. if there exist positive constants δ and β such that:*

$$\int_t^{t+\beta} D(\tau)d\tau \geq \delta > 0 \tag{7.84}$$

then:

$$\lim_{t\to\infty} (\xi_2 - \hat{\xi}_2) = 0 \tag{7.85}$$

Proof: the proof of the theorem is immediate if one observes that, from (7.77), (7.79), (7.80), (7.83), the dynamics of the estimation error is equal to:

$$\frac{d(\xi_2 - \hat{\xi}_2)}{dt} = -D(\xi_2 - \hat{\xi}_2) \tag{7.86}$$

QED

Remark #1: the persistence-of-excitation condition on D simply requires that D is not equal to zero for too long. This condition is clearly easily fulfilled in fedbatch and continuous reactors.

Remark #2: the general formulation of the estimation algorithm (7.82), (7.83) with the introduction of the left inverse allows for larger flexibility in the use of the asymptotic observer since it permits the possible presence of a number of measured variables larger than M to be taken into account. Let us illustrate this in the following example.

Example: anaerobic digestion process. Let us consider the simplified two step anaerobic digestion reaction network:

$$S_1 \longrightarrow X_1 + S_2 + P_1 \qquad \text{(acidogenesis)} \qquad (7.87)$$

$$S_2 \longrightarrow X_2 + P_1 + P_2 \qquad \text{(methanisation)} \qquad (7.88)$$

where S_1, S_2, X_1, X_2, P_1 and P_2 represent the organic matter, the volatile fatty acids, the acidogenic bacteria, the methanogenic bacteria, the carbon dioxide, and the methane, respectively. The associated general dynamical model is given by the following vectors and matrices:

$$\xi = \begin{bmatrix} S_1 \\ S_2 \\ X_1 \\ X_2 \\ P_1 \\ P_2 \end{bmatrix}, \; Y = \begin{bmatrix} -\frac{1}{Y_1} & 0 \\ Y_3 & -\frac{1}{Y_2} \\ 1 & 0 \\ 0 & 1 \\ Y_4 & Y_5 \\ 0 & Y_6 \end{bmatrix}, \; F = \begin{bmatrix} DS_{in} \\ 0 \\ 0 \\ 0 \\ 0 \\ 0 \end{bmatrix}, \; Q = \begin{bmatrix} 0 \\ 0 \\ 0 \\ 0 \\ Q_1 \\ Q_2 \end{bmatrix} \quad (7.89)$$

$$\rho = \begin{bmatrix} \rho_1 \\ \rho_2 \end{bmatrix} = \begin{bmatrix} \mu_1 X_1 \\ \mu_2 X_2 \end{bmatrix} \qquad (7.90)$$

Let us first define one state transformation ζ, e.g.:

$$\xi_a = \begin{bmatrix} X_1 \\ X_2 \end{bmatrix}, \qquad \xi_b = \begin{bmatrix} S_1 \\ S_2 \\ P_1 \\ P_2 \end{bmatrix} \qquad (7.91)$$

with:

$$Y_a = \begin{bmatrix} 1 & 0 \\ 0 & 1 \end{bmatrix}, \qquad Y_b = \begin{bmatrix} -\frac{1}{Y_1} & 0 \\ Y_3 & -\frac{1}{Y_2} \\ Y_4 & Y_5 \\ 0 & Y_6 \end{bmatrix} \qquad (7.92)$$

Therefore if C_b is chosen as an identity matrix ($C_b = I_4$), then C_a is equal to:

$$C_a = -Y_b Y_a^{-1} = \begin{bmatrix} \frac{1}{Y_1} & 0 \\ -Y_3 & \frac{1}{Y_2} \\ -Y_4 & -Y_5 \\ 0 & -Y_6 \end{bmatrix} \qquad (7.93)$$

The dynamics of ζ are here equal to:

$$\frac{d}{dt} \begin{bmatrix} \zeta_1 \\ \zeta_2 \\ \zeta_3 \\ \zeta_4 \end{bmatrix} = -D \begin{bmatrix} \zeta_1 \\ \zeta_2 \\ \zeta_3 \\ \zeta_4 \end{bmatrix} + \begin{bmatrix} DS_{in} \\ 0 \\ -Q_1 \\ -Q_2 \end{bmatrix} \qquad (7.94)$$

Note that since $F_a = Q_a = 0$, the dynamics of ζ are also independent of the yield coefficients. This has been called a *nice partition* in Bastin and Dochain [14].

Case #1: S_2 and P_2 are measured on-line

Then A_1 and A_2 are equal to:

$$A_1 = \begin{bmatrix} 0 & 0 \\ 1 & 0 \\ 0 & 0 \\ 0 & 1 \end{bmatrix}, \quad A_2 = \begin{bmatrix} 1 & \frac{1}{Y_1} & 0 & 0 \\ 0 & -Y_3 & \frac{1}{Y_2} & 0 \\ 0 & -Y_4 & -Y_5 & 1 \\ 0 & 0 & -Y_6 & 0 \end{bmatrix} \tag{7.95}$$

Therefore, S_1, X_1, X_2 and P_1 can be estimated by using the asymptotic observer via the dynamical equation of ζ (7.94) as follows:

$$\hat{S}_1 = \hat{\zeta}_1 + \frac{1}{Y_1 Y_3}(S_2 + \frac{P_2}{Y_6} - \frac{\hat{\zeta}_4}{Y_6} - \hat{\zeta}_2) \tag{7.96}$$

$$\hat{X}_1 = \frac{1}{Y_3}(S_2 + \frac{P_2}{Y_6} - \frac{\hat{\zeta}_4}{Y_6} - \hat{\zeta}_2) \tag{7.97}$$

$$\hat{X}_2 = \frac{1}{Y_6}(P_2 - \hat{\zeta}_4) \tag{7.98}$$

$$\hat{P}_1 = \hat{\zeta}_3 + \frac{Y_4}{Y_3}S_2 - \frac{Y_4}{Y_3}\hat{\zeta}_2 + \frac{Y_4 + Y_3 Y_5}{Y_3 Y_6}(P_2 - \hat{\zeta}_2) \tag{7.99}$$

Case #2: S_1 and X_1 are measured on-line

This choice is interesting because it corresponds to a wrong choice of measured variables, since S_1 and X_1 are not independent. Indeed the matrices A_1 and A_2 are then equal to:

$$A_1 = \begin{bmatrix} \frac{1}{Y_1} & 1 \\ 0 & Y_3 \\ 0 & -Y_4 \\ 0 & 0 \end{bmatrix}, \quad A_2 = \begin{bmatrix} 0 & 0 & 0 & 0 \\ 1 & \frac{1}{Y_2} & 0 & 0 \\ 0 & -Y_5 & 1 & 0 \\ 0 & -Y_6 & 0 & 1 \end{bmatrix} \tag{7.100}$$

The matrix A_2 is obviously not invertible (the first row is equal to zero!).

This example shows that not any choice of M measured components is valid for the implementation of the asymptotic observer: the submatrix Y_1 must be full rank, i.e. the measured components must be independent or the measured components have to take part in all the reactions (at least one in each reaction) in order to avoid a submatrix Y_1 with (a) column(s) only filled with zeros.

Case #3: S_2, P_1 and P_2 are measured on-line

Now the number of measured variables is larger than rank(Y). In this case, the vectors ξ_1 and ξ_2 are equal to:

$$\xi_1 = \begin{bmatrix} S_2 \\ P_1 \\ P_2 \end{bmatrix}, \quad \xi_2 = \begin{bmatrix} S_1 \\ X_1 \\ X_2 \end{bmatrix} \tag{7.101}$$

and the matrices A_1 and A_2 are then defined as follows:

$$A_1 = \begin{bmatrix} 0 & 0 & 0 \\ 1 & 0 & 0 \\ 0 & 1 & 0 \\ 0 & 0 & 1 \end{bmatrix}, \quad A_2 = \begin{bmatrix} 1 & \frac{1}{Y_1} & 0 \\ 0 & -Y_3 & \frac{1}{Y_2} \\ 0 & -Y_4 & -Y_5 \\ 0 & 0 & -Y_6 \end{bmatrix} \tag{7.102}$$

A_2 is not a square matrix anymore. Then we can choose, as a left inverse, its left pseudo-inverse:

$$A_2^+ = (A_2^T A_2)^{-1} A_2^T \tag{7.103}$$

which takes the following form in our example:

$$A_2^+ = \begin{bmatrix} 0 & 1 & 0 & 0 \\ 0 & 0 & 1 & 0 \\ 0 & 0 & 0 & 1 \end{bmatrix} \tag{7.104}$$

Implementation aspects: choice of the sampling period.   The practical computer implementation of the asymptotic observer (7.79)(7.80) or (7.82)(7.83) requires that it be rewritten in a discrete-time form. This can be done simply by replacing the time derivative of ζ by a finite difference (using a first order Euler approximation):

$$\frac{d\hat{\zeta}}{dt} \longrightarrow \frac{\hat{\zeta}_{t+1} - \hat{\zeta}_t}{T} \tag{7.105}$$

where T is the sampling period and t and $t + 1$ are time indices. The asymptotic observer is then written as follows for the general case $p \geq M$:

$$\hat{\zeta}_{t+1} = \hat{\zeta}_t - T D_t \hat{\zeta}_t + T C_a (F_{at} - Q_{at}) + C_b (F_{bt} - Q_{bt}) \tag{7.106}$$

$$\hat{\xi}_{2,t+1} = A_2^+ [\hat{\zeta}_{t+1} - A_1 \xi_{1,t+1}] \tag{7.107}$$

For the discrete-time equation, the value of the sampling period plays a role in the stability. In fact, if the dilution rate D is bounded as follows:

$$0 \le D(t) \le D_{max} \qquad (7.108)$$

then equation (7.106) will remain stable as long as T is smaller than $2/D_{max}$:

$$T \le \frac{2}{D_{max}} \qquad (7.109)$$

Equation (7.109) gives a condition for the stability of the discrete-time version of the asymptotic observer. But even if the sampling period T is chosen so as to fulfill condition (7.109), large values of T may introduce oscillations in the estimation. As a matter of fact, if we assume that D is constant, the dynamics of ζ are characterised by a (discrete-time) pole equal to $(1 - TD)$. It will be negative if T is larger than $1/D$ $(T > 1/D)$ and then corresponds to oscillating (underdamped) dynamics for ζ. Therefore, in order to avoid (undesirable) oscillations, condition (7.109) can be replaced by:

$$T \le \frac{1}{D_{max}} \qquad (7.110)$$

Remark: note that different sampling periods may be used for the computation of the variables ζ and the calculation of the observation of ξ_2 , e.g. a "fast" computation of the variables ζ and a slower computation of the estimated values of ξ_2. This choice may depend on the measurement sampling interval which may be different from one variable to another.

Application to an anaerobic digestion process. Let us consider the on-line estimation of S_1, X_1 and X_2 from the measurements of methane P_2 and volatile fatty acids S_2. This corresponds to the case #1 hereabove[18].

Note that in practice the methane gas is a low solubility product, and therefore the model reduction approach discussed in Chapter 2 (Section 2.9) applies:

$$P_2 = 0, \ \frac{dP_2}{dt} = 0 \qquad (7.111)$$

Therefore the asymptotic observer specialises as follows:

$$\frac{d\hat{\zeta}_1}{dt} = -D\hat{\zeta}_1 + DS_{in} \qquad (7.112)$$

$$\frac{d\hat{\zeta}_2}{dt} = -D\hat{\zeta}_2 \qquad (7.113)$$

---

[18]We did not consider the on-line estimation of CO_2 (P_1) which is indeed directly obtained either via the gas flow rate and CO_2 percentage measurement, or via the gas flow rate and the CH_4 percentage measurement under the assumption that the outflow gas is composed almost exclusively by CO_2 and CH_4.

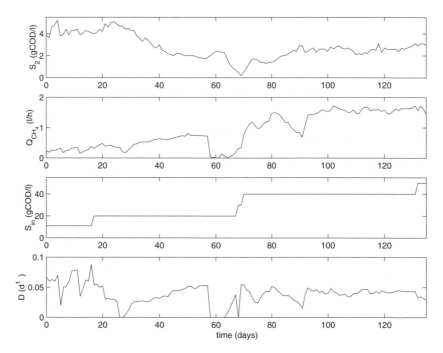

FIG. 7.5. On-line data of the asymptotic observer for the anaerobic digestion process.

$$\frac{d\hat{\zeta}_4}{dt} = -D\hat{\zeta}_4 - Q_2 \tag{7.114}$$

$$\hat{S}_1 = \hat{\zeta}_1 + \frac{1}{Y_1 Y_3}\left(S_2 - \frac{\hat{\zeta}_4}{Y_6} - \hat{\zeta}_2\right) \tag{7.115}$$

$$\hat{X}_1 = \frac{1}{Y_3}\left(S_2 - \frac{\hat{\zeta}_4}{Y_6} - \hat{\zeta}_2\right) \tag{7.116}$$

$$\hat{X}_2 = -\frac{1}{Y_6}\hat{\zeta}_4 \tag{7.117}$$

The asymptotic observer has been implemented on a 60 liter pilot CSTR of the Unit of Bioengineering, Université Catholique de Louvain, Belgium ([14]). The values of the yield coefficients are equal to:

$$Y_1 = 0.3125, \quad Y_2 = 0.035, \quad Y_3 = 5.7, \quad Y_6 = 27.3 \tag{7.118}$$

Figure 7.5 shows the data used for the on-line estimation (volatile fatty acids S_2, methane gas flow rate Q_2 denoted Q_{CH_4} in the figure), influent organic matter

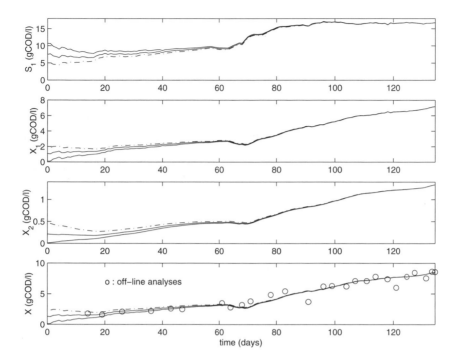

FIG. 7.6. Estimation results of the asymptotic observer of the anaerobic digestion process
(-: $\hat{\zeta}_1(0) = 11$, $\hat{\zeta}_2(0) = 4$, $\hat{\zeta}_4(0) = -6$; - -: $\hat{\zeta}_1(0) = 11$, $\hat{\zeta}_2(0) = 3.9$, $\hat{\zeta}_4(0) = 0$; - . -: $\hat{\zeta}_1(0) =$
11, $\hat{\zeta}_2(0) = 6.65$, $\hat{\zeta}_4(0) = -13.65$).

concentration S_{in}, and dilution rate D. The estimation results are given in Figure
7.6. Different initial conditions for the auxiliary variables ζ have been considered:

$$- : \hat{\zeta}_1(0) = 11, \ \hat{\zeta}_2(0) = 4, \ \hat{\zeta}_4(0) = -6$$
$$-- : \hat{\zeta}_1(0) = 11, \ \hat{\zeta}_2(0) = 3.9, \ \hat{\zeta}_4(0) = 0$$
$$-.- : \hat{\zeta}_1(0) = 11, \ \hat{\zeta}_2(0) = 6.65, \ \hat{\zeta}_4(0) = -13.65$$

Note that the asymptotic observer converges after an initial transient to the same
estimation profile. The convergence time is approximately equal to 60 days, which
corresponds to three times the residence time (or equivalently the inverse of the
dilution rate, see Figure 7.5).

    One of the major difficulties that we had to face here was the validation of
the on-line estimates provided by the asymptotic observer, since no measurement,
even indirect, of the different (acidogenic and methanogenic) bacterial populations
is available. So we have been using COD data for the validation, more precisely
the difference between the total COD and the soluble COD, which are known
to represent fairly well the active biomass population in the process, i.e. $X_1 +$

X_2 ([14]). The last figure in Figure 7.6 compares these data with the sum of the estimates $\hat{X}_1 + \hat{X}_2$ for the three different initial conditions.

7.4.2   Asymptotic Observers for Multi-Tank Reactors

The design of the above asymptotic observer can be easily extended to multi-reactor processes, since the definition of the transformation ζ is the same. The main question is the stability of the dynamics of the auxiliary variables ζ (see Chapter 2, Section 2.8):

$$\frac{d\zeta}{dt} = -(C_b D_{bb} + C_a D_{ab})C_b^{-1}\zeta + C_a(F_a - Q_a) + C_b(F_b - Q_b)$$

$$+[(C_b D_{bb} + C_a D_{ab})C_b^{-1} - C_b D_{ba} - C_a D_{aa}]\xi_a \qquad (7.119)$$

with:

$$D = \begin{bmatrix} D_{aa} & D_{ab} \\ D_{ba} & D_{bb} \end{bmatrix} \qquad (7.120)$$

The stability of (7.119) is determined by the matrix:

$$-(C_b D_{bb} + C_a D_{ab})C_b^{-1} \qquad (7.121)$$

We shall not develop this point in detail here (see e.g. [56]). Rather we shall concentrate on one example (the basic model of the activated sludge process), and discuss its stability.

Let us first recall the equation of the basic model of the activated sludge process, written as follows in the General Dynamical Model format:

$$\xi = \begin{bmatrix} S \\ S_O \\ X \\ X_R \end{bmatrix}, \ Y = \begin{bmatrix} -\frac{1}{Y_S} \\ -\frac{1}{Y_O} \\ 1 \\ 0 \end{bmatrix}, \ F = \begin{bmatrix} D_{in} S_{in} \\ D_{in} S_{O,in} + k_L a(S_O^* - S_O) \\ 0 \\ 0 \end{bmatrix} \qquad (7.122)$$

$$\rho = \mu X, \ Q = 0, \ D = \begin{bmatrix} D_1 & 0 & 0 & 0 \\ 0 & D_1 & 0 & 0 \\ 0 & 0 & D_1 & -D_2 \\ 0 & 0 & -D_3 & D_4 \end{bmatrix} \qquad (7.123)$$

with the following definitions for D_{in}, D_1, D_2, D_3, and D_4:

$$D_{in} = \frac{F_{in}}{V}, \ D_2 = \frac{F_R}{V}, \ D_1 = D_{in} + D_2, \ D_3 = \frac{F_{in} + F_R}{V_S}, \ D_4 = \frac{F_R + F_W}{V_S}$$

$$(7.124)$$

Let us consider, as in Section 2.8.3, the following state partition:

$$\xi_a = X, \ \xi_b = \begin{bmatrix} S \\ S_O \\ X_R \end{bmatrix} \tag{7.125}$$

This means that the different dilution matrices are equal to:

$$D_{aa} = D_1, \ D_{ab} = \begin{bmatrix} 0 & 0 & -D_2 \end{bmatrix} \tag{7.126}$$

$$D_{ba} = \begin{bmatrix} 0 \\ 0 \\ -D_3 \end{bmatrix}, \ D_{bb} = \begin{bmatrix} D_1 & 0 & 0 \\ 0 & D_1 & 0 \\ 0 & 0 & D_4 \end{bmatrix} \tag{7.127}$$

If we select $C_b = I$, the matrix C_a is equal to:

$$C_a = \begin{bmatrix} -\frac{1}{Y_S} \\ -\frac{1}{Y_O} \\ 0 \end{bmatrix} \tag{7.128}$$

and the auxiliary variable ζ is written as follows:

$$\zeta = \begin{bmatrix} S + \frac{1}{Y_S}X \\ S_O + \frac{1}{Y_O}X \\ X_R \end{bmatrix} \tag{7.129}$$

The stability of the dynamics of the auxiliary ζ (and the underlying asymptotic observer) is determined by the matrix (7.121) which specialises here as follows:

$$\begin{bmatrix} -D_1 & 0 & -\frac{D_2}{Y_S} \\ 0 & -D_1 & -\frac{D_2}{Y_O} \\ 0 & 0 & -D_4 \end{bmatrix} \tag{7.130}$$

Its eigenvalues are negative if D_1 and D_4 are positive, i.e. if

$$F_{in} + F_R > 0 \text{ and } F_R + F_W > 0 \tag{7.131}$$

which holds for instance if the sludge is being recycled ($F_R > 0$).

Let us now consider the problem of estimating the concentrations of organic matter S, and of the biomass X and X_R both in the aerator and in the settler, from on-line data of dissolved oxygen S_O. This means that ξ_1 and ξ_2 are defined as follows:

$$\xi_1 = S_O, \ \xi_2 = \begin{bmatrix} S \\ X \\ X_R \end{bmatrix} \tag{7.132}$$

The asymptotic observer is then written as follows:

$$\frac{d\hat{\zeta}_1}{dt} = -\frac{F_{in} + F_R}{V}\hat{\zeta}_1 + \frac{F_{in}}{V}S_{in} + \frac{F_R}{VY_S}\zeta_3 \tag{7.133}$$

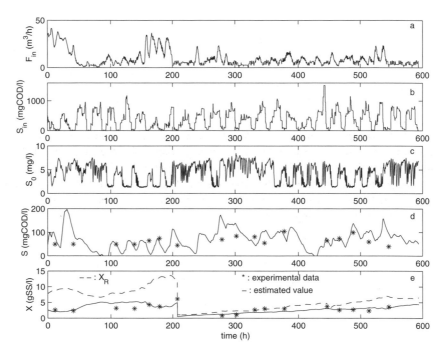

FIG. 7.7. Estimation results of the asymptotic observer of an activated sludge process.

$$\frac{d\hat{\zeta}_2}{dt} = -\frac{F_{in} + F_R}{V}\hat{\zeta}_2 + k_L a(S_O^* - S_O) + \frac{F_R}{VY_O}\hat{\zeta}_3 \qquad (7.134)$$

$$\frac{d\hat{\zeta}_3}{dt} = -\frac{F_W + F_R}{V_S}\hat{\zeta}_3 + \frac{F_{in} + F_R}{V_S}(\hat{\zeta}_2 - S_O)Y_O \qquad (7.135)$$

$$\hat{S} = (\hat{\zeta}_1 - Y_O\hat{\zeta}_2 + Y_O S_O)Y_S \qquad (7.136)$$

$$\hat{X} = (\hat{\zeta}_2 - S_O)Y_O \qquad (7.137)$$

$$\hat{X}_R = \hat{\zeta}_3 \qquad (7.138)$$

The implementation of the asymptotic observer further requires on-line data of the influent flow rate F_{in}, of the influent organic matter concentration S_{in}, and of the recycle and waste flow rates F_R and F_W. And besides the values of the yield coefficients Y_S and Y_O, we also need the values of the mass transfer coefficient $k_L a$ and of the saturation concentration S_O^*.

Figure 7.7 shows the application of the asymptotic observer to a wastewater treatment plant (the Maria Middelares treatment plant, Gent, Belgium)(see also [76]). The available data were the on-line data of dissolved oxygen and influent COD (obtained from an on-line respirometer), some off-line measurements of the

COD in the effluent and of the suspended solids in the aerator (yet no precise measurement of the recycle flow rate and only time points (t = 208 h and 449 h) when discontinuous sludge wastage was carried out). The state estimation is a challenging question if we consider the measurement constraints, and the oversimplification of the model dynamics. Figure 7.7 shows the estimation results. Figures a, b and c show the data of the influent flow rate F_{in}, the influent substrate concentration S_{in} and the dissolved oxygen S_O, respectively, while Figures d and e compare the estimation results given by the asymptotic observer with the off-line data for S and X, respectively. The computed value of X_R is also shown (in dotted lines) in Figure d. The recycle flow rate F_R was kept constant at the design value of 8 m^3/h. The volumes are: $V = 380\ m^3$, $V_S = 265\ m^3$; the determination of S_O^* and $k_L a$ gave the following result:

$$S_O^* = 10\ mg/l,\ k_L a = 1.7\ h^{-1} \tag{7.139}$$

We note that although the reaction scheme is very simple and that some of the data are not very precise, the reaction scheme is fairly well validated.

7.4.3 Asymptotic Observers for Tubular Bioreactors

Let us now discuss the extension and application of the asymptotic observer design to non completely mixed reactors.

Let us modify the first assumption introduced in Section 7.4.1:

1bis. p (= M) components are measured on-line *along the reactor*.

and introduce a sixth one:

6. The axial mass dispersion coefficient D_{am} and the reactor section A are known.

Recall that we had considered in Section 2.8.2 (Chapter 2) $C_b = I$ and that the vector of the fixed component ξ_{fi} has been put in ξ_b:

$$\xi_b = \begin{bmatrix} \xi_{bf} \\ \xi_{fi} \end{bmatrix} \tag{7.140}$$

Then we can rewrite the auxiliary variable ζ as follows:

$$\zeta = \begin{bmatrix} \zeta_{fl} \\ \zeta_{fi} \end{bmatrix} = \begin{bmatrix} \xi_{bf} \\ \xi_{fi} \end{bmatrix} + \begin{bmatrix} C_{af} \\ C_{ae} \end{bmatrix} \xi_a \tag{7.141}$$

The dynamics of ζ can then be written as follows:

$$\frac{\partial \zeta_{fl}}{\partial t} = -\frac{F_{in}}{A} \frac{\partial \zeta_{fl}}{\partial z} + D_{am} \frac{\partial^2 \zeta_{fl}}{\partial z^2} \tag{7.142}$$

$$\frac{\partial \zeta_{fi}}{\partial t} = -\frac{F_{in}}{A} C_{ae} \frac{\partial \xi_a}{\partial z} + D_{am} C_{ae} \frac{\partial^2 \xi_a}{\partial z^2} \qquad (7.143)$$

Assume that the measured components are put in the vector ξ_a (i.e. we consider a particular choice for ξ_1 and ξ_2 ($\xi_1 = \xi_a, \xi_2 = \xi_b$); this is done only to simplify the approach but, of course, other choices are possible). Then by using the same arguments as above and the equations (7.141), (7.142) and (7.143), we obtain the following asymptotic observer for tubular bioreactors:

$$\frac{\partial \hat{\zeta}_{fl}}{\partial t} = -\frac{F_{in}}{A} \frac{\partial \hat{\zeta}_{fl}}{\partial z} + D_{am} \frac{\partial^2 \hat{\zeta}_{fl}}{\partial z^2} \qquad (7.144)$$

$$\frac{\partial \hat{\zeta}_{fi}}{\partial t} = -\frac{F_{in}}{A} C_{ae} \frac{\partial \xi_a}{\partial z} + D_{am} C_{ae} \frac{\partial^2 \xi_a}{\partial z^2} \qquad (7.145)$$

$$\hat{\xi}_2 = \hat{\zeta} - \left[\frac{C_{af}}{C_{ae}} \right] \xi_1 \qquad (7.146)$$

There remain two key questions with the above asymptotic observer:

1. The above observer is written under the form of ("infinite dimensional") partial derivative equations (PDEs). These are not very easy or convenient to handle in practical control and monitoring applications. Moreover, it is assumed that p components are available for on-line measurement *along the reactor*. In line with a number of works on the subject (e.g. [97], [136]), we propose to "reduce" the above PDE equations to a finite number of ordinary differential equations (ODEs) at a finite number of positions along the reactor and to consider the reduced equations for the practical application of the observer. This will the object of the next section.

2. The second key question is to know whether the proposed observer is reliable, i.e. under which the conditions it will give estimates that converge to their true values: this will be addressed on page 284.

Practical implementation of the asymptotic observer. Let us reduce the PDEs of the asymptotic observer (7.144) to a finite number of ODEs. We shall not discuss the choice of one reduction method here (see e.g. [277], [211] for this topic); we shall only assume that the user has chosen one method for approximating the PDEs (e.g. finite differences, orthogonal collocation,...) and that the reduced model is a fairly good representation of the PDE asymptotic observer (7.144). Whatever the reduction method, the partial derivatives of the variables ζ_i with respect to the space variable z are approximated by a weighted sum of ζ_i at a finite number of positions z_j ($j = 0$ to q, where 0 and q hold for the input and output of the reactor, respectively) along the reactor:

$$\frac{\partial^k}{\partial z^k}\begin{bmatrix}\varsigma_i(z=z_1)\\\varsigma_i(z=z_2)\\\vdots\\\varsigma_i(z=z_q)\end{bmatrix}\cong[\tilde{c}_k\mid\tilde{C}_k]\begin{bmatrix}\varsigma_i(z=z_0)\\\varsigma_i(z=z_1)\\\vdots\\\varsigma_i(z=z_q)\end{bmatrix},\ k=1,2 \qquad (7.147)$$

$$dim(\tilde{c}_k)=q\times 1,\ dim(\tilde{C}_k)=q\times q \qquad (7.148)$$

If we consider a finite difference approximation, the matrices $\tilde{C}_k$ (k = 1, 2) are equal to:

$$\tilde{C}_1=\frac{1}{\Delta z}\begin{bmatrix}1&0&0&\cdots&0&0\\-1&1&0&\cdots&\vdots&\vdots\\0&-1&1&\cdots&\vdots&\vdots\\0&0&-1&\cdots&\vdots&\vdots\\\vdots&\vdots&\vdots&\cdots&1&0\\0&0&0&\cdots&-1&1\end{bmatrix},\ \tilde{c}_1=\frac{1}{\Delta z}\begin{bmatrix}1\\0\\\vdots\\0\end{bmatrix} \qquad (7.149)$$

$$\tilde{C}_2=\frac{1}{(\Delta z)^2}\begin{bmatrix}-2&1&0&\cdots&0&0\\1&-2&1&\cdots&\vdots&\vdots\\0&1&-2&\cdots&\vdots&\vdots\\0&0&1&\cdots&\vdots&\vdots\\\vdots&\vdots&\vdots&\cdots&-2&1\\0&0&0&\cdots&1&-2\end{bmatrix},\ \tilde{c}_2=\frac{1}{(\Delta z)^2}\begin{bmatrix}1\\0\\\vdots\\0\end{bmatrix} \qquad (7.150)$$

where Δz is the spatial discretisation step.

The reduction procedure results in a rewriting of the auxiliary variables ς_f and of the asymptotic observer equations under ODEs at each position z_j (j = 1 to q). Then the dynamics of the auxiliary variables ς_f and of the asymptotic observer become:

Dynamics of ς_f:

$$\frac{d\varsigma_{rl}}{dt}=[D_{am}C_2-\frac{F}{A}C_1]\varsigma_{rl}+[D_{am}c_2-\frac{F}{A}c_1](\xi_{2,in}+C_{af}\xi_{1,in}) \qquad (7.151)$$

$$\frac{d\varsigma_{ri}}{dt}=[D_{am}C_2-\frac{F}{A}C_1]C_{ae}\xi_a+[D_{am}c_2-\frac{F}{A}c_1]C_{ae}\xi_{1,in} \qquad (7.152)$$

Asymptotic Observer:

$$\frac{d\hat{\zeta}_{rl}}{dt} = [D_{am}C_2 - \frac{F}{A}C_1]\hat{\zeta}_{rl} + [D_{am}c_2 - \frac{F}{A}c_1](\xi_{2,in} + C_{af}\xi_{1,in}) \quad (7.153)$$

$$\frac{d\hat{\zeta}_{ri}}{dt} = [D_{am}C_2 - \frac{F}{A}C_1]C_{ae}\xi_a + [D_{am}c_2 - \frac{F}{A}c_1]C_{ae}\xi_{1,in} \quad (7.154)$$

with:

$$\zeta_r = \begin{bmatrix} \zeta_{rl} \\ \zeta_{ri} \end{bmatrix} = \begin{bmatrix} \zeta_{r,1}(z = z_1) \\ \vdots \\ \zeta_{r,1}(z = z_q) \\ \vdots \\ \zeta_{r,R}(z = z_1) \\ \vdots \\ \zeta_{r,R}(z = z_q) \end{bmatrix} \quad (7.155)$$

$$C_k = \begin{bmatrix} \tilde{C}_k & 0 & \cdots & 0 \\ 0 & \tilde{C}_k & \cdots & \vdots \\ \vdots & \vdots & \cdots & 0 \\ 0 & 0 & \cdots & \tilde{C}_k \end{bmatrix}, \quad c_k = \begin{bmatrix} \tilde{c}_k \\ \tilde{c}_k \\ \vdots \\ \tilde{c}_k \end{bmatrix}, \quad k = 1, 2 \quad (7.156)$$

One important feature of the above equations (7.151)(7.153) is that the tubular reactor is approximated by a stirred multi-tank reactor. In case of a finite difference approximation, the equations represent a cascade of stirred tank reactors; but with other approximation methods (e.g. collocation methods), the model exhibits interconnections between each of the stirred tank reactors of the ODE model since the entries of the matrices $\tilde{C}_k$ ($k = 1, 2$) are generally different from zero [155].

Stability properties of the asymptotic observer. Let us start by analysing the stability of the PDE asymptotic observer.

If we define the estimation error e on the auxiliary variables ζ as follows:

$$e = \zeta - \hat{\zeta} = \begin{bmatrix} \zeta_{fl} - \hat{\zeta}_{fl} \\ \zeta_{fi} - \hat{\zeta}_{fi} \end{bmatrix} = \begin{bmatrix} e_{fl} \\ e_{fi} \end{bmatrix} \quad (7.157)$$

the dynamics of e are given by the following equations:

$$\frac{\partial e_{fl}}{\partial t} = -\frac{F_{in}}{A}\frac{\partial e_{fl}}{\partial z} + D_{am}\frac{\partial^2 e_{fl}}{\partial z^2}, \quad e_{fl}(t = 0, z) = e_{fl0}(z) \quad (7.158)$$

$$\frac{\partial e_{fi}}{\partial t} = 0, \qquad\qquad\qquad e_{fi}(t=0,z) = e_{fi0}(z) \quad (7.159)$$

The analysis of the second equation (7.159) shows that under the assumption presented above, any initial bias will remain (this is exactly the same as for the batch reactor in stirred tank reactors).

Let us now analyse the first error equation (7.158). First note that it is a linear equation. If D_{am} is different from zero (parabolic equation), it is characterised (see e.g. [285]) by the following simple, real, negative eigenvalues λ_n, $n \geq 1$:

$$\lambda_n = -\frac{s_n^2 + u^2}{4D_{am}} < -\frac{u^2}{4D_{am}} < 0, \quad \text{for all } n \geq 1, \qquad (7.160)$$

where $\{s_n : n \geq 1\}$ is the set of all the solutions to the equation (called the resolvent equation):

$$\tan(\frac{L}{2D_{am}}s) = \frac{2us}{s^2 - u^2}, \quad s > 0, \qquad (7.161)$$

with $u = F_{in}/A$ and L the length of the reactor.

If D_{am} is equal to zero (hyperbolic equation, plug flow reactor), the solution of the equation (7.158) is of the following form:

$$\begin{aligned} e_{fl}(t,z) &= e_{fl0}(z - ut), \quad \text{if } t < z/u \\ &= 0 \qquad\qquad\quad \text{if } t \geq z/u \end{aligned}$$

Therefore the equation (7.158) is asymptotically stable, and the asymptotic observer for the components ξ_f is convergent.

Let us concentrate on the stability properties of the reduced form (7.153) of the asymptotic observer, to see in which conditions the above stability result is correctly transferred to the discretised version of the asymptotic observer. For obvious reasons, we shall only concentrate on the error dynamics (7.158). If we define the observation error:

$$e = \zeta_r - \hat{\zeta}_r \qquad\qquad\qquad (7.162)$$

then the dynamics of the observation error e is readily obtained from (7.151)(7.153):

$$\frac{de}{dt} = [D_{am}C_2 - \frac{F}{A}]e \qquad\qquad (7.163)$$

Therefore the stability depends on the state matrix:

$$D_{ma}C_2 - \frac{F}{A}C_1 \qquad\qquad (7.164)$$

Because of the diagonal structure of the matrices C_1 and C_2, the stability of the above state matrix (7.164) depends on the stability of each submatrix:

$$D_{ma}\tilde{C}_2 - \frac{F}{A}\tilde{C}_1 \tag{7.165}$$

Therefore it follows that the asymptotic observer (7.153) will be asymptotically stable, i.e.:

$$\lim_{t\to\infty} \hat{\xi}_r = \xi_r \tag{7.166}$$

if the eigenvalues of the matrix $D_{ma}\tilde{C}_2 - \frac{F}{A}\tilde{C}_1$ are stable. Note that the stability of the asymptotic observer only depends on the axial mass transfer and not on the kinetics. In other words, the reactor may be unstable (due to kinetics like the Haldane kinetics, see Section 2.2.6) while the asymptotic observer is asymptotically stable (because of stable hydrodynamics).

Note also that it is routine to check the stability of the matrix (7.165) by using any matrix computation program. An interesting particular case is the finite difference approximation of a fixed bed reactor without dispersion ($D_{am} = 0$). Indeed, the stability then simply depends on the matrix $\frac{F}{A}C_1$, and it is straightforward to check that the matrix $\frac{F}{A}C_1$ (see equation (7.149)) is stable as long as F is positive. The stability properties of the hydrodynamics approximated via orthogonal collocation are analysed in Lefévre *et al.* [155].

Example.   Let us consider a plug flow reactor with fixed biomass and with the following reaction scheme:

$$S \longrightarrow X + P \tag{7.167}$$

$$X \longrightarrow X_d \tag{7.168}$$

with X_d the dead biomass. The dynamics of the process are given by the following equations:

$$\frac{\partial S}{\partial t} = -\frac{F_{in}}{A}\frac{\partial S}{\partial z} - \frac{1}{Y_1}\mu X \tag{7.169}$$

$$\frac{\partial P}{\partial t} = -\frac{F_{in}}{A}\frac{\partial P}{\partial z} + Y_2\mu X \tag{7.170}$$

$$\frac{\partial X}{\partial t} = \mu X - k_d X \tag{7.171}$$

$$\frac{\partial X_d}{\partial t} = -\frac{F_{in}}{A}\frac{\partial X_d}{\partial z} + k_d X \tag{7.172}$$

Assume first that S and X_d are available for on-line measurement (while X and P are not). Assume also that the specific growth rate μ and the death/detachment coefficient k_d are unknown. Then the asymptotic observer can be derived as follows. The auxiliary variables ζ are derived from equation (7.141) and are, for instance, equal to:

$$\zeta_1 = X + X_d + Y_1 S \tag{7.173}$$

$$\zeta_2 = P + Y_1 Y_2 S \tag{7.174}$$

The asymptotic observer equations are then written as follows:

$$\frac{\partial \hat{\zeta}_1}{\partial t} = -\frac{F_{in}}{A} \frac{\partial (X_d + Y_1 S)}{\partial z} \tag{7.175}$$

$$\frac{\partial \hat{\zeta}_2}{\partial t} = -\frac{F_{in}}{A} \frac{\partial \hat{\zeta}_2}{\partial z} \tag{7.176}$$

$$\hat{X} = \hat{\zeta}_1 - X_d - Y_1 S \tag{7.177}$$

$$\hat{P} = \hat{\zeta}_2 - Y_1 Y_2 S \tag{7.178}$$

Another option is when the death coefficient k_d is known. Then only one measurement is necessary, e.g. the substrate concentration S. There are now three entries for the auxiliary variable ζ, e.g.:

$$\zeta_1 = X + Y_1 S \tag{7.179}$$

$$\zeta_2 = P + Y_1 Y_2 S \tag{7.180}$$

$$\zeta_3 = X_d \tag{7.181}$$

Then the asymptotic observer specialises as follows:

$$\frac{\partial \hat{\zeta}_1}{\partial t} = -\frac{F_{in}}{A} Y_1 \frac{\partial S}{\partial z} - k_d(\hat{\zeta}_1 - Y_1 S) \tag{7.182}$$

$$\frac{\partial \hat{\zeta}_2}{\partial t} = -\frac{F_{in}}{A} \frac{\partial \hat{\zeta}_2}{\partial z} \tag{7.183}$$

$$\frac{\partial \hat{\zeta}_3}{\partial t} = -\frac{F_{in}}{A} \frac{\partial \hat{\zeta}_3}{\partial z} + k_d(\hat{\zeta}_1 - Y_1 S) \tag{7.184}$$

$$\hat{X} = \hat{\zeta}_1 - X_d - Y_1 S \tag{7.185}$$

$$\hat{P} = \hat{\zeta}_2 - Y_1 Y_2 S \tag{7.186}$$

$$\hat{X}_d = \hat{\zeta}_3 \tag{7.187}$$

7.4.4 Asymptotic Observers as a Tool for Model Selection

Here above we have seen that asymptotic observers are independent of the process kinetics. In this section an approach is suggested on how to use asymptotic observers for model selection, more precisely to model and validate reaction schemes independently of the reaction kinetics [76]. We shall first introduce the modelling concept and procedure based on the asymptotic observer. The results will be illustrated with one process: a detoxification reactor.

Model selection procedure. The basic idea of the approach is to use the asymptotic observer not for on-line estimation but for model building, more precisely for modelling the reaction network part of the dynamics independently of the kinetics. It is based on the convergence properties of the observer: because of these, we know that the asymptotic observer is able to give a reliable replica of the dynamics of the process components if the reaction scheme is correct (i.e. representative of the key process reactions) and only if the values of the yield coefficients are known (and not the reaction kinetics).

Assume that a set of experimental data of the concentrations and (when appropriate) gaseous outflow rates of the key process components (whose number is equal to N) as well as of the hydrodynamics (mainly, the flow rates) are available. The model selection procedure is the following.

1. Select a plausible reaction network involving the measured components.
2. Build the dynamical model in the format of the General Dynamical Model.
3. Select M (= number of a priori selected reactions) components and build the asymptotic observer to estimate the $N - M$ remaining components.
4. Try to fit the estimated values of the $N-M$ components to their experimental values by selecting appropriate values of the yield coefficients.
5. The final step is usual: either the fitting is satisfactory and you stop, or it is not and you go back to the first step (select another reaction scheme candidate, possibly with another set of process components).

Illustrative case study: a detoxification process.   Soil decontamination has become an important matter in wastewater treatment in recent years, resulting for instance in the increasing use and development of biological detoxification processes [125]. For instance, polychlorinated aliphatic compounds [175], [33], recalcitrant compounds, and halogenated aromatic compounds [40] extracted from the contaminated soil may be treated by microbial mixed populations in strict anaerobic conditions. In this context, a packed bed reactor has been developed and successfully applied to the anaerobic destruction of a mixture of toxic and recalcitrant molecules (chlorinated aliphatics) on a laboratory scale [175], [33].

The anaerobic detoxification is a complex biochemical network involving different microorganism populations and biochemical reactions. A four-reaction model is often considered to describe it [178] (see Figure 7.8). Here again the constraints on the available data for modelling are quite hard: only off-line measurements of hydrogen, acetate, toxics and non-toxics at the reactor output, and of the co-substrate in the influent are available. A plausible reaction network is then:

$$S \longrightarrow H_2 + V A_2 \tag{7.188}$$

$$H_2 + T \longrightarrow N_T \tag{7.189}$$

where S, H_2, $V A_2$, T and N_T represent the co-substrate (ammonium citrate), the hydrogen, the acetate, the toxics and the non-toxics, respectively. Different sets of data corresponding to different experiments are available. Because of the high

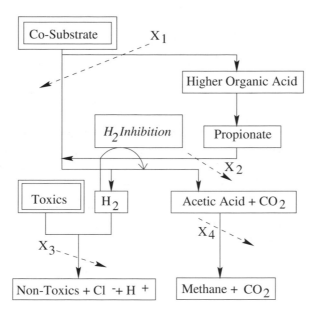

FIG. 7.8. Scheme of the four-reaction model of the anaerobic detoxification process.

recirculation rate (usually 18 d^{-1}), the reactor is run under rather homogeneous conditions, and the transport time delay between the reactor input and its output is negligible with regard to the measurement frequency (the lowest measurement frequency is 3-4 days for the acetate). This validates the use of a stirred tank reactor model. Therefore the dynamics of the process in the General Dynamical Model formalism are as follows:

$$
\xi = \begin{bmatrix} S \\ VA_2 \\ H_2 \\ T \\ N_T \end{bmatrix}, \quad Y = \begin{bmatrix} -1 & 0 \\ Y_1 & 0 \\ Y_2 & -\frac{1}{Y_3} \\ 0 & -\frac{1}{Y_4} \\ 0 & 1 \end{bmatrix}, \quad F = \begin{bmatrix} DS_{in} \\ 0 \\ 0 \\ DT_{in} \\ DN_{in} \end{bmatrix}, \quad \rho = \begin{bmatrix} \rho_1 \\ \rho_2 \end{bmatrix}, \quad Q = O
$$

(7.190)

Let us consider the state partition:

$$
\xi_1 = \begin{bmatrix} S \\ H_2 \end{bmatrix}, \quad \xi_2 = \begin{bmatrix} VA_2 \\ T \\ N_T \end{bmatrix}
$$

(7.191)

The asymptotic observer equations follow from the above model:

$$
\frac{d\hat{\xi}_1}{dt} = -D\hat{\xi}_1 - k_1 DS_{in}
$$

(7.192)

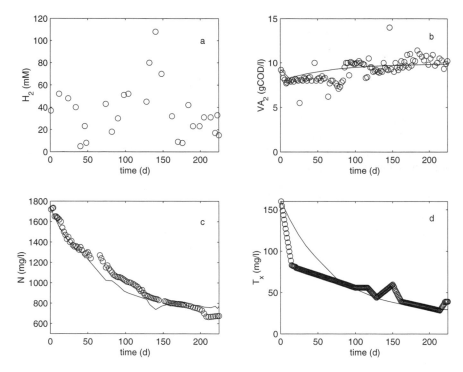

FIG. 7.9. Experimental data and estimation results of the detoxification process (Bioengineering unit, Louvain-la-Neuve, Belgium) (o: experimental data, -: asymptotic observer results).

$$\frac{d\hat{\zeta}_2}{dt} = -D\hat{\zeta}_2 - \frac{Y_2Y_3}{Y_4}DS_{in} + DT_{in} \qquad (7.193)$$

$$\frac{d\hat{\zeta}_3}{dt} = -D\hat{\zeta}_3 + Y_2Y_3DS_{in} + DBN_{in} \qquad (7.194)$$

$$V\hat{A}_2 = \hat{\zeta}_1 + Y_1S \qquad (7.195)$$

$$\hat{T} = \hat{\zeta}_2 + \frac{Y_2Y_3}{Y_4}S + \frac{Y_3}{Y_4}H_2 \qquad (7.196)$$

$$\hat{N}_T = \hat{\zeta}_3 - Y_2Y_3S - Y_3H_2 \qquad (7.197)$$

The asymptotic observer-based modelling procedure has been applied on three sets of data with similar results. Since the co-substrate is not measured at the reactor output and COD estimation indicates its complete utilisation, its value has been assumed in rough approximation to be negligible. Figure 7.9 shows one set of results for which $D = 0.014$ d^{-1}, $S_{in} = 0.3$ gCOD/l, $N_{in} = 457$ mg/l, $T_{in} = 25$ mg/l. Figure 7.9.a gives the data of H_2 and Figures 7.9.b, c and d compare the experimental data of VA_2, T and N with their estimates provided by the above asymptotic observer for the following values of the yield coefficients: $Y_1 = 33$, $Y_2 = 800$, $Y_3 = 1.25$, $Y_4 = 5000$. Note the fairly good validation of the reaction scheme.

7.5 Observers for Processes with Badly Known Kinetics

So far, we have considered (classical) observers when the kinetics are perfectly known, and asymptotic observers when the kinetics are assumed to be unknown. We shall now introduce an intermediate class of observers, i.e. observers of processes for which the structure ("models") of the kinetics are known but with badly or unknown parameter values [77].

7.5.1 State Observer with the Unknown Parameters as Design Parameters

The first approach that we propose in this section is based on the following idea: why not use the badly known kinetic parameters as extra design observer parameters in order to guarantee (at least) zero steady state observation errors for the unmeasured variables?

Let us denote the badly known parameters by θ. The process dynamics (7.25) can be rewritten as follows:

$$\frac{dx}{dt} = f(x, u, \theta) \tag{7.198}$$

If we consider that the output vector consists of state variables (as it is often the case in (bio)processes), we can define a state partition with the measured variables ($y = x_1$) and the unmeasured variables, x_2:

$$\frac{dx_1}{dt} = f_1(x, u, \theta) \tag{7.199}$$

$$\frac{dx_2}{dt} = f_2(x, u, \theta) \tag{7.200}$$

The observer design remains basically the same, but now we choose θ such that $x_2 - \hat{x}_2$ is equal to zero in steady state, i.e.:

$$\theta : (x_2(\theta) - \hat{x}_2(\theta))_{ss} = 0 \tag{7.201}$$

Let us illustrate this idea on the simple microbial growth example (Section 7.3.4). The objective is to select one of the kinetic parameters used in the observer such that the estimation error is equal to zero in steady state. Let us first consider that the badly known parameter is the saturation constant K_S (this means that we use $\tilde{K}_S$ and μ_{max} in the observer equations). Then the dynamics of the observation errors is here equal to:

$$\frac{d}{dt} \begin{bmatrix} e_S \\ e_X \end{bmatrix} = \begin{bmatrix} -D - \frac{1}{Y_1}\tilde{\alpha}X - k_1 & -\frac{1}{Y_1}\tilde{\alpha}\hat{S} \\ \tilde{\alpha} - k_2 & \tilde{\alpha}\hat{S} - D \end{bmatrix} \begin{bmatrix} e_S \\ e_X \end{bmatrix}$$

$$+ \begin{bmatrix} -\frac{1}{Y_1}SX \\ SX \end{bmatrix} e_\alpha \qquad (7.202)$$

with:

$$\tilde{\alpha} = \frac{\mu_{max}}{\tilde{K}_S + \hat{S}}, \quad e_\alpha = \mu_{max}[\frac{1}{K_S + S} - \frac{1}{\tilde{K}_S + \hat{S}}] \qquad (7.203)$$

e_S and e_X are the estimation errors on S and X. As a matter of illustration, if we consider the ELO, the gains k_1 and k_2 would be chosen from equations (7.72)(7.73). Let us set de_S/dt, de_X/dt to zero. Then e_X in steady state, $\bar{e}_X$, is given by the following expression:

$$\bar{e}_X = \frac{(\frac{1}{Y}k_2 + k_1 + D)SXe_\alpha}{(D - \tilde{\alpha}\hat{S})(D + k_1 + \frac{1}{Y_1}\tilde{\alpha}X) + \frac{1}{Y_1}\tilde{\alpha}\hat{S}(\tilde{\alpha}X - k_2)} \qquad (7.204)$$

We note that the steady state error $\bar{e}_X$ will be equal to zero if $\frac{1}{Y_1}k_2 + k_1 + D = 0$. This gives the following expression for $\tilde{K}_S$ from (7.72) (7.73):

$$\tilde{K}_S = -\hat{S} + \frac{\mu_{max}D\hat{S}}{D^2 - (\lambda_1 + \lambda_2)D + \lambda_1\lambda_2} \qquad (7.205)$$

By using a similar approach, if μ_{max} is assumed to be badly known instead of K_S, the value of $\tilde{\mu}_{max}$ in the observer that guarantees zero steady state error for the estimation of X is given by the following relationship:

$$\tilde{\mu}_{max} = \frac{(D^2 - (\lambda_1 + \lambda_2)D + \lambda_1\lambda_2)(K_S + \hat{S})}{D\hat{S}} \qquad (7.206)$$

Figures 7.10 and 7.11 illustrate the behaviour of the ELO with $\tilde{K}_S$ given by equation (7.205) for $\lambda_1 = \lambda_2 = -0.2$. Figure 7.10 shows the convergence of the observer when $\hat{X}(0) = 0$. Figure 7.11 illustrates the performance of the observer in presence of a square wave of the influent concentration S_{in} (between 5 and 6 g/l, variations at time t = 40, 80, 120 and 160 h). The process is assumed to be initially in steady state ($X(0) = 2.054$ g/l, $S(0) = 0.893$ g/l); the observer has been initialised with the correct value for $\hat{S}$, and a wrong value for $\hat{X}$ (= 1 g/l). Note that, as expected, the estimate of X (dotted line) converges to the "true" simulated value. This is obviously done at the price of a biased estimate of the measured variable S. Note that in Figure 7.11 the estimate of S rapidly converges to its biased value (around 0.68 g/l).

Our numerical experience with different kinetic models and estimation problems tell us that the gains of the observer have to be selected not too large in order to avoid strange transient behaviour of the observer. So far we have not been able to obtain satisfactorily (i.e. under realistic conditions) theoretical stability and convergence for this approach.

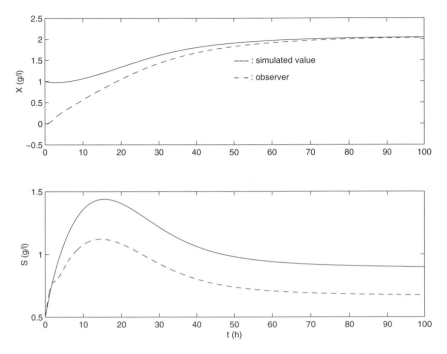

FIG. 7.10. Extended Luenberger observer with K_S computed from (7.205)(initial transient)(-: "true" simulated values; - -: estimates).

7.5.2 *Adaptive State Observer*

The second approach that we propose in order to handle (kinetic) parameter uncertainty in the state observation of (bio)chemical processes is the design of adaptive observers, i.e. observers that also estimate the badly known parameters (e.g. [58], [14]). One of the original features of the present adaptive observers is to consider a nominal (default) process model, i.e. a model with nominal values of the badly known parameters.

For simplicity we consider here that the badly known parameters are such that the process models are linear in these parameters. Then we can write the right hand side of (7.198) as follows:

$$f(x, u, \theta) = \bar{f}(x, u, \bar{\theta}) + b(x, u)\Delta\theta \qquad (7.207)$$

From the above equations and the observer equations (7.27), the adaptive observer is readily obtained (by considering the badly known parameters (here $\Delta\theta$) as unmeasured states with dynamics equal to zero):

$$\frac{d\hat{x}}{dt} = \bar{f}(\hat{x}, u, \bar{\theta}) + b(\hat{x}, u)\widehat{\Delta\theta} + K(y - \hat{y}) \qquad (7.208)$$

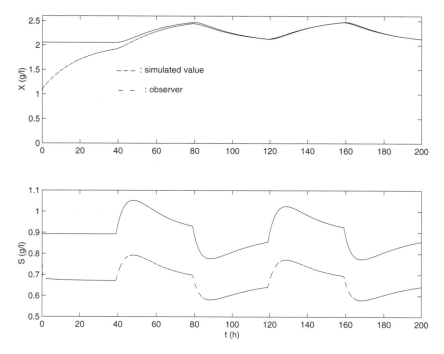

FIG. 7.11. Extended Luenberger observer with K_S computed from (7.205)(square wave S_{in})(-: "true" simulated values; - -: estimates).

$$\frac{d\widehat{\Delta\theta}}{dt} = \Gamma(y - \hat{y}) \qquad (7.209)$$

Some important remarks are probably necessary at this point. Although the design of adaptive observers has been a very active research field in the 1980s (see e.g. [150], [70], [183], [167], [14]), the theoretical analysis of adaptive observers may become easily highly complex. Moreover our experience shows that their application to processes is indeed so far quite disappointing because they are very difficult to tune properly in practice. That's why we have limited our study here to the following items:

1. to base the design on a model with nominal values of the badly known parameters;
2. to consider here only models linear in the badly known parameters;
3. to allow only one badly known parameter per measured variable;
4. to consider what is called in the adaptive control nomenclature as the "averaging" (see e.g. [165]) as a key tuning rule for the adaptive observer.

Item 1 is important to increase the flexibility on the observer dynamics. Item 2 is only an a priori choice that simplifies the approach. Item 3 is probably the most

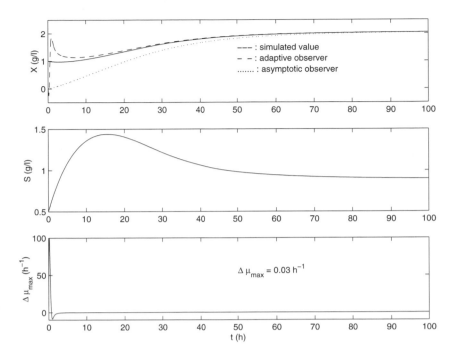

FIG. 7.12. Adaptive observer (initial transient).

essential assumption in order to have some guarantee for the successful implementation of the observer, and is rather easy to understand from a practical viewpoint: how can one expect to handle properly uncertainty on different kinetics and distinguish between them if there is not a sufficient amount of information about them, in particular one independent information per independent unknown. Finally, item 4 is also a key issue in the implementation of the adaptive observer: one important underlying idea of averaging is that the dynamics of the parameter adaptation has to be slower than the dynamics of the rest of the observer.

In the rest of the section we shall consider the same example as before (simple microbial growth, Section 7.3.4) on which we shall perform the theoretical analysis of the adaptive observer and illustrate its performance in numerical simulation. Assume that μ_{max} is the badly known parameter with:

$$\mu_{max} = \bar{\mu}_{max} + \Delta\mu_{max} \qquad (7.210)$$

where $\bar{\mu}_{max}$ is a nominal value of μ_{max}. Then equation (7.207) specialises as follows:

$$\frac{dS}{dt} = -\frac{1}{Y_1}\bar{\mu}_{max}\frac{SX}{K_S + S} + DS_{in} - DS - \frac{1}{Y_1}\frac{SX}{K_S + S}\Delta\mu_{max} \qquad (7.211)$$

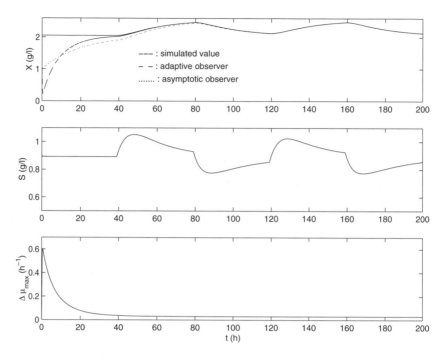

FIG. 7.13. Adaptive observer (square wave S_{in}).

$$\frac{dX}{dt} = \bar{\mu}_{max}\frac{SX}{K_S + S} - DX + \frac{SX}{K_S + S}\Delta\mu_{max} \qquad (7.212)$$

At this point we consider an asymptotic observer for estimating the value of X that appears in the regressor (i.e. the multiplicative term) of $\Delta\mu_{max}$. Its introduction is fully motivated by the theoretical analysis that will be performed below. The asymptotic observer is based on the state transformation: $\zeta = S + \frac{1}{Y}X$. Its dynamical equations are readily derived from the model equations (7.43),(7.44):

$$\frac{d\zeta}{dt} = -D\zeta + DS_{in} \qquad (7.213)$$

$$X_e = Y_1(\zeta - S) \qquad (7.214)$$

where X_e denotes the estimate of X given by the asymptotic observer (we use a different notation in order to avoid the confusion with the estimate of X ($\hat{X}$) given by the adaptive observer). The adaptive observer is then equal to:

$$\frac{d\hat{S}}{dt} = -\frac{1}{Y_1}\bar{\mu}_{max}\frac{S\hat{X}}{K_S + S} + DS_{in} - DS - \frac{1}{Y_1}\frac{SX_e}{K_S + S}\widehat{\Delta\mu_{max}}$$
$$+K_1(S - \hat{S}) \qquad (7.215)$$

$$\frac{d\hat{X}}{dt} = \bar{\mu}_{max}\frac{S\hat{X}}{K_S + S} - D\hat{X} + \frac{SX_e}{K_S + S}\widehat{\Delta\mu_{max}} - K_2(S - \hat{S}) \quad (7.216)$$

$$\frac{d\widehat{\Delta\mu_{max}}}{dt} = -\gamma(S - \hat{S}) \quad (7.217)$$

The analysis of the theoretical stability and convergence properties of the adaptive observer is based on the estimation error dynamics. If we define:

$$e_S = S - \hat{S}, \; e_X = X - \hat{X}, \; e_\mu = \Delta\mu_{max} - \widehat{\Delta\mu_{max}} \quad (7.218)$$

these are written as follows:

$$\frac{de}{dt} = Ae + Be_{Xe} \quad (7.219)$$

with:

$$A = \begin{bmatrix} -K_1 & -\frac{1}{Y_1}\frac{\bar{\mu}_{max}S}{K_S+S} & -\frac{1}{Y_1}\frac{SX_e}{K_S+S} \\ K_2 & \frac{\bar{\mu}_{max}S}{K_S+S} & \frac{SX_e}{K_S+S} - D \\ \gamma & 0 & 0 \end{bmatrix}$$

$$e = \begin{bmatrix} e_S \\ e_X \\ e_\mu \end{bmatrix}, \; B = \begin{bmatrix} -\frac{1}{Y_1}\frac{\Delta\mu_{max}S}{K_S+S} \\ \frac{\Delta\mu_{max}S}{K_S+S} \\ 0 \end{bmatrix}, \; e_{Xe} = X - X_e$$

From the theory of the asymptotic observers, we know that e_{Xe} will tend asymptotically to zero, $\lim_{t\to\infty} e_{Xe} = 0$. We also know (see e.g. [14]) that the matrix B is bounded. Similarly, if K_1, K_2 and γ are bounded, then A is bounded. Therefore the error dynamics are that of a time-varying system with an input, e_{Xe}, going asymptotically to zero. If the state matrix A is asymptotically stable, we can use a classical stability result (e.g. [284], p.55) to state that the estimation errors will tend asymptotically to zero. Let us check now that A is a stable matrix. Its characteristic polynomial $\det(\lambda I - A)$ is equal to:

$$\lambda^3 + \lambda^2(K_1 + D - \frac{\bar{\mu}_{max}S}{K_S + S}) + \lambda(K_1(D - \frac{\bar{\mu}_{max}S}{K_S + S})$$
$$+ K_2\frac{1}{Y}\frac{\bar{\mu}_{max}S}{K_S + S} + \gamma\frac{1}{Y}\frac{SX_e}{K_S + S}) + \gamma\frac{1}{Y}D\frac{SX_e}{K_S + S}$$

It is then routine to check that A is asymptotically stable via a proper choice of the gains K_1, K_2, and γ, e.g. if we assign the dynamics of the observer with the three eigenvalues λ_1, λ_2, λ_3. Figure 7.12 illustrates the performance of the observer under the same initial and operating conditions as in Figure 7.10, but with a 10% error on μ_{max}. The gains have been chosen primarily to assign the dynamics

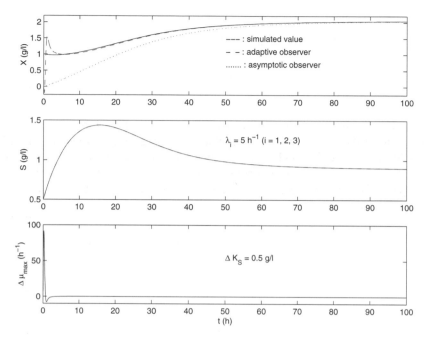

FIG. 7.14. Adaptive observer (10% error on K_S).

with $\lambda_1 = \lambda_2 = \lambda_3 = 5 \text{ h}^{-1}$, then γ has been reduced by a factor 20 to follow the averaging recommendations (the factor is somewhat arbitrary (10 would also have been good too for instance), it just gives a rough idea of a typical value for γ). Figure 7.13 illustrates the performance of the adaptive observer in the presence of a square wave influent substrate concentration.

Note that the adaptive observer (which assumes the knowledge of the kinetics model) is able to converge faster than the asymptotic observer (which ignores the kinetics model).

It is also worth noting that the adaptive observer (7.215)(7.216)(7.217) is capable of handling nonlinear uncertainties like those on K_S, as it illustrated in Figure 7.14, which presents a numerical simulation performed under the same conditions as in Figure 7.12, but with a 10% error on K_S. The adaptive observer compensates the error on K_S by a bias on the estimate of μ_{max}.

7.5.3   Generalisation

The generalisation of both proposed observers is based on the General Dynamical Model:

$$\frac{d\xi}{dt} = -D\xi + Y\rho + F - Q \tag{7.220}$$

If we consider that the measured outputs y are process components, then the part of the General Dynamical Model associated to y is written as follows:

$$\frac{dy}{dt} = -Dy + Y_y\rho + F_y - Q_y \qquad (7.221)$$

where Y_y, F_y and Q_y gather the rows of Y, F and Q related to y. The key assumptions at this point are to consider that the number of measured components is (at least) equal to the number of uncertain kinetics, and that the measured components are independent (this basically requires that the submatrix Y_y is full rank). The generalisation consists then of introducing the following state transformation (see also [191]):

$$\Psi = Y_y^{-1}y \qquad (7.222)$$

This transformation decouples all the kinetic terms and associates each of them to only one variable Ψ. Indeed equation (7.221) can be rewritten as follows:

$$\frac{d\Psi}{dt} = -D\Psi + \rho + Y_y^{-1}(F_y - Q_y) \qquad (7.223)$$

It is then straightforward to extend the design of both observers introduced in the above sections to each decoupled system (Ψ_i, ρ_i).

7.6   On-line Parameter Estimation

So far we have considered the estimation of state variables, i.e. state observation. In this section, we address the problem of estimating the reaction rates from on-line knowledge of the state variables (knowledge available either from measurements or from state estimation). It has been decided to restrict the section to the development, analysis and implementation of estimators of kinetic parameters. Our motivation for doing so comes basically from our experience that shows the kinetic parameters are in practice difficult to evaluate precisely via kinetic models and are often the primary and key source of uncertainty of bioprocess models (these difficulties has also motivated the writing of the preceding chapters, particularly Chapter 5 dedicated to experiment design for parameter estimation). It is also important to note that the approaches presented here do not exclusively apply to the estimation of kinetic parameters, but can also be used for the estimation of any other process parameters, like yield coefficients or transfer coefficients. Examples of estimation of such parameters can be found e.g. [14].

Two approaches will be considered: the observer-based estimator and the recursive least squares estimator.

7.6.1   *The Observer-Based Estimator*

The underlying idea of the observer-based estimator is (as its name suggests) to consider the classical observer structure already introduced in Section 7.3. Yet here we go a step further by assimilating the unknown parameter as unknown states for which the dynamics are equal to zero. This means that if we consider that the dynamics of the process are given by the following equations:

$$\frac{dx}{dt} = f(x, u, \theta) \tag{7.224}$$

we assume that the unknown parameters θ have the following "dynamics":

$$\frac{d\theta}{dt} = 0 \tag{7.225}$$

and we consider both equations (7.224)(7.225) for the design of the observer-based estimator.

Let us now proceed further in the design of the estimator. Let us first consider the General Dynamical Model:

$$\frac{d\xi}{dt} = Y\rho(\xi) - D\xi - Q + F \tag{7.226}$$

We assume that:

a) the matrix Y of yield coefficients is known,
b) the dilution rate D, the feed rates F and the gaseous outflow rates Q are measured on-line,
c) the vector of state variables ξ is known either by measurement or by estimation using an asymptotic observer (as described in Section 7.4).

We further assume that the vector $\rho(\xi)$ of reaction rates is partially unknown and written as follows:

$$\rho(\xi) = \begin{bmatrix} H(\xi)r(\xi) \\ h(\xi) \end{bmatrix} \tag{7.227}$$

where $H(\xi)$ is a diagonal matrix of known functions of the state and $r(\xi)$ a vector of unknown functions of ξ with $\dim r(\xi) = n_u$.

The known reaction rates are given by vector $h(\xi)$ with $\dim h(\xi) = M - n_u$.

Using equation (7.227), the general dynamical model is rewritten as:

$$\frac{d\xi}{dt} = Y_u H(\xi)r(\xi) + Y_k h(\xi) - D\xi - Q + F \tag{7.228}$$

where Y_u and Y_k are matrices of yield coefficients associated with the unknown and known reaction rates respectively.

The observer-based estimator is written as follows:

$$\frac{d\hat{\xi}}{dt} = Y_u H(\xi)\hat{r} + Y_k h(\xi) - D\xi - Q + F - \Omega(\xi - \hat{\xi}) \tag{7.229}$$

$$\frac{d\hat{r}}{dt} = [Y_u H(\xi)]^T \, \Gamma(\xi - \hat{\xi}) \tag{7.230}$$

As in the classical observer, the update of the parameter vector $\hat{r}$ is driven by the deviation term $(\xi - \hat{\xi})$ which reflects the mismatch between $\hat{r}$ and r.

The matrices Ω and Γ are tuning parameters for adjusting the rate of convergence of the algorithm [14]. A common choice is:

$$\Omega = diag(-\omega_i), \quad \Gamma = diag(-\gamma_j), \quad \omega_i, \gamma_j > 0 \tag{7.231}$$

With this choice, the stability of the estimator is satisfied.

The tuning procedure may be simplified if the state equations are first decoupled using the transformation (7.222) already considered in Section 7.5.3 [191]:

$$\Psi = Y_u^{-1}\xi \tag{7.232}$$

The dynamical equations of the new variables Ψ are readily obtained from (7.228):

$$\frac{d\Psi}{dt} = H(\xi)r(\xi) + Y_u^{-1}Y_k h(\xi) - D\Psi + Y_u^{-1}(F - Q) \tag{7.233}$$

Note that each entry Ψ_i of Ψ is a linear combination of the entries ξ_i of ξ. Applying the estimation algorithm to the transformed state equations yields:

$$\frac{d\hat{\Psi}}{dt} = H\hat{r} + Y_u^{-1}Y_k h - D\Psi + Y_u^{-1}(F - Q) - \Omega(\Psi - \hat{\Psi}) \tag{7.234}$$

$$\frac{d\hat{r}}{dt} = H\Gamma(\Psi - \hat{\Psi}) \tag{7.235}$$

Recall that H is a diagonal matrix. Let us gather each variable Ψ_i with its related parameter r_i and rearrange the entries of the vector $[\Psi, r]^T$ in the following order in a vector ζ:

$$\zeta = \begin{bmatrix} \Psi_1 \\ r_1 \\ \Psi_2 \\ r_2 \\ \dots \\ \Psi_p \\ r_p \end{bmatrix} \tag{7.236}$$

Basic tuning rule. Let us first define the estimation error:

$$e = \zeta - \hat{\zeta} \tag{7.237}$$

The estimation error dynamics are readily derived from equations (7.233) (7.234) and (7.235):

$$\frac{de}{dt} = Ae + b \qquad (7.238)$$

with a block diagonal matrix A with 2×2 blocks:

$$A = diag\{A_i\}, \quad A_i = \begin{bmatrix} -\omega_i & h_i(\xi) \\ -\gamma_i & 0 \end{bmatrix}, \quad i = 1 \ to \ p \qquad (7.239)$$

and b equal to:

$$b = \begin{bmatrix} 0 \\ \frac{dr_1}{dt} \\ 0 \\ \frac{dr_2}{dt} \\ \dots \\ 0 \\ \frac{dr_p}{dt} \end{bmatrix} \qquad (7.240)$$

The characteristic equation of the matrix A, $det(\lambda I - A)$, is equal to:

$$det(\lambda I - A) = \prod_{i=1}^{p} (\lambda^2 + \omega_i \lambda + \gamma_i h_i(\xi)) \qquad (7.241)$$

The key idea of the tuning rule consists of choosing each γ_i inversely proportional to the corresponding term $h_i(\xi)$:

$$\gamma_i = \frac{\bar{\gamma}_i}{h_i(\xi)}, \quad \bar{\gamma}_i > 0, \quad i = 1 \ to \ p \qquad (7.242)$$

With the choice above, the characteristic equation (7.241) is rewritten as follows:

$$det(\lambda I - A) = \prod_{i=1}^{p} (\lambda^2 + \omega_i \lambda + \bar{\gamma}_i) \qquad (7.243)$$

and the observer-based estimator dynamics are now independent of the state variables. Such a choice corresponds to a Lyapunov transformation (see [192]). It is obviously valid for values of $h_i(\xi) \neq 0$: this condition is usually met easily in (bio)process applications, as it will be illustrated in the following section.

The values of the design parameters can then be set to arbitrarily fix the estimator's dynamics for each unknown parameter r_i. Since the estimator reduces, via the transformations, to a set of independent second-order linear systems, the classical rules for assigning the dynamics of second-order linear systems apply straightforwardly here. The reader is therefore referred to the classical automatic control textbooks for further information on the subject. However the following basic guidelines are suggested.

One important guideline is to choose real poles:

$$\omega_i^2 - 4\bar{\gamma}_i \geq 0 \qquad (7.244)$$

The objective of this is to avoid oscillations in the estimation of the parameters that do not correspond to any physical phenomenon related to the estimated reaction rates.

Pomerleau and Perrier ([198]) suggest choosing double poles, i.e.:

$$\bar{\gamma}_i = \frac{\omega_i^2}{4} \qquad (7.245)$$

The tuning of the estimation algorithm reduces to the choice of one design parameter, ω_i, per estimated parameter. This allows having a design procedure that has the double advantage of being simple (one design parameter) and flexible (each parameter estimation can be tuned differently if needed, e.g. if the time variations of the parameters are different).

As an alternative, Oliveira *et al.* ([186]) propose to choose complex poles with a damping factor equal to 0.7 in order to increase the speed of convergence of the estimator with a reduced overshoot. (Generally speaking, the damping factor can be freely chosen; the choice may then depend on the type and nature of the application, of the time variations of the parameter to be estimated, and of the noise on the measured data. This means that then there are, in this approach, two design parameters per estimated parameter).

So far, we have suggested that it is possible to assign arbitrarily the dynamics of the estimator. However in presence of noisy data, it appears that indeed a compromise has to be made between a fast estimator convergence and a good noise rejection. A detailed and somewhat involved study is performed in Bastin and Dochain [14] (pp. 162-72) to analyse the performance of the observer-based estimator both in theory and in numerical simulation in the presence of bounded noisy data in the particular case of the estimation of the specific growth rate of a simple microbial growth process. The theoretical optimisation analysis is based on the evaluation of the asymptotic properties of the estimator and results in the following optimal value for ω_1:

$$\omega_{1,opt} = 2\sqrt{\frac{M_1}{\alpha(M_2^2 + Y_1 S_{max} M_2)}}, \qquad 0 < \alpha < 1 \qquad (7.246)$$

where M_1 and M_2 are the upper bounds on the time derivative of μ and on the measurement noise, respectively, and S_{max} the maximum value of the influent substrate concentration S_{in}.

This result is probably rather conservative because it is based on upper bounds for the measurement noise, the time variation of μ and the influent substrate concentration, but it is qualitatively confirmed by numerical simulation studies which

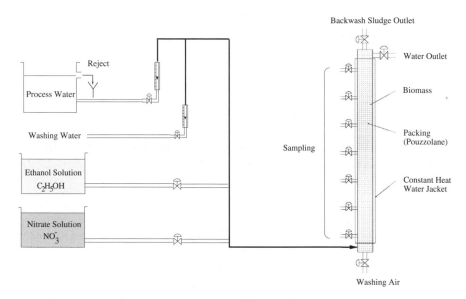

FIG. 7.15. Experimental set-up of the denitrification process.

also give a value of the design parameters that minimises the estimation error. Because it is conservative, the theoretical optimum has to be interpreted with care, but since it is qualitatively correct, our suggestion (also based on our practical experience), in presence of noisy measurements, is to perform numerical simulations with a plausible reaction rate model and noise in order to get a first initial guess for the design parameter values. These can then be adjusted when applied to the real process.

Application to a denitrification process.   Let us consider the application to the observer-based estimator to a denitrification process. The process considered here is a pilot-scale denitrifying biofilter [34], [36]. The denitrification process under study is a submerged granular fixed bed biofilter (see Figure 7.15); it is packed with pouzzolane, a very porous volcanic material which retains a big amount of microorganisms even after back-washing. Thus the initial biomass concentration is often very close to the maximum active biomass concentration. The substrates to be removed can move freely along the reactor. The inside temperature is kept almost constant via a double insulation jacket. An ethanol solution is mixed with the feeding nitrate in a tank at the inlet of the reactor. Eight measurement points are distributed at every thirty centimeters along the column for the measurement of nitrate and nitrite concentrations.

The denitrification process is defined as being the reduction of nitrates (electron acceptor) NO_3^- into gaseous nitrogen N_2 by using organic carbon (electron donor) with the production of intermediate compounds, namely nitrites NO_2^-. In the denitrification, two sequential biological reactions are considered in the model, denitratation (1) and denitritation (2):

$$NO3^- \xrightarrow{(1)} NO2^- \xrightarrow{(2)} N_2$$

If we assume that the liquid and solid mediums are in completely mixed conditions and that the biomass is fixed, the dynamics of the process are given by the following equations:

$$\frac{dS_{NO_3}}{dt} = -DS_{NO_3} + DS_{NO_3,in} - \frac{1}{Y_1}\mu_1 X \tag{7.247}$$

$$\frac{dS_{NO_2}}{dt} = -DS_{NO_2} + Y_3\mu_1 X - \frac{1}{Y_2}\mu_2 X \tag{7.248}$$

$$\frac{dS_C}{dt} = -DS_C - DS_{C,in} - \frac{1}{Y_4}\mu_1 X - \frac{1}{Y_5}\mu_2 X \tag{7.249}$$

$$\frac{dX}{dt} = (\mu_1 + \mu_2 - k_d) X \tag{7.250}$$

where S_C is the (external) source of carbon (ethanol) for both biochemical reactions, and k_d is the biomass death coefficient.

As a matter of illustration of the flexibility of the algorithm, let us consider that we consider the following rewriting of the specific growth rates μ_1 and μ_2:

$$\mu_1 = r_1 S_{NO_3}, \quad \mu_2 = r_2 S_{NO_2} \tag{7.251}$$

which simply expresses an explicit dependence between the specific growth rates and one of their limiting substrates. Then:

$$r = \begin{bmatrix} r_1 \\ r_2 \end{bmatrix}, \quad H = \begin{bmatrix} S_{NO_3}X & 0 \\ 0 & S_{NO_2}X \end{bmatrix} \tag{7.252}$$

In the present experiments, the concentrations of nitrate S_{NO_3} and nitrite S_{NO_2} are accessible for on-line measurement. The auxiliary variables Ψ are obtained from the inverse of the yield coefficient matrix Y associated to these two variables:

$$Y = \begin{bmatrix} -\dfrac{1}{Y_1} & 0 \\ Y_3 & -\dfrac{1}{Y_2} \end{bmatrix} \tag{7.253}$$

The auxiliary variables are then defined as follows:

$$\Psi_1 = -Y_1 S_{NO_3}, \quad \Psi_2 = -Y_2 S_{NO_2} - Y_1 Y_2 Y_3 S_{NO_3} \qquad (7.254)$$

The observer-based estimator is written as follows:

$$\frac{d\hat{\Psi}_1}{dt} = -D\Psi_1 - DY_1 S_{NO_{3,in}} + \hat{r}_1 S_{NO_3} X + \omega_1(\Psi_1 - \hat{\Psi}_1) \qquad (7.255)$$

$$\frac{d\hat{r}_1}{dt} = \frac{\omega_1^2}{4 S_{NO_3} X}(\Psi_1 - \hat{\Psi}_1) \qquad (7.256)$$

$$\frac{d\hat{\Psi}_2}{dt} = -D\Psi_2 - DY_2(Y_1 Y_3 S_{NO_{3,in}} + S_{NO_{2,in}}) + \hat{r}_2 S_{NO_2} X$$

$$+\omega_2(\Psi_2 - \hat{\Psi}_2) \qquad (7.257)$$

$$\frac{d\hat{r}_2}{dt} = \frac{\omega_2^2}{4 S_{NO_2} X}(\Psi_2 - \hat{\Psi}_2) \qquad (7.258)$$

Figure 7.16 shows the estimation results of r_1 and r_2 based on real-life data. The (known) parameters of the model are equal to:

$$Y_1 = 1.05, \quad Y_2 = 0.67, \quad Y_3 = 1.05, \quad Y_4 = 0.33, \quad Y_5 = 0.22, \quad k_d = 0.01 h^{-1}$$

The flow rate was equal to 32.5 l/h. The influent concentrations are given in Figure 7.16. The observer-based estimator was initialised as follows:

$$\omega_1 = 0.5 \, h^{-1}, \quad \omega_2 = 0.5 \, h^{-1}, \quad \hat{\Psi}_1(0) = \Psi(0), \quad \hat{\Psi}_2(0) = \Psi_2(0) \ (7.259)$$

$$\hat{r}_1(0) = 0.0001 \, m^3/g/h, \quad \hat{r}_2(0) = 0.0002 \, m^3/g/h \qquad (7.260)$$

Validation of the estimation has been performed on data of ethanol S_C by comparing real-life data with values computed on the basis of the estimated values of r_1 and r_2 and the mass balance equation of S_C, i.e.:

$$\frac{d\hat{S}_C}{dt} = -D\hat{S}_C - D S_{C,in} - \frac{1}{Y_4}\hat{r}_1 S_{NO_3} X - \frac{1}{Y_5}\hat{r}_2 S_{NO_2} X \qquad (7.261)$$

The values are given in Figure 7.16 (bottom right).

7.6.2 The Recursive Least Squares Estimator

Since the model is linear in the unknown parameters, an alternative estimation algorithm may be the least squares algorithm in recursive form, or the recursive least squares (RLS) scheme. We have presented in Chapter 6 the non recursive version of the least squares estimation algorithm.

Let us consider the following discrete-time equation linear in the (unknown) parameter θ:

$$y_t = \phi_t^T \theta_t \qquad (7.262)$$

where t is the time index. The RLS scheme with forgetting factor γ based on the above equation (7.262) is indeed written as follows (see e.g. [151] for a detailed approach to derive the recursive form from the batch ("off-line") version):

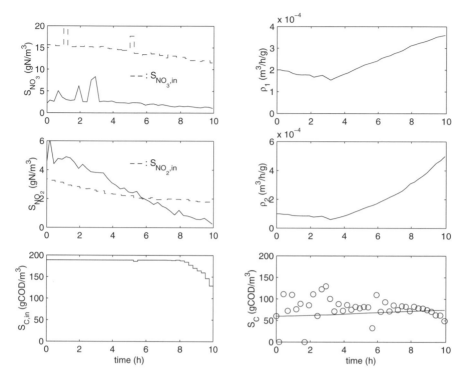

FIG. 7.16. Observer-based estimator for a denitrifying biofilter

$$\hat{\theta}_{t+1} = \hat{\theta}_t + g_t \phi_t (y_t - \phi_t^T \hat{\theta}_t) \qquad (7.263)$$

$$g_t = \frac{g_{t-1}}{\gamma} \left[I - \phi_t (\gamma I + \phi_t^T g_{t-1} \phi_t)^{-1} \phi_t^T g_{t-1} \right] \qquad g_0 > 0, \ 0 < \gamma \le 1$$

$$(7.264)$$

where I is the identity matrix, and g_t is the gain of the estimator. The role of the forgetting factor is to follow possible time variations of the unknown parameter θ by giving a weight to recent data that is higher than that given to old data. $\gamma = 1$ means that an equal weight is given to all data, and there is then no forgetting effect. The forgetting factor is typically (but not necessarily) chosen between 0.9 and 0.99 in practical applications.

Let us see how to apply the RLS scheme to the estimation of kinetic parameters on the basis of the General Dynamical Model. If we consider an Euler approximation for the time derivative:

$$\frac{d\xi}{dt} = \frac{\xi_{t+1} - \xi_t}{\Delta t} \qquad (7.265)$$

with Δt the sampling period, the version used for the observer-based estimator (7.228) is written in discrete-time as follows:

$$\xi_{t+1} = \xi_t + \Delta t(Y_u H(\xi_t)r(\xi_t) + Y_k h(\xi_t) - D_t\xi_t - Q_t + F_t) \qquad (7.266)$$

The RLS scheme can be immediately applied by considering that:

$$y_t = \xi_{t+1} - \xi_t - \Delta t(Y_k h(\xi_t) - D_t\xi_t - Q_t + F_t) \qquad (7.267)$$

$$\phi_t^T = \Delta t Y_u H(\xi_t), \ \theta_t = r(\xi_t) \qquad (7.268)$$

The selection of the initial gain g_0 is typically an important choice for starting correctly the RLS estimator. Obviously it must be positive definite. Usually the literature suggests starting with a diagonal matrix with large values. However in practice too large values may result in very large transients in the starting phase. Our recommendation for RLS estimation with forgetting factor is to select initial values for entries of the gain matrix that have the same order of magnitude as those reached in steady state.

This can be easily done when estimating only one parameter. Indeed the gain matrix g_t (as well the regressor ϕ_t) reduces to a scalar and is written as follows:

$$g_t = \frac{g_{t-1}}{\gamma + g_{t-1}\phi_t^2} \qquad (7.269)$$

In steady state , $g_t = g_{t-1}$, and therefore we can draw the following expression for g_t:

$$g_t = \frac{1 - \gamma}{\phi_t^2} \qquad (7.270)$$

Therefore our recommendation is then to select the initial value g_0 close to $\dfrac{1 - \gamma}{\phi_0^2}$.

Application to an anaerobic digestion process.   Let us consider the anaerobic digestion process already introduced in Section 7.4 and Figures 7.5 and 7.6.

Here again, methane P_2 and volatile fatty acids S_2 are available measurements. Let us use them to estimate the specific growth rates μ_1 and μ_2 of the acidogenesis and of the methanisation, respectively.

In discrete-time, and by considering the low solubility assumption on methane, the mass balance equations for methane and volatile fatty acids are written as follows:

$$Q_{2t} = Y_6 \mu_{2t} X_{2t} \qquad (7.271)$$

$$S_{2,t+1} = S_{2t} - \Delta t D_t S_{2t} + \frac{1}{Y_2}\mu_{1t}\Delta t X_{1t} - Y_3\mu_{2t}\Delta t X_{2t} \qquad (7.272)$$

Because μ_2 appears in both equations while μ_1 appears only in the second, we have here designed the estimators in cascade. At each time step, we proceed as follows:

1. estimation of μ_2 on the basis of equation (7.271);

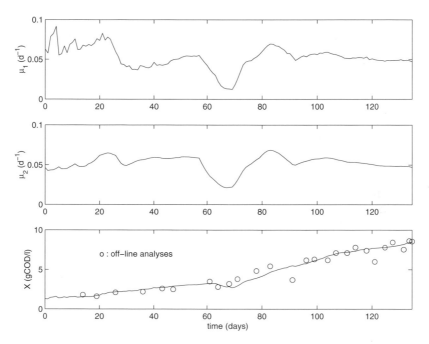

FIG. 7.17. On-line estimation of the specific growth rates in an anaerobic digestion process with a recursive least squares algorithm.

2. estimation of μ_1 on the basis of equation (7.272) and with the estimate of μ_2 given at step 1

The estimation results are given in Figure 7.17. The values of the biomass concentrations X_1 and X_2 are provided by the asymptotic observer of Section 7.4. The RLS estimators have been initialised as follows:

$$\text{estimation of } \mu_2 : y_t = Q_{2t}, \ \phi_t = Y_6 X_{2t}, \ \gamma = 0.9, \ g_0 = 0.01 \quad (7.273)$$

$$\text{estimation of } \mu_1 : y_t = S_{2,t+1} - S_{2t} + \Delta t \, D_t S_{2t} + Y_3 \hat{\mu}_{2t} \Delta t \, X_{2t} \quad (7.274)$$

$$\phi_t = \frac{1}{Y_2} \Delta t \, X_{1t}, \ \gamma = 0.9, \ g_0 = 0.01 \quad (7.275)$$

The estimates $\hat{\mu}_1$ and $\hat{\mu}_2$ have been validated by considering, as in Section 7.4 and Figure 7.6, the off-line data of total and soluble COD to calculate the biomass concentration. These have been compared to values computed on the basis of the estimates of μ_1 and μ_2 with the mass balance equation for $X_1 + X_2$:

$$\frac{d(\widehat{X_1 + X_2})}{dt} = -D(\widehat{X_1 + X_2}) + \hat{\mu}_1 X_1 + \hat{\mu}_2 X_2 \quad (7.276)$$

7.7 Conclusions

This chapter has been dedicated to the design of software sensors to estimate concentrations (state observers) and kinetic parameters (parameter estimation). We have started the chapter with theoretical notions of state observability, which appeared to be important to evaluate the possibility to apply and design state observers. The state observers have been presented in three sections: classical observers, asymptotic observers, and "intermediate" observers. The last section has been concerned with the design of recursive parameter estimation algorithms. Two approaches have been presented: the observer-based estimator, and the recursive least squares estimator. All the approaches proposed in the present chapter have been mathematically analysed. Their performances have been illustrated via numerical simulations and/or real-life results. And we have proposed several guidelines to help the user to implement them.

  At this point, we would also like to draw attention to the following conclusions concerning the application of the software sensors introduced here.

1. As it has been already explained, the parameter estimation section has been restricted to the estimation of kinetic parameters. Our motivation for doing so comes basically from our experience that shows that kinetic parameters are in practice difficult to evaluate precisely via kinetic models and are often the primary and key source of uncertainty of bioprocess models (these difficulties have also motivated the writing of the preceding chapters, particularly Chapter 5 dedicated to experiment design for parameter estimation). It is also important to note that the approaches presented here do not exclusively apply to the estimation of kinetic parameters, but can also be used for the estimation of any other process parameters, like yield coefficients or transfer coefficients. Examples of estimations of such parameters can be found e.g. [14].

2. Today the estimation of states and parameters in nonlinear systems remains generally speaking an open question. This explains why it is an active research area. In the present chapter, we have followed a line in which we have privileged three approaches (classical observers, asymptotic observers, and "intermediate" observers). The reason for doing so is twofold. First, this sequence had the pedagogical advantage to present approaches that are dedicated to solve three types of problems (state observation with known model parameters, with unknown model structure, and with badly known parameters, respectively). Secondly, the proposed approaches present a high degree of reliability due to our experience in implementing them (as illustrated in the numerical and real-life examples). But this choice does not mean that we consider that other approaches must be rejected or are generally speaking worse than those presented here. We even have the feeling that we are giving here an instantaneous photograph of the estimation of

state and parameter estimation in WWTP, and that the situation may change rapidly with possibly new emerging approaches.

3. Yet, our experience shows that today the asymptotic observer represents the best state observers that can be applied to bioprocesses in general and to WWTP in particular. This argument is largely based to the large number of applications that we have seen over the last 20 years. The deciding factor in favour of the asymptotic observers is the following. The asymptotic observer is indeed a reliable (because of its theoretical stability and convergence properties) replica of the reaction scheme and of the hydraulics that have been chosen by the user to represent the dynamics of the process. Therefore, if for any reason the asymptotic observer does not work well, the diagnosis is usually easy: it is either due to a bad selection of the reaction scheme or of the hydrodynamics, or to wrong values of the yield coefficients, or finally to on-line measurements that may appear to be unreliable.

APPENDIX A

Glossary

A.1 Model Constituents

constant: model constituent, whose value is constant throughout all possible applications of the model.

forcing function: function used as model input.

input/output model: model that describes system behavior as being a function of only present input and past inputs and outputs.

model: abstraction of reality.

model structure: the relations between inputs, outputs and eventually states formulated as equations.

observation equation: equation in state-space model, that relates the state variables to the outputs (sometimes also denoted as output equation).

parameter: model constituent, whose value needs to be determined for each specific application of the model.

state: present situation of the system as described by the model.

state variable: model constituent in state-space models, acting as mediator between inputs and outputs and used for a descriptive representation of the system.

state-space model: model that includes a descriptive representation of the system by means of an additional set of state variables.

state-transition equation: function that relates the future state of the system to the present state and inputs.

transfer function: same as input/output function.

variables: inputs, outputs and eventually state variables in model equations.

A.2 Model Attributes

adaptive: model that interacts with the real system and changes the values of its inputs or state variables depending on past output values.

aggregated: model that contains state variables that represent functional classes of different constituents (e.g. organisms) or that simplifies the spatial configuration of a system by lumping it together (cf. segregated).

black-box: model that describes the observed behaviour of the corresponding subsystem without being based on the mechanisms of this subsystem (cf. white-box).

complex: a relative attribute that values whether the model contains more state variables, parameters, forcing functions, etc. or an attribute that qualifies that (irrespective of the number of variables) there exists chaotic solutions of the model equations (cf. simple).

continuous space: the model resolves the spatial domain of the system continuously (cf. discrete space).

continuous time: the model resolves the time axis continuously; the time evolution is usually described by differential equations (cf. discrete time).

conceptual: a model that contains a description of the ideas/hypothesis on system behaviour without giving a mathematical formulation.

deterministic: the time evolution of the model solution is uniquely determined by the initial state (for state-space models) and the time evolution of inputs (cf. stochastic).

discrete space: the model approximates the spatial domain of the system by a number of mixed compartments (cf. continuous space).

discrete time: the model divides the time axis into periods of finite length and the output or the state of the model in the next period is given as an algebraic equation depending on the old inputs or states (cf. continuous time).

distributed parameter: model with more than one independent variable, i.e. model behaviour is governed by partial differential equation in time and space.

dynamic: the model describes the time evolution of a system; a solution of the model may anyway be in steady-state (cf. static, steady-state).

empirical: the model equations are not based on generally accepted laws but are just of a descriptive nature (cf. phenomenological, mechanistic).

grey-box: the model consists of submodels that partly are based on mechanistic, partly on phenomenological descriptions (cf. black-box, white-box).

heuristic: model not based on rigorous development but on rules of thumb, feeling, qualitative reasoning.

linear: model equations are linear in input variables (for input-output models) or in state variables (for state-space models).

lumped: equal to aggregated.

mechanistic: model with the goal of describing the mechanisms that lead to the observed behaviour (cf. phenomenological).

nonlinear: model equations are nonlinear in input variables (for input-output models) or in input and state variables (for state-space models).

phenomenological: describing the observed phenomena without representing the mechanisms governing the behaviour (cf. mechanistic).

physical: model that is based on a description with physical, chemical or biological laws; sometimes also used for small scale reproductions of a system made in physical materials.

reduced order: model of reduced complexity obtained by direct deduction (e.g. by aggregation/lumping) from a more complex basic model.

reductionist: hierarchical description of a system by resolving it in subsystems that again are resolved in sub-subsystems until a description level is reached at which a satisfying description is possible without empirical assumptions (in the ideal case down to a description that is based on natural laws).

segregated: model that separates variables in more functional classes (cf. aggregated).

semi-physical: equal to grey-box .

simple: relative attribute that describes that the model equations contain only few state variables, parameters, forcing functions, etc. and the solutions show simple behaviour (periodic or quasi-periodic; cf. complex).

static: the model only describes the steady-state solution of a system (cf. dynamic).

steady state: the model only describes the steady-state solution of a system (cf. dynamic).

stochastic: the time evolution of the model contains random elements (cf. deterministic).

time-invariant: the way the model processes input to output does not change with time.

transparent: equal to white-box.

white-box: model that describes a system by one or several submodels (white-boxes) that describe the observed behaviour of the corresponding subsystem by describing the relevant mechanisms of this subsystem (cf. black-box).

A.3   Terms of Model Building

calibration: the same as parameter estimation but not necessarily by using statistical methods.

corroboration: the same as validation; the term introduced by Popper [200] makes it clearer that the correctness of the model cannot be proved and that each successful test only increases the belief that the model is correct (cf. confirmation, falsification, validation, verification).

confirmation: the same as validation; the term makes it clearer that the correctness of the model cannot be proved and that each successful test only increases the belief that the model is correct (cf. corroboration, falsification, validation, verification).

falsification: demonstrating the invalidity of a model by showing that the model results deviate significantly from the measurements confirmation, corroboration, validation and verification are fail trials of falsification (cf. confirmation, corroboration, validation, verification).

frame definition: selection of which components of a system are to be described and specification of classes of models to be included in the model structure selection process and specification of the experimental conditions for use of the model (experimental frame).

identifiability analysis (structural, practical): evaluation of the uniqueness of the estimates of model parameters from measured data. Structural (theoretical, a priori) identifiability analysis assesses the uniqueness of parameter estimates from ideal data for a given experimental frame, practical (a posteriori) identifiability analysis assesses the accuracy with which parameters can be estimated with a given data set. In the latter case identifiability is not an objective property, but it depends on the required accuracy.

model building: the process of finding an adequate model of a system by (iteratively) processing the following model building steps: Problem formulation, prior knowledge collection, system identification and model testing (see Figure 1.1).

model reduction: simplification of an existing model in order to improve its identifiability without loosing the description of the most important phenomena.

model (structure) selection: selecting out of a given set of model structures the structure that makes an optimal (as simple as possible but as complicated as required for the intended purpose) description of measured data possible.

optimal experimental design: using a (preliminary) model of a system in order to plan an experiment that maximises the possible gain in information.

parameter estimation (batch, recursive): process of finding parameter values that lead to an optimal agreement of model results with measured data by using statistical methods. Time series of data can be used as a whole (batch estimation) or data points from within a moving data window can be used. In

the latter case the parameters become time-dependent and the algorithm can
be implemented to modify the previous estimate by considering the omitted
and the new data points (recursive estimation).

regression (linear, nonlinear): the same as parameter estimation, however usually used for the special case of algebraically given linear or nonlinear model
equations and using the (weighted) least squares technique goodness-of-fit
criterion.

simulation (interactive, real-time, Monte Carlo): calculating the solution of a
model for given values of the parameters, inputs and intial values (usually
numerically). Interactive simulations are processed on a computer which allows the user to interact with the program (stop, change parameter values,
etc.). Monte Carlo simulation is a method to propagate probability distributions of parameters, inputs and initial values to probability distributions
of model results by performing a lot of simulations with parameter values
sampled randomly from the probability distribution of the parameters, inputs
and initial values.

system identification: finding a model to solve a given problem by (iteratively)
processing the following identification steps: Frame definition, model structure selection, parameter estimation, model diagnosis. (see Figure 1.1).

structure characterisation: the same as model (structure) selection.

uncertainty analysis: estimating the uncertainty of model predictions and
analysing the sources of uncertainty.

validation: test of a model with a data set not used for identification; note that
such tests only increase the belief in the correctness of the model, it is not
possible to prove that the model is correct (cf. confirmation, corroboration,
falsification, verification).

verification: the same as validation (cf. confirmation, corroboration, falsification,
validation).

APPENDIX B

Nomenclature

a_i, i = 1 to n	: Coefficients of the polynomial A(q)	
a	: Inhibition constant (logistic model)	[l/g]
a, b	: Constants (weir flow rate model)	
a, b, c	: Constants (μ(pH))	
A	: State matrix	
A	: Reactor cross-section	[m^2]
A_L	: Liquid phase cross-section	[m^2]
A_S	: Solid phase cross-section	[m^2]
A_{se}	: Settler cross-section	[m^2]
$A(q)$	: Polynomial in q (denominator of a transfer function)	
b	: Death coefficient	[/h]
b_i, i = 1 to m	: Coefficients of the polynomial B(q)	
B	: Input matrix	
$B(q)$	: Polynomial in q (numerator of a transfer function)	
C	: Component concentration	[g/l]
C	: Joint experimental design criterion for SC and PE	
C_i	: Measurement error covariance matrix	
D	: Dilution rate	[/h]
D	: Discriminative power of an experiment	
D_{ma}	: Axial mass dispersion (diffusion) coefficient	[m^2/s]

DO	: Dissolved oxygen	
e	: Observation or estimation error	
E	: Parameter estimation accuracy of an experiment	
E	: Activation energy	[J/mol]
E	: Expectation operator	
f	: State function	
F	: Hydraulic flow rate	[l/h]
F	: Feed rate vector	[g/l/h]
F	: Fisher Information Matrix	
F_l	: Solid flux	[g/m^2/h]
g	: Gravity constant	[m/s^2]
h	: Output function	
$H(s)$	: Transfer function	
h, H	: Height	[m]
J	: Objective functional	
J_{opt}	: Minimal value of the objective functional	
k_0	: Kinetic constant	
k_r, k_s	: Rate constants	[/h]
$k_L a$	: Mass transfer coefficient	[/h]
K	: Gain matrix (of an observer)	
K_c	: Contois model constant	
K_i	: Inhibition constant	[g/l]
K_p	: Product inhibition constant	[g/l]
K_s	: Affinity (or saturation, or Michaelis-Menten) constant	[g/l]
L	: Length	[m]
L	: Output matrix	
L_f	: Lie derivative along the vector field f	
m_s	: Maintenance coefficient	[/h]
n	: Expansion index	
n	: Number of data in moving window regression	
N	: Number of weirs	
N	: Number of measured data	
O	: Observability matrix	
OUR	: Oxygen uptake rate	[mgO_2/l.min]
OTR	: Oxygen transfer rate	[mgO_2/l.min]
p_i	: Number of parameters of model i	
p_L	: Liquid phase pressure	[N/m^2]
p_S	: Solid phase pressure	[N/m^2]
P	: Product concentration	[g/l]
Q	: Gaseous outflow rate (vector)	[g/l/h]
Q_i	: Weighting matrix	
$r(f)$	: Reliability of an inflection point f	

$r_\epsilon(\tau)$	: Covariance with lag τ	
R	: Ideal gas constant	[J/mol/K]
R	: Riccati matrix	
s	: Laplace variable	
s^2	: Residual mean square	
S	: Substrate concentration	[g/l]
S_O	: Oxygen concentration	[g/l]
S_O^*	: Oxygen saturation concentration	[g/l]
t	: Time	[min]
T	: Sampling period	[min]
T	: Temperature	[K]
t_{puls}	: Time of pulse addition	[min]
u	: Fluid superficial velocity	[m/h]
$u(t)$	: Input vector	
u_0	: Fluid superficial velocity in absence of solid particles	[m/h]
u_S	: Particles' velocity	[m/h]
U_T	: Terminal settling velocity of the particles	[m/h]
V	: Parameter estimation covariance matrix	
V	: Volume	[m^3]
W	: Weighing matrix (Kalman observer)	
w_D	: Weight attributed to model discrimination	
w_E	: Weight attributed to parameter estimation	
x	: State vector	
X	: Concentration of biomass	[g/l]
X_d	: Concentration of dead/detached biomass	[g/l]
Y	: Yield coefficient (matrix)	
y	: Measurement vector	
$\hat{y}_i(\theta)$	: Model prediction vector for model i	
z	: Space	[m]

B.1   Greek Letters

α_i	: Reaction order	
δ	: Small parameter (singular perturbation)	
ϵ	: Void fraction	
λ	: Local state isomorphism (structural identifiability)	
λ_{max}	: Largest eigenvalue of F (in absolute values)	
λ	: Eigenvalue	
λ_{min}	: Smallest eigenvalue of F (in absolute values)	
Γ	: Gain matrix (observer-based estimator)	
μ_{max}	: Maximum specific growth rate	[/h]
μ	: Specific growth rate	[/h]

ν	: Specific production rate	[/h]
ν	: Settling velocity	[/h]
ν	: Maximum settling velocity	[/h]
η	: Non-growth associated specific production rate	[/h]
ρ	: Reaction rate	[/h]
ρ_L	: Liquid phase density	[g/m^3]
ρ_S	: Solid phase density	[g/m^3]
ϕ	: Conversion rate	[/h]
Ψ	: State transformation (state observer and observer-based estimator)	
$\sigma(\theta_i)$	: Standard error of parameter i	
θ_i	: Parameter i	
χ_i	: Experimental conditions for experiment i	
ξ	: Vector of the bioprocess component concentrations	[g/l]
Ω	: Gain matrix (observer-based estimator)	

B.2   Indices

B, A	: Autotrophic bacteria
B, H	: Heterotrophic bacteria
end	: Endogenous
ex	: Exogenous
in	: Influent
min	: Minimum
max	: Maximum
ND	: Organic nitrogen
NH	: Ammonia nitrogen
NO	: Nitrate nitrogen
out	: Effluent
R	: Recycle
W	: Waste

B.3   Abbreviations

AIC	: Akaike's Information Criterion
$BFGS$	: Broyden, Fletcher, Goldfarb and Shanno
$CSTR$	: Continuous stirred tank reactor
det	: Determinant (of a matrix)
DFP	: Davidson-Fletcher-Powell
DUD	: doesn't use derivatives
GA	: Genetic algorithm
GIC	: General Information Criterion

$iidN$	: independent and identically distributed normally
MAP	: Maximum a posteriori
$MCSM$	: Monte Carlo set-membership
ODE	: Ordinary differential equation
OED/PE	: Optimal experimental design for parameter estimation
OED/SC	: Optimal experimental design for structure characterization
OLS	: Ordinary least squares
PDE	: Partial differential equation
PDF	: Probability density function
PE	: Parameter estimation
$PRBS$	: Pseudo random binary signal
RSM	: Response surface methodology
SC	: Structure characterization
SCE	: Shuffled complex evolution
SSR_i	: Sum of squared residuals of model i
tr	: Trace (of a matrix)
WLS	: Weighted least squares
$WWTP$	: Wastewater treatment process (or plant)

References

[1] U. Abeling and C.F. Seyfried C.F. Anaerobic-aerobic treatment of high-strength ammonium wastewater - nitrogen removal via nitrite. *Wat. Sci. Tech.*, 26(5-6):1007–1015, 1992.

[2] S. Aborhey and D. Williamson. State and parameter estimation of microbial gowth processes. *Automatica*, 14:493–498, 1978.

[3] P. Agrawal and H.C. Lim. Analyses of various control schemes for continuous bioreactors. *Adv. Biochem. Eng./Biotechnol., A. Fiechter (Ed.)*, 30:61–90, 1984.

[4] S. Aiba, M. Shoda, and M. Nagatani. Kinetics of product inhibition in alcohol fermentation. *Biotechnol. Bioeng.*, 10:40–50, 1968.

[5] A. Aivasidis, H. Hochscherf, G. Rottmann, T. Hagen, M.T. Mertens, G. Reiners, and C. Wandrey. Neuere konzepte zur prozessaberwachung und -regelung bei der biologischen stickstoffelimination. *AWT*, 5:48–55, 1992.

[6] H. Akaike. A new look at the statistical model identification. *IEEE Trans. Aut. Control*, 19:716–723, 1974.

[7] J.G. Alvarez and J.G. Alvarez. Analysis and control of fermentation processes by optimal and geometric methods. *Proc. ACC*, 2:1112–1117, 1988.

[8] J.F. Andrews. A mathematical model for the continuous culture of microorganisms utilizing inhibiting substrates. *Biotechnol. Bioeng.*, 10:707–723, 1968.

[9] G.L. Atkins. The use of non-parametric methods for fitting equations to biological data. In G.C. Vansteenkiste and P.C. Young, editors, *Modelling and Data Analysis in Biotechnology and Medical Engineering*, pages 209–217. North-Holland, Amsterdam, 1983.

[10] D. Baetens, L.H. Hosten, B. Petersen, S. Van Volsem, and P.A. Vanrolleghem. Optimal experimental design for the calibration of models of phosphorous removing activated sludge systems. *Journal A*, 41(3):65–73, 2000.

[11] B. Balmelle, K.M. Nguyen, B. Capdeville, J.C. Cornier, and A. Deguin. Study of factors controlling nitrite build-up in biological processes for water nitrification. *Wat. Sci. Tech.*, 26(5-6):1017–1025, 1992.

[12] M. Baltes, R. Schneider, C. Sturm, and M. Reuss. Optimal experimental design for parameter estimation in unstructured growth models. *Biotechnol. Prog.*, 10:480–488, 1994.

[13] Y. Bard. *Nonlinear Parameter Estimation*. Academic Press, London, 1974.

[14] G. Bastin and D. Dochain. *On-line Estimation and Adaptive Control of Bioreactors*. Elsevier, Amsterdam, 1990.

[15] G. Bastin, D. Dochain, M. Haest, M. Installé, and Ph. Opdenacker. Modelling and adaptive control of a continuous anaerobic fermentation process. In A. Halme, editor, *Modelling and Control of Biotechnical Processes*, pages 299–306. Pergamon Press, Oxford, 1982.

[16] G. Bastin and J. Levine. On state reachability of reaction systems. *Proc. 29th CDC*, pages 2819–2824, 1990.

[17] D.M. Bates and D.G. Watts. Relative curvature measures of nonlinearity. *J. Roy. Stat. Soc.*, B42:1–25, 1980.

[18] E.M.L. Beale. Confidence regions in nonlinear estimation. *J. Roy. Stat. Soc.*, B22:41–88, 1960.

[19] J.V. Beck and K.J. Arnold. *Parameter Estimation in Engineering and Science*. John Wiley, New York, 1977.

[20] M.B. Beck. System identification and control. In G.G. Patry and D. Chapman, editors, *Dynamical Modelling and Expert Systems in Wastewater Engineering*, pages 261–323. Lewis Publishers, Chelsea, Michigan, 1989.

[21] M.B. Beck. Uncertainty, system identification and the prediction of water quality. In M.B. Beck and G. van Straten, editors, *Uncertainty and Forecasting of Water Quality*, pages 3–68. Springer Verlag, Berlin, 1983.

[22] M.B. Beck. Identification, estimation and control of biological waste-water treatment processes. *IEE Proc. (Part D)*, 133(5):254–264, 1986.

[23] M.B. Beck. Water quality modeling: A review of the analysis of uncertainty. *Water Res. Res.*, 23:1393–1442, 1987.

[24] M.B. Beck and P.C. Young. An introduction to system identification, parameter and state estimation. *Computer Appl. in Ferment. Technol.: Modelling and Control Biotechnol. Proc., Fish N., Fox R. and Thornhill N. (Eds)*, Elsevier, London:129–158, 1989.

[25] C.M. Bender and S.A. Orszag. *Advanced Mathematical Methods for Scientists and Engineers*. McGraw-Hill, Auckland, 1984.

[26] O. Bernard, Z. Hadj-Sadok, and D. Dochain. Advanced monitoring and control of anaerobic treatment plants: II - Dynamical model development and

identification. *Proc. Watermatex 2000*, 3.57–3.64, 2000.

[27] O. Bernard, Z. Hadj-Sadok, and D. Dochain. Dynamical model development and parameter identification for an anaerobic wastewater treatment process. *Biotechnol. Bioeng.*, to appear, 2001.

[28] K. Beven and A. Binley. The future of distributed models: Model calibration and uncertainty prediction. *Hydrol. Proc.*, 6:279–298, 1992.

[29] R. Binot, T. Bol, H. Naveau, and E.J. Nyns. Biomethanation by immobilized fluidised cells. *Wat. Sci. Tech.*, 15:103–115, 1983.

[30] B. Bonnet, D. Dochain, and J.Ph. Steyer. Dynamical modelling of an anaerobic digestion fluidised bed reactor. *Wat. Sci. Tech.*, 36(5):285–292, 1997.

[31] C. Bonvillain, D. Benyamina, M. Schaegger, A. Pauss, O. Bernard, and D. Dochain. Modelling of an extensive wastewater treatment plant (lagoon), based on two-year intensive follow-up. *Proc. Advanced Wastewater Treatment, Recycling and Reuse (AWT98)*, pages 535–541, 1998.

[32] B. Bouaziz and D. Dochain. Control analysis of fixed bed reactors: a singular perturbation approach. *Proc. ECC'93*, pages 1741–1745, 1993.

[33] J.B. Boucquey, P. Renard, P. Amerlynck, P. Modesto Filho, S.N. Agathos, H.P. Naveau, and E.J. Nyns. High-rate continuous biodegradation of concentrated chlorinated aliphatics by a durable enrichment of methanogenic origin under carrier-dependent conditions. *Biotechnol. Bioeng.*, 47:298–307, 1995.

[34] S. Bourrel. *Estimation et Commande d'un Procédé à Paramètres Répartis Utilisé pour le Traitement Biologique de l'Eau à Potabiliser*. PhD thesis, Université Paul Sabatier, Toulouse, France, 1996.

[35] S. V. Bourrel, J.P. Babary, S. Julien, M.T. Nihtila, and D. Dochain. Modelling and identification of a fixed-bed denitrification bioreactor. *Syst. Anal. Model. Simul.*, 30:289–309, 1998.

[36] S. V. Bourrel, D. Dochain, J.P. Babary, and I. Queinnec. Modelling, identification and control of a denitrifying biofilter. *J. Process Control*, 10:73–91, 2000.

[37] J.R. Bowen, A. Acrivos, and A.K. Oppenheim. Singular perturbation refinement to quasi-steady-state approximation in chemical kinetics. *Chem. Eng. Sci.*, 18:177–181, 1963.

[38] G. Box and G. Jenkins. *Time Series Analysis, Forecasting and Control*. Holden-Day, 1976.

[39] G.E.P. Box and W.J. Hill. Discrimination among mechanistic models. *Technometrics*, 9:57–71, 1967.

[40] S.A. Boyd and D.R. Shelton. Anaerobic biodegradation of chlorophenols in fresh and acclimated sludge. *Appl. Environ. Microbiol.*, 47:272–277, 1984.

[41] W.C. Boyle and P.M. Berthouex. Biological wastewater treatment model building: fits and misfits. *Biotechnol. Bioeng.*, 16:1139–1159, 1974.

[42] R.P. Brent. *Algorithms for Minimization without Derivatives*. Prentice-Hall, Englewood Cliffs, NJ, 1973.

[43] G.E. Briggs and J.B.S. Haldane. *Biochem. J.*, 242:3973, 1925.

[44] H. Brouwer, A. Klapwijk, and K.J. Keesman. Identification of activated sludge and wastewater characteristics using respirometric batch experiments. *Wat. Res.*, 32:1240–1254, 1998.

[45] C.G. Broyden. A new method of solving nonlinear simultaneous equations. *Comput. J.*, 12:94–99, 1969.

[46] K.F. Cacossa and D.A. Vaccari. Calibration of a compressive gravity thickening model from a single batch settling curve. *Wat. Sci. Tech.*, 30(8):107–116, 1994.

[47] G. Caminal, F.J. Lafuente, J. Lopez-Santin, M. Poch, and C. Sola. Application of the extended Kalman filter to identification of enzymatic deactivation. *Biotechnol. Bioeng.*, 24:366–369, 1987.

[48] C.W. Carroll. The created response surface technique for optimizing nonlinear, restrained, systems. *Operations Res.*, 9:169–184, 1961.

[49] J. Carstensen, P. Harremoës, and R. Strube. Software sensors based on the grey-box modelling approach. *Wat. Sci. Tech.*, 33(1):117–126, 1996.

[50] J. Carstensen, H. Madsen, N.K. Poulsen, and M.K. Nielsen. Identification of wastewater treatment processes for nutrient removal on a full-scale WWTP by statistical methods. *Wat. Res*, 28:2055–2066, 1994.

[51] J. Carstensen, P.A. Vanrolleghem, W. Rauch, and P. Reichert. Terminology and methodology in modelling for water quality management - a review. *Wat. Sci. Tech.*, 36(5):157–168, 1997.

[52] J.L. Casti. *Dynamical Systems and Their Application - Linear Theory*. Academic Press, New York, 1977.

[53] H. Caswell. The validation problem. In B.C. Patten, editor, *System Analysis and Simulation in Ecology, Vol.IV*, pages 313–325. Academic Press, New York, 1976.

[54] H.T. Chang and B.E. Rittmann. Predicting bed dynamics in three-phase, fluidized-bed biofilm reactors. *Wat. Sci. Tech.*, 29(10-11):231–241, 1994.

[55] M.J. Chappell, K.R. Godfrey, and S. Vajda. Global identifiability of the parameters of nonlinear systems with specified inputs: a comparison of methods. *Math. Biosci.*, 102:41–73, 1990.

[56] L. Chen. *Modelling, Identifiability and Control of Complex Biotechnological Processes*. PhD thesis, Université Catholique de Louvain, Belgium, 1992.

[57] L. Chen and G. Bastin. Structural identifiability of the yield coefficients in bioprocess models when the reaction rates are unknown. *Math. Biosci.*, 132:35–67, 1996.

[58] Y.H. Chen. Adaptive robust observers for non-linear uncertain systems. *Int. J. Systems Sci.*, 21:803–814, 1990.

[59] S.H. Cho, F. Colin, M. Sardin, and C. Prost. Settling velocity model of activated sludge. *Wat. Res.*, 27:1237–1242, 1993.

[60] C.K. Chui and G. Chen. *Linear Systems and Optimal Control*. Springer Verlag,

Berlin, 1988.

[61] D. Contois. Kinetics of bacterial growth relationship between population density and specific growth rate of continuous culture. *J. Gen. Microbiol.*, 21:845–864, 1959.

[62] A. Cornish-Bowden. Parameter estimating procedures for the michaelis-menten model: Reply to Tseng and Hsu. *J. Theor. Biol.*, 153:437–330, 1991.

[63] P.V. Danckwerts. Continuous flow systems. Distribution of residence times. *Chem. Eng. Sci.*, 2:1–13, 1953.

[64] B. De Clercq, F. Coen, B. Vanderhaegen and P.A. Vanrolleghem. Calibrating simple models for mixing and flow propagation in waste water treatment plants. *Wat. Sci. Tech.*, 39(4):61–69, 1999.

[65] J.G. de Gooijer, B. Abraham, A. Gould, and L. Robinson. Methods for determining the order of an autoregressive-moving average process: A survey. *Int. Statist. Rev.*, 53:301–329, 1985.

[66] P. de Larminat and Y. Thomas. *Automatique des Systèmes Linéaires. 2. Identification.* Flammarion Sciences, Paris, 1977.

[67] D. Defour, D. Derycke, J. Liessens, and P. Pipyn. Field experience with different systems for biomass accumulation in anaerobic reactor technology. *Wat. Sci. Tech.*, 30(12):181–191, 1994.

[68] B. De heyder, P.A. Vanrolleghem, H. Van Langenhove, and W. Verstraete. Kinetic characterisation of mass transfer limited degradation of a poorly water soluble gas in batch experiments. *Biotechnol. Bioeng.*, 55:511–519, 1997.

[69] R.I. Dick and B.B. Ewing. Evaluation of the activated sludge thickening theories. *J. Sanit. Engng. Div. Am. Soc. Civ. Eng.*, 93:9–29, 1967.

[70] D. Dochain. *On-line Parameter Estimation, Adaptive State Estimation and Adaptive Control of Fermentation Processes.* PhD. Thesis, Université Catholique de Louvain, Belgium, 1986.

[71] D. Dochain and L. Chen. Local observability and controllability of stirred tank reactors. *J. Process Control*, 2:139–144, 1992.

[72] D. Dochain and B. Bouaziz. Approximation of the dynamical model of fixed bed reactors via a singular perturbation approach. *Proc. IMACS Int. Symp. MIM-S2'93*, pages 34–39, 1993.

[73] D. Dochain. *Contribution to the Analysis and Control of Distributed Parameter Systems with Application to (Bio)chemical Processes and Robotics.* Thèse d'Aggrégation de l'Enseignement Supérieur, Université Catholique de Louvain, Belgium, 1994.

[74] D. Dochain, P.A. Vanrolleghem, and M. Van Daele. Structural identifiability of biokinetic models of activated sludge respiration. *Wat. Res.*, 29:2571–2578, 1995.

[75] D. Dochain. Modelling the dynamics of settlers: a basic problem. *Proc. FAB'97, Med. Fac. Landbouww. Univ. Gent*, 62:1617–1624, 1997.

[76] D. Dochain, S. Agathos, and P. Vanrolleghem. Asymptotic observers as a tool for

modelling process dynamics. *Wat. Sci. Tech.*, 36(5):259–268, 1997.

[77] D. Dochain and M. Pengov. State observers for processes with uncertain kinetics. *Proc. ADCHEM2000*, pages 171–176, 2000.

[78] D. Dochain. *Automatique des Bioprocédés*. Traité IC2, Section Systèmes Automatisés, Hermès, Paris, 2001.

[79] D. Draper. Assessment and propagation of model uncertainty (with discussion). *J. Roy. Stat. Soc.*, B57(1):45–97, 1995.

[80] N.R. Draper and H. Smith. *Applied Regression Analysis*. John Wiley, New York, 1981.

[81] Q. Duan, S. Sorooshian, and V. Gupta. Effective and efficient global optimisation for conceptual rainfall-runoff models. *Water Res. Res.*, 28:1015–1031, 1992.

[82] F. Ehlinger, Y. Escoffier, J.P. Couderc, J.P. Leyris, and R. Moletta. Development of an automatic control system for monitoring an anaerobic fluidized-bed. *Wat. Sci. Tech.*, 29(10-11):289–295, 1994.

[83] G.A. Ekama, J.L. Barnard, F.W. Günthert, P. Krebs, J.A. McCorquodale, D.S. Parker, and E.J. Wahlberg. *Secondary Settling Tanks: Theory, Modelling, Design and Operation*. IWA (Scientific and Technical Report n6), London, 1997.

[84] B. Eramo, R. Gavasci, A. Misiti, and P. Viotti. Validation of a multisubstrate mathematical model for the simulation of the denitrification process in fluidized bed biofilm reactors. *Wat. Sci. Tech.*, 29(10-11):401–408, 1994.

[85] S. Feyo de Azevedo, M.A. Romero-Ogawa, and A.P. Wardle. Modelling of tubular fixed-bed catalytic reactors: a brief review. *Trans. I.Chem.E.*, 68(Part A):2–8, 1990.

[86] M. Fjeld, O.A. Asbjornsen, and K.J. Aström. Reaction invariants and their importance in the analysis of eigenvectors, state observability and controllability of the continuous stirred tank reactor. *Chem. Eng. Sci.*, 29:1917–1926, 1974.

[87] R. Fletcher. *Practical Methods of Optimization*. John Wiley, New York, 1987.

[88] H.S. Fogler. *Elements of Chemical Reaction Engineering*. Prentice-Hall, Englewood Cliffs, NJ, 1986.

[89] P.U. Foscolo and L.G. Gibilaro. Fluid dynamic stability of fluidised suspensions: the particle bed model. *Chem. Eng. Sci.*, 42:1489–1500, 1987.

[90] G.F. Franklin, J.D. Powell, and A. Emami-Naeini. *Feedback Control of Dynamic Systems (3rd ed.)*. Addison-Wesley, Reading, Mass., 1995.

[91] A.G. Frederickson and H.M. Tsuchiya. Microbial kinetics and dynamics. In L. Lapidus and N.R. Amundson, editors, *Chemical Reactor Theory*, chapter 7. Prentice-Hall, Englewood Cliffs, NJ, 1977.

[92] J.L. Fripiat, T. Bol, R. Binot, H. Naveau, and E.J. Nyns. A strategy for the evaluation of methane production from different types of substrate biomass. *In R. Buvet, M.F. Fox and D.J. Picker (eds.), Biomethane, Production and Uses*,

Roger Bowskill Ltd, Exeter, UK:95–105, 1984.

[93] G.F. Froment and K.B. Bischoff. *Chemical Reactor Analysis and Design*. John Wiley, New York, 1990.

[94] M.J. Fuente, C. De Prada, and P. Vega. Optimization tools in a continuous dynamical simulation language. In W. Krug and A. Lehmann, editors, *Simulation and AI in Computer-aided Techniques*, pages 475–479. Society for Computer Simulation, San Diego, 1992.

[95] G.R. Gavalas. *Nonlinear Differential Equations of Chemically Reacting Systems*. Springer Verlag, Berlin, 1968.

[96] K.R. Gegenfurtner. Praxis: Brent's algorithm for function minimization. *Behavior Research Methods, Instruments & Computers*, 24:560–564, 1992.

[97] C. Georgakis, R. Aris, and R. Amundson. Studies in the control of tubular reactors - I. General considerations. *Chem. Eng. Sci.*, 32:1359–1369, 1977.

[98] K. Gernaey, P. Vanrolleghem, and W. Verstraete. On-line estimation of kinetic parameters of NH_4^+ oxidizing bacteria in activated sludge samples using titration in-sensor-experiments. *Wat. Res.*, 32:71–80, 1998.

[99] K.R. Godfrey and J.J. Di Stefano III. Identifiability of model parameters. In P. Young, editor, *Identification and System Parameter Estimation*, pages 89–114. Pergamon Press, Oxford, 1985.

[100] K.R. Godfrey and J.J. Di Stefano III. Identifiability of model parameters. In E. Walter, editor, *Identifiability of Parametric Models*, pages 1–19. Pergamon Press, Oxford, 1987.

[101] D.E. Goldberg. *Genetic algorithms in Search, Optimization and Machine Learning*. Addison-Wesley, Menlo Park, California, 1989.

[102] R.F. Goncalves, L. Le Grand, and T. Rogalla. Biological phosphorus uptake in submerged biofilters with nitrogen removal. *Wat. Sci. Tech.*, 29(10-11):135–143, 1994.

[103] G.C. Goodwin. Identification: Experiment design. In M. Singh, editor, *Systems and Control Encyclopedia*, Volume 4, pages 2257–2264. Pergamon Press, Oxford, 1987.

[104] G.C. Goodwin, M. Gevers, and B. Ninness. Quantifying the error in estimated transfer functions with application to model order selection. *IEEE Trans. Autom. Control*, 37:913–928, 1992.

[105] L.A. Gould. *Chemical Process Control: Theory and Applications*. Addison-Wesley, Reading, Mass., 1969.

[106] S. Grégoire, D. Dochain, A. Pauss, and M. Schaegger. Identification of a dynamical model of a waste stabilisation pond. *Proc. Watermatex 2000*, pages 1.51–1.58, 2000.

[107] J.B.S. Haldane. *Enzymes*. Longmans, London, 1930.

[108] R.P. Hamalainen, A. Halme, and A. Gyllenberg. A control model for activated sludge wastewater treatment process. *Proc. 6th IFAC World Congress, Boston, Paper 61:6*, 1975.

[109] J. Hamilton, R. Jain, P. Antoniou, S.A. Svoronos, B. Koopman, and G. Lyberatos. Modeling and pilot-scale verification for predenitrification process. *J. Environ. Eng.*, 118:38–55, 1992.

[110] E.J. Hannan. The estimation of the order of an ARMA process. *Ann. Statist.*, 8:1071–1081, 1980.

[111] P. Harremoës, A.G. Capodaglio, B.G. Hellstrom, M. Henze, K.N. Jensen, A. Lynggaard-Jensen, R. Otterpohl, and H. Soeberg. Wastewater treatment plants under transient loading - performance, modelling and control. *Wat. Sci. Tech.*, 27(12):71–115, 1993.

[112] P. Harremoës and J. Carstensen. Deterministic versus stochastic interpretation of continuously monitored sewer systems. *European Water Pollution Control*, 4(5):117–126, 1994.

[113] A. Harvey. *Forecasting, Structural Time Series Models and the Kalman Filter.* Cambridge University Press, Reading, Mass., 1989.

[114] F.G. Heineken, H.M. Tsuchiya, and R. Aris. On the mathematical status of the pseudo-steady state hypothesis of biochemical kinetics. *Math. Biosci.*, 1:95–113, 1967.

[115] A. Heitzer, H.-P..E. Kohler, Reichert P., and G. Hamer. Utility of phenomenological models for describing temperature dependence of bacterial growth. *Appl. Environ. Microbiol.*, 57:2656–2665, 1991.

[116] C. Hellinga, P.A. Vanrolleghem, M.C.M. van Loosdrecht, and J.J. Heijnen. The potentials of off-gas analysis for monitoring and control of waste water treatment plants. *Wat. Sci. Tech.*, 33(1):13–23, 1996.

[117] C. Hellinga, M.C.M. van Loosdrecht and J.J. Heijnen. Model based design of a novel process for ammonia removal from concentrated flows. *Math. Comp. Mod.Dyn. Syst.*, 5:351–371, 1999.

[118] M. Henze. Characterization of wastewater for modelling of activated sludge processes. *Wat. Sci. Tech.*, 25(6):1–15, 1992.

[119] M. Henze, R. Dupont, P. Grau, and A. de la Sota. Rising sludge in secondary settlers due to denitrification. *Wat. Res.*, 27:231–236, 1993.

[120] M. Henze, C.P.L. Grady Jr, W. Gujer, G.v.R. Marais, and T. Matsuo. Activated sludge model no. 1. Technical report, IAWPRC, Scientific and Technical Reports No. 1, 1986.

[121] M. Henze, W. Mino, W. Gujer, T. Matsuo, M.C. Wentzel, G.v.R. Marais, and T. Matsuo. Activated sludge model no. 2. Technical report, IAWQ Scientific and Technical Reports No. 3, IAWQ, London, 1995.

[122] H. Hertz, A. Krogh, and R. Palmer. *Introduction to the Theory of Neural Computation.* Addison-Wesley, Reading, Mass., 1991.

[123] P.D.H. Hill. A review of experimental design procedures for regression model discrimination. *Technometrics*, 20:15–21, 1978.

[124] W.J. Hill, W.G. Hunter, and D.W. Wichern. A joint design criterion for the dual problem of model discrimination and parameter estimation. *Technometrics*,

10:145–160, 1968.

[125] C. Holliger, A.J.M. Stams, and A.J.B. Zehnder. Anaerobic degradation of recal-citrant compounds. *Anaerobic Digestion, E.R. Hall and P.N. Hobson (eds)*. Pergamon Press, Oxford, 1988.

[126] A. Holmberg. On the practical identifiability of microbial growth models incor-porating Michaelis-Menten type nonlinearities. *Math. Biosci.*, 62:23–43, 1982.

[127] A. Holmberg and J. Ranta. Procedures for parameter and state estimation of microbial growth process models. *Automatica*, 18:181–193, 1982.

[128] J. Holst, U. Holst, H. Madsen, and H. Melgaard. Validation of grey box models. *Proc. IFAC Symposium on Adaptive Control and Signal Processing*, pages 407–414, 1992.

[129] G.M. Hornberger and R.C. Spear. An approach to the analysis of behavior and sensitivity in environmental systems. In M.B. Beck and G. van Straten, editors, *Uncertainty and Forecasting of Water Quality*, pages 101–116. Springer-Verlag, Heidelberg, 1983.

[130] W.G. Hunter and A.M. Reiner. Designs for discriminating between two rival models. *Technometrics*, 7:307–323, 1965.

[131] K.F. Janning, P. Harremoës, and M. Nielsen. Evaluating and modelling the kinet-ics in a full scale submerged denitrification filter. *Wat. Sci. Tech.*, 32(8):115–123, 1995.

[132] H.H. Jean and L.H. Fan. On the model equations of Gibilaro and Foscolo with corrected buoyancy force. *Powder Technology*, 72:201–205, 1992.

[133] U. Jeppsson. *Modelling Aspects of Wastewater Treatment Processes*. PhD thesis, Lund Institute of Technology, Sweden, 1996.

[134] U. Jeppsson and G. Olsson. Reduced order models for on-line parameter identifi-cation of the activated sludge process. *Wat. Sci. Tech.*, 28(11-12):173–183, 1993.

[135] D.B. Johnson and P.M. Berthouex. Using multiresponse data to estimate bioki-netic parameters. *Biotechnol. Bioeng.*, 17:571–583, 1975.

[136] S.B. Jorgensen. Fixed bed reactor dynamics and control - A review. *Proc. IFAC Control of Distillation Columns and Chemical Reactors*. Pergamon, pages 11–24, 1986.

[137] S.E. Jørgensen. *Integration of Ecosystem Theories: A Pattern*. Kluwer, Dor-drecht, 1992.

[138] S. Julien, J.P. Babary, and P. Lessard. Theoretical and practical identifiability of a reduced order model in an activated sludge process doing nitrification and denitrification. *Wat. Sci. Tech.*, 37(12):309–368, 1998.

[139] S. Julien, J.P. Babary, and P. Lessard. Identifiability and identification of an acti-vated sludge process model. *Syst. Anal. Model. Sim.*, 37:481–499, 2000.

[140] R.L. Kashyap. Inconsistency of the AIC rule for estimating the order of autore-gressive models. *IEEE Trans. Autom. Control*, 25:996–998, 1980.

[141] K.J. Keesman. Membership-set estimation using random-scanning and principal component analysis. *Math. Comp. Simul.*, 32:535–543, 1990.

[142] K.J. Keesman, H. Spanjers, and G. van Straten. Analysis of endogenous process behavior in activated sludge. *Biotechnol. Bioeng.*, 57:155–163, 1998.

[143] K.J. Keesman and G. van Straten. Modified set theoretic identification of an ill-defined water quality system from poor data. In B. Beck, editor, *System Analysis in Water Quality Management*, pages 297–308. Pergamon Press, Oxford, 1987.

[144] K.J. Keesman and G. van Straten. Set-membership approach to identification and prediction of lake eutrophication. *Water Res. Res.*, 26:2643–2652, 1990.

[145] R. Khanna and J.H. Seinfeld. Mathematical modelling of packed bed reactors: numerical solutions and control model development. *Adv. Chem. Eng.*, 13:113–191, 1987.

[146] A.B. Koehler and E.S. Murphree. A comparison of the Akaike and Schwarz criteria for selecting model order. *Appl. Statist.*, 37:187–195, 1988.

[147] H.-P. Kohler, A. Heitzer, and G. Hamer. Improved unstructured model describing temperature dependence of bacterial maximum specific growth rates. *Proceedings EERO/EFB International Symposium Environmental Biotechnology ISEB. Ostend, Belgium, April 22-25 1991*, Part II:511–514, 1991.

[148] P. Kokotovic, H.K. Khalil, and J. O'Reilly. *Singular Perturbation Methods in Control: Analysis and Design.* Academic Press, London, 1986.

[149] Z. Kong, P.A. Vanrolleghem, and W. Verstraete. Automated respiration inhibition kinetics analysis (ARIKA) with a respirographic biosensor. *Wat. Sci. Tech.*, 30(4):275–284, 1994.

[150] G. Kresseilmeier. Adaptive observers with exponential rate of convergence. *IEEE Trans. Autom. Control*, 22:2–8, 1977.

[151] H. Kwakernaak and R. Sivan. *Linear Optimal Control Systems.* John Wiley, New York, 1972.

[152] E.B. Lee and L. Markus. *Foundations of Optimal Control Theory.* John Wiley, New York, 1967.

[153] S.H. Lee, P. Tsobanakis, J.A. Phillips, and C. Georgakis. Issues in the optimization, estimation and control of fed-batch bioreactors using tendency models. *Proc. ICCAFT 5/IFAC-BIO 2, Keystone, Colorado*, 1992.

[154] T.T. Lee, F.Y. Wang, and R.B. Newell. Distributed parameter approach to the dynamics of complex biological processes. *AIChE J.*, 45:2245–2268, 1999.

[155] L. Lefèvre, D. Dochain, S. Feyo de Azevedo, and A. Magnus. Analysis of the orthogonal collocation method when applied to the numerical integration of chemical reactor models. *Computers and Chemical Engineering*, 24 (12):2571–2588, 2000.

[156] J.A. Lennox, Z. Yuan and J. Harmand. A systematic approach to error isolation in computerized wastewater simulation models. *Wat. Sci. Tech.*, 43(7):367–376, 2001.

[157] H. Linhart and W. Zucchini. *Model Selection*. John Wiley, New York, 1986.

[158] L. Ljung. Asymptotic behavior of the extended Kalman filter as a parameter estimator for linear systems. *IEEE Trans. Autom. Control*, 24:36–50, 1979.

[159] L. Ljung. *System Identification - Theory for the User*. Prentice-Hall, Englewood Cliffs, NJ, 1999.

[160] J.R. Lobry. *Re-évaluation du Modèle de Croissance de Monod. Effet des Antibiotiques sur l'Energie de Maintenance*. PhD thesis, Université Claude Bernard, Lyon, France, 1991.

[161] J.R. Lobry and J.P. Flandrois. Comparison of estimates of Monod's growth model parameters from the same data set. *Binary*, 3:20–23, 1991.

[162] J.R. Lobry, L. Rosso, and J.P. Flandrois. A Fortran subroutine for the determination of parameter confidence limits in non-linear models. *Binary*, 3:86–93, 1991.

[163] A. Lübbert. Characterization of bioreactors. In K. Schügerl, editor, *Biotechnology, A Multi-Volume Comprehensive Treatise, Vol.4, Measuring, Modelling and Control*, pages 107–148. VCH, Weinheim, 1991.

[164] R. Luedeking and E. Piret. A kinetic study of the lactic acid fermentation batch process at controlled pH. *J. Bioch. Microb. Technol. Eng.*, I (4):393–412, 1959.

[165] I. Mareels and J.W. Polderman. *Adaptive Systems. An Introduction*. Birkhauser, Boston, 1996.

[166] A.T. Marino, J.J. Di Stefano III, and E.M. Landaw. Dimsum: An expert system for multiexponential model discrimination. *Am. J. Physiol.*, 262:E546–E556, 1992.

[167] R. Marino. Adaptive observers for single output nonlinear systems. *IEEE Trans. Autom. Control*, 35:1054–1058, 1990.

[168] D.W. Marquardt. An algorithm for least squares estimation of nonlinear parameters. *J. Soc. Ind. Appl. Math.*, 11:431–441, 1963.

[169] S. Marsili-Libelli. Optimal control of the activated sludge process. *Trans. Inst. Meas. Control*, 6:146–152, 1984.

[170] S. Marsili-Libelli. Modelling, identification and control of the activated sludge process. *Adv. Biochem. Eng./Biotechn.*, 38:90–148, 1989.

[171] S. Marsili-Libelli. Parameter estimation of ecological models. *Ecol. Modelling*, 62:233–258, 1992.

[172] H. Melcer, W.J. Parker, and B.E. Rittmann. Modeling of volatile organic contaminants in trickling filter systems. *Wat. Sci. Tech.*, 31(1):95–104, 1995.

[173] L. Michaelis and M.L. Menten. Die Kinetic der Invertinwirkung. *Biochemische Zeitschrift*, 49:334–369, 1913.

[174] M. Milanese and A. Vicino. Optimal estimation theory for dynamic systems with set membership uncertainty: an overview. *Automatica*, 27:997–1009, 1991.

[175] P. Modesto Filho, P. Amerlynck, E.J. Nyns, and H.P. Naveau. Acclimation of methanogenic consortium to polychlorinated compounds in a fixed-film sta-

tionary bed reactor. *Appl. Microb. Biot.*, 1992.

[176] J. Monod. *Recherches sur la Croissance des Cultures Bactériennes*. Hermann, Paris, 1942.

[177] A. Moser. *Bioprocess Technology*. Springer-Verlag, New York, 1988.

[178] F.E. Mosey. Mathematical modelling of the anaerobic digestion process: regulatory mechanisms for the formation of short-chain volatile acids from glucose. *Wat. Sci. Tech.*, 15(8/9):209–232, 1983.

[179] A. Munack. Optimal feeding strategy for identification of Monod-type models by fed-batch experiments. *Proc. Comp. Appl. in Ferm. Techn.: Mod. and Control of Biotech. Proc.*, pages 195–204, 1989.

[180] A. Munack. Optimization of sampling. In K. Schugerl, editor, *Biotechnology, a Multi-volume Comprehensive Treatise*, volume 4, pages 251–264. VCH, Weinheim, 1991.

[181] A. Munack. Some improvements in the identification of bioprocesses. In M.N. Karim and G. Stephanopoulos, editors, *Modelling and Control of Biotechnical Processes*, pages 89–94. Pergamon Press, Oxford, 1992.

[182] A. Munack and C. Posten. Design of optimal dynamical experiments for parameter estimation. *Proc. ACC*, pages 2010–2016, 1989.

[183] K.S. Narendra and A.M. Annaswamy. *Stable Adaptive Systems*. Prentice-Hall, Englewood Cliffs, NJ, 1989.

[184] J.A. Nelder and R. Mead. A simplex method for function minimization. *Comp. J.*, 7:308–313, 1964.

[185] J.P. Norton. *An Introduction to Identification*. Academic Press, London, 1986.

[186] R. Oliveira, E.C. Ferreira, F. Oliveira, and S. Feyo de Azevedo. A study on the convergence of observer-based kinetics estimators in stirred tank bioreactors. *J. Process Control*, 6:367–371, 1996.

[187] G. Olsson and B. Newell. *Wastewater Treatment Systems. Modelling, Diagnosis and Control*. IWA Publishing, London, 1999.

[188] G. Olsson and J.P. Stephenson. The propagation of hydraulic disturbances and flow rate reconstruction in activated sludge plants. *Env. Technol. Letters*, 6:536–545, 1985.

[189] R.V. O'Neill and R.H. Gardner. Sources of uncertainty in ecological models. In G.S. Innis and R.V. O'Neill, editors, *Methodology in Systems Modelling and Simulation*, pages 447–463. North-Holland, Amsterdam, 1979.

[190] S.L. Ong. A comparison of estimates of kinetic constants for a suspended growth treatment system from various linear transformations. *J. Water Pollut. Control Fed.*, 62:894–900, 1990.

[191] M. Perrier, S. Feyo de Azevedo, E.C. Ferreira, and D. Dochain. Tuning of observer-based estimators: theory and application to the on-line estimation of kinetic parameters. *Control Eng. Practice*, 8 (4):377–388, 2000.

[192] M. Perrier and D. Dochain. Evaluation of control strategies for anaerobic digestion processes. *Int. J. Adaptive Cont. Signal Proc.*, 7:309–321, 1993.

[193] B. Petersen. *Calibration, Identifiability and Optimal Experimental Design of Activated Sludge Models*. PhD thesis, Ghent University, Belgium, 2000.

[194] B. Petersen, K. Gernaey, and P.A. Vanrolleghem. Improved theoretical identifiability of model parameters by combined respirometric-titrimetric measurements. A generalization of results. *Proc. 3rd IMACS Symposium on Mathematical Modelling MATHMOD, I. Troch and F. Breitenecker (Eds)*, 2:639–642, 2000.

[195] B. Petersen, K. Gernaey, and P.A. Vanrolleghem. Practical identifiability of model parameters by combined respirometric-titrimetric measurements. *Wat. Sci. Tech.*, 43(7):347–356, 2001.

[196] E.E. Petersen. *Chemical Reaction Analysis*. Prentice-Hall, Englewood Cliffs, NJ, 1965.

[197] H. Pohjanpalo. System identifiability based on the power series expansion of the solution. *Math. Biosci.*, 41:21–33, 1978.

[198] Y. Pomerleau and M. Perrier. Estimation of multiple specific growth rates in bioprocesses. *AIChE J.*, 36:207–215, 1990.

[199] D. Poncelet, H. Naveau, E.J. Nyns, and D. Dochain. Transient response of a solid-liquid model biological fluidised bed to a step change in fluid superficial velocity. *J. Chem. Tech. Biotechnol.*, 48:439–452, 1990.

[200] K.R. Popper. *The Logic of Scientific Discovery*. Hutchinson, London, 1980.

[201] C. Posten and A. Munack. On-line application of parameter estimation accuracy to biotechnical processes. *Proc. ACC*, pages 2181–2186, 1990.

[202] M.J.D. Powell. An efficient method for finding the minimum of a function of several variables without calculating derivatives. *Comp. J.*, 7:155–162, 1964.

[203] W.H. Press, B.P. Flannery, S.A. Teukolsky, and W.T. Vetterling. *Numerical Recipes: The Art of Scientific Computing*. Cambridge University Press, Cambridge, UK, 1986.

[204] W.L. Price. A controlled random search procedure for global optimization. *Comp. J.*, 20:367–370, 1979.

[205] I. Queinnec and D. Dochain. Modelling and simulation of the steady-state of secondary settlers in wastewater treatment plants. *Wat. Sci. Tech.*, 43(7):39-46, 2001.

[206] A. Raksanyi, Y. Lecourtier, E. Walter, and A. Venot. Identifiability and distinguishability testing via computer algebra. *Math. Biosci.*, 77:245–266, 1985.

[207] M.L. Ralston and R.I. Jennrich. DUD, a derivative-free algorithm for nonlinear least squares. *Technometrics*, 20:7–14, 1978.

[208] D.A. Ratkowsky. *Nonlinear Regression Modeling - A Unified Practical Approach*. Marcel Dekker, Basel, Switzerland, 1983.

[209] D.A. Ratkowsky. A suitable parameterization of the Michaelis-Menten enzyme reaction. *Biochem. J.*, 240:357–360, 1986.

[210] D.A. Ratkowsky, T. Ross, T.A. McMeekin, and J. Olley. Comparison of Arrhenius-type and Belehradek-type models for prediction of bacterial

growth in foods. *J. Appl. Bacteriol.*, 71:452–459, 1991.

[211] W.H. Ray and D.G. Lainiotis. *Distributed Parameter Systems, Identification, Estimation and Control.* Marcel Dekker, New York, 1978.

[212] K.H. Reckhow and S.C. Chapra. Confirmation of water quality models. *Ecol. Modelling*, 20:113–133, 1983.

[213] P. Reichert. *Concepts Underlying a Computer Program for the Identification and Simulation of Aquatic Systems.* PhD thesis, Swiss Federal Institute of Environmental Science and Technology (EAWAG), Dübendorf, Switzerland, 1994.

[214] P. Reichert and M. Omlin. On the usefulness of overparameterized ecological models. *Ecol. Modelling*, 95:289–299, 1997.

[215] P. Reichert and P.A. Vanrolleghem. Identifiability and uncertainty analysis of the river water quality model no. 1 (RWQM1). *Wat. Sci. Tech.* 43(7):329–338, 2001.

[216] J.F. Richardson and W.N. Zaki. Sedimentation and fluidisation. Part I. *Trans. Inst. Chem. Eng.*, 32:35–53, 1954.

[217] R.G. Riefler, D.P. Ahlfeld, and B.F. Smets. Respirometric assay for biofilm kinetics estimation: Parameter identifiability and retrievability. *Biotechnol. Bioeng.*, 57:35–45, 1998.

[218] J.A. Robinson. Determining microbial parameters using nonlinear regression analysis: Advantages and limitations in microbial ecology. *Adv. Microb. Ecol.*, 8:61–114, 1985.

[219] M.R. Rose. *Quantitative Ecological Theory - An Introduction to Basic Models.* Croom Helm, London, 1987.

[220] H.H. Rosenbrock. An automated method for finding the greatest or least value of a function. *Comput. J.*, 3:175–184, 1960.

[221] P.J. Rousseeuw and A.M. Leroy. *Robust Regression and Outlier Detection.* John Wiley, New York, 1987.

[222] A. Rozzi. Modelling and control of anaerobic digestion processes. *Trans. Inst. Meas. Control*, 6:153–159, 1984.

[223] A. Rozzi. Anaerobic process control by bicarbonate monitoring. *Env. Technol. Letters*, 6:594–601, 1985.

[224] E. Sarner. Removal of dissolved and particulate organic matter in high-rate trickling filters. *Wat. Res.*, 15:671–678, 1981.

[225] K. Schmidt and S.H. Isaacs. An evolutionary algorithm for initial state and parameter estimation in complex biochemical models. In *Preprints 6th IFAC Conference on Computer Applications in Biotechnology - CAB6. Garmisch-Partenkirchen, Germany, May 14-17 1995*, pages 239–242. 1995.

[226] M. Schuetze. *Integrated simulation and optimum control of the urban wastewater system.* PhD thesis, Imperial College, London, UK, 1998.

[227] K. Schugerl. Biofluidization: application of the fluidization technique in biotechnology. *Can. J. Chem. Eng.*, 67:178–184, 1989.

[228] G. Schwarz. Estimating the dimension of a model. *Ann. Statist.*, 6:461–464, 1978.

[229] Z. Shi and K. Shimizu. Neuro-fuzzy control of bioreactor systems with pattern recognition. *J. Ferment. Bioeng.*, 74:39–45, 1992.

[230] H. Siegrist, P. Krebs, R. Bahler, I. Purtschert I., C. Rtck, and R. Rufer. Denitrification in secondary clarifiers. *Wat. Sci. Tech.*, 31(2):205–214, 1995.

[231] J. Singer and D.C.G. Lewis. *Water and Water Engineering*, pages 105–111, 1966.

[232] T. Söderström and P. Stoica. *Model validation and model structure determination*. Prentice-Hall, Englewood Cliffs, NJ, 1989.

[233] U. Sollfrank and W. Gujer. Characterisation of domestic wastewater for mathematical modelling of the activated sludge process. *Wat. Sci. Tech.*, 23(4-6):1057–1066, 1991.

[234] E.D. Sontag. *Mathematical Control Theory. Deterministic Finite Dimensional Systems (2nd ed.)*. Texts in Applied Mathematics, 6, Springer Verlag, New York, 1998.

[235] S. Sorooshian, H.V. Gupta, and L.A. Bastidas. Calibration of hydrologic models using multi-objectives and visualization techniques. Final report project EAR-9418147. Technical report, University of Arizona, Tucson, USA, 1998.

[236] H. Spanjers and K. Keesman. Identification of wastewater biodegradation kinetics. In *Proceedings 3rd IEEE Conference on Control Applications, Session on Environmental Process Control. Glasgow, UK, August 24-26 1994*, pages 1011–1016, 1994.

[237] W. Spendley, G.R. Hext, and F.R. Himswokin. Sequential application of simplex designs in optimisation and evolutionary operation. *Technometrics*, 4:441–461, 1962.

[238] J.A. Spriet. Structure characterization - An overview. In H.A. Barker and P.C. Young, editors, *Identification and System Parameter Estimation*, pages 749–756. Pergamon Press, Oxford, 1985.

[239] J.A. Spriet and P. Herman. Simulation study of structure characterisation methods. *IMACS 1983*, pages 452–459, 1983.

[240] J.A. Spriet and G.C. Vansteenkiste. *Computer-aided Modelling and Simulation*. Academic Press, London, 1982.

[241] H. Stehfest. An operational dynamic model for the final clarifier. *Trans. Inst. Meas. Control*, 6:160–164, 1984.

[242] G. Stephanopoulos and K.-Y. San. Studies on on-line bioreactor identification. *Biotechnol. Bioeng.*, 26:1176–1188, 1984.

[243] P.M. Sutton and P.N. Mishra. Activated carbon based biological fluidized beds for contaminated water and wastewater treatment: A state-of-the-art review. *Wat. Sci. Tech.*, 29(10-11):309–317, 1994.

[244] L. Szalai, P. Krebs, and W. Rodi. Simulation of flow in circular clarifiers with and without swirl. *J. Hydraulic Eng.*, 120:4–17, 1993.

[245] I. Takács, G.G. Patry, and D. Nolasco. A dynamic model of the clarification-thickening process. *Wat. Res.*, 25(10):1263–1271, 1991.

[246] H. te Braake, R. Babuska, E. van Can, and C. Hellinga. Predictive control in biotechnology using fuzzy and neural models. In J. VanImpe, P.A. Vanrolleghem, and D. Iserentant, editors, *Advanced Instrumentation, Data Interpretation and Control of Biotechnological Processes*. Kluwer, Dordrecht, 437–464, 1998.

[247] G. Tessier. Croissance des populations bactériennes et quantités d'aliments disponibles. *Rev. Sci.*, 80–209, 1942.

[248] M.L. Thompson and M.A. Kramer. Modelling chemical processes using prior knowledge and neural networks. *AIChE J.*, 40:1328–1340, 1994.

[249] L. Tijhuis, W.A.J. van Benthum, M.C.M. van Loosdrecht, and J.J. Heijnen. Solids retention time in spherical biofilms in a biofilm airlift suspension reactor. *Biotechnol. Bioeng.*, 44:867–879, 1994.

[250] H. Topiwala and C.G. Sinclair. Temperature relationship in continuous culture. *Biotechnol. Bioeng.*, 13:795–813, 1971.

[251] S. Tseng and J.-P. Hsu. A comparison of the parameter estimating procedures for the Michaelis-Menten model. *J. Theor. Biol.*, 143:457–464, 1990.

[252] H. Tulleken. *Grey-box Modelling and Identification Topics*. PhD thesis, Delft University of Technology, Holland, 1992.

[253] A. Urrutikoetxea and J.L. Garcia-Heras. Secondary settling in activated sludge: a lab-scale dynamic model of thickening. *Proc. FAB, Med. Fac. Landbouww. Univ. Gent*, 59/4a:2025–2036, 1994.

[254] S. Vajda, K.R. Godfrey, and H. Rabitz. Similarity transformation approach to identifiability analysis of nonlinear compartmental models. *Math. Biosci.*, 93:217–248, 1989.

[255] V. Van Breusegem and G. Bastin. Reduced order modelling of reaction systems: a singular perturbation approach. *Proc. 30th IEEE Conf. Decision and Control*, 1049–1054, 1992.

[256] H.L. Vangheluwe, F. Claeys, S. Kops, F. Coen, and G.C. Vansteenkiste. A modelling and simulation environment for wastewater treatment plant design. *Proc. European Simulation Symposium 1996*, Genoa, Italy:90–97, 1996.

[257] P. Van Overschee and B. De Moor. N4SID: Two subspace algorithms for the identification of combined deterministic-stochastic systems. *Automatica*, 30:75–93, 1994.

[258] P.A. Vanrolleghem, M. Van Daele, and D. Dochain. Practical identifiability of a biokinetic model of activated sludge respiration. *Wat. Res.*, 29:2561–2570, 1995.

[259] P.A. Vanrolleghem, M. Van Daele, P. Van Overschee, and G.C. Vansteenkiste. Model structure characterization of nonlinear wastewater treatment systems. *Proc. 10th IFAC Conf. Syst. Ident.*, 1:279–284, 1994.

[260] P.A. Vanrolleghem, A. Vanderhasselt, P. Krebs, and P. Reichert. Identification of a second-order one-dimensional clarifier model from on-line data. *Technical report, Ghent University*, 1997.

[261] P.A. Vanrolleghem and M. Van Daele. Optimal experimental design for structure characterization of biodegradation models: On-line implementation in a respirographic biosensor. *Wat. Sci. Tech.*, 30(4):243–253, 1994.

[262] P.A. Vanrolleghem and J.F. Van Impe. On the use of structured compartment models for the activated sludge biodegradation process. *Proceedings Workshop Modelling, Monitoring and Control of the Activated Sludge Process. Med. Fac. Landbouww. Rijksuniv. Gent*, 57:2215–2228, 1992.

[263] P.A. Vanrolleghem, J.F. Van Impe, J. Vandewalle, and W. Verstraete. Advanced monitoring and control of the activated sludge process: On-line estimation of crucial biological variables in a structured model with the RODTOX biosensor. In M.N. Karim and G. Stephanopoulos, editors, *Modelling and Control of Biotechnical Processes*, pages 355–358. Pergamon Press, Oxford, 1992.

[264] P.A. Vanrolleghem and K.J. Keesman. Identification of biodegradation models under model and data uncertainty. *Wat. Sci. Tech.*, 33(2):91–105, 1996.

[265] P.A. Vanrolleghem and W. Verstraete. On-line monitoring equipment for wastewater treatment processes: State of the art. *Proc. TI-KVIV Studiedag Optimalisatie van Waterzuiveringsinstallaties door Proceskontrole en -sturing*, pages 1–22, 1993.

[266] P.A. Vanrolleghem and W. Verstraete. Simultaneous biokinetic characterization of heterotrophic and nitrifying populations of activated sludge with an on-line respirographic biosensor. *Wat. Sci. Tech.*, 28(11-12):377–387, 1993.

[267] G.C. Vansteenkiste and J.A. Spriet. Modelling ill-defined systems. In F.E. Cellier, editor, *Progress in Modelling and Simulation*, pages 11–38. Academic Press, London, 1982.

[268] G. van Straten. Analytical methods for parameter-space delimitation and application to shallow-lake phytoplankton-dynamics modeling. *Appl. Math. Comput.*, 17:459–482, 1985.

[269] L. Van Vooren. *Buffer capacity based multipurpose hard- and software sensor for environmental applications*. PhD thesis, Ghent University, Belgium, 2000.

[270] A. Varma and R. Aris. Stirred pots and empty tubes. In L. Lapidus and N.R. Amundson, editors, *Chemical Reactor Theory: A Review.*, pages 79–154. Prentice-Hall, Englewood Cliffs, NJ, 1977.

[271] R. Verhulst. Notice sur la loi que la population suit dans son accroissement. *Corr. Math. et Phys., A. Quetelet (Ed.)*, t. X:113, 1838.

[272] L.K. Vermeersch, B. Kroes, and P.A. Vanrolleghem. Feature based model identification of nonlinear biotechnological processes. *Ecol. Modelling*, 75:629–640, 1994.

[273] K.J. Versyck, J.E. Claes, and J.F. Van Impe. Practical identification of un-

structured growth kinetics by application of optimal experimental design. *Biotechnol. Prog.*, 13:524–531, 1997.

[274] K.J. Versyck and J.F. Van Impe. Optimal design of system identification experiments for bioprocesses. *Journal A*, 41(2):25–34, 2000.

[275] C. Vialas. *Modélisation et Contribution à la Conception d'un Procédé Biotechnologique*. PhD thesis, ENSIEG, Grenoble, France, 1984.

[276] C. Vialas, A. Cheruy, and S. Gentil. An experimental approach to improve the Monod model identification. In A. Johnson, editor, *Modelling and Control of Biotechnological Processes*, pages 155–160. Pergamon, Oxford, 1986.

[277] J. Villadsen and M.L. Michelsen. *Solution of Differential Equation Models by Polynomial Approximation*. Prentice-Hall, Englewood Cliffs, NJ, 1978.

[278] Z.Z. Vitasovic. *An Integrated Control Strategy for the Activated Sludge Process*. PhD thesis, Rice University, Houston, USA, 1986.

[279] E. Walter. *Identifiability of State Space Models*. Springer-Verlag, Berlin, 1982.

[280] O. Wanner, J. Kappeler, and W. Gujer. Calibration of an activated sludge model based on human expertise and on a mathematical optimization technique - A comparison. *Wat. Sci. Tech.*, 25(6):141–148, 1992.

[281] R.W. Watts, S.A. Svoronos, and B. Koopman. One-dimensional modeling of secondary clarifiers using a concentration and feed velocity-dependent dispersion coefficient. *Wat. Res.*, 30:2112–2124, 1996.

[282] S.R. Weijers and P.A. Vanrolleghem. A procedure for selecting the most important parameters in calibrating the activated sludge model No.1 with full-scale plant data. *Wat. Sci. Tech.*, 36(5):69–79, 1997.

[283] T. Wik, A. Mattsson, E. Hansson, and C. Niklasson. Nitrification in a tertiary trickling filter at high hydraulic loads - pilot plant operation and mathematical modelling. *Wat. Sci. Tech.*, 32(8):185–192, 1995.

[284] J.L. Willems. *Stability Theory of Dynamical Systems*. Nelson, London, 1970.

[285] J. Winkin, D. Dochain, and Ph. Ligarius. Dynamical analysis of distributed parameter tubular reactors. *Automatica*, 36:349–361, 2000.

[286] L. Yang and J.E. Alleman. Investigation of batchwise nitrite build-up by an enriched nitrification culture. *Wat. Sci. Tech.*, 26(5-6):997–1005, 1992.

[287] Y.J. Yoo, J. Hong, and R.T. Hatch. Sequential estimation of states and kinetic parameters and optimization of fermentation processes. *Proc. ACC*, 2:866–871, 1985.

[288] P.C. Young. The validity of models for badly defined systems. In M.B. Beck and G. van Straten, editors, *Uncertainty and Forecasting of Water Quality*, pages 69–100. Springer-Verlag, Heidelberg, 1983.

[289] I. Zambettakis. *Simplification des Systèmes à Paramétres Distribués Multiéchelles. Application à la Commande*. PhD thesis, Université des Sciences et Techniques de Lille, France, 1987.

[290] E. Zauderer. *Partial Differential Equations of Applied Mathematics*. John Wiley, New York, 1989.

[291] M.H. Zwietering, J.T. De Koos, B.E. Hasenack, J.C. De Wit, and K. Van 'T Riet. Modelling of bacterial growth as a function of temperature. *Appl. Environ. Microbiol.*, 57:1094–1101, 1991.

Index